"Ted and Jill Heske's Fuzzy Logic for Real World Design represents a practical guide for both the working engineer as well as for the computer novice. They start "from scratch" by introducing the fundamental concepts of fuzzy logic in a way that will have you thinking in fuzzy variables and rules before you realize how much you've already learned. The chapters on Fuzzy Engineering will appeal to the control engineer who is considering applying fuzzy techniques to current projects. The chapters on object-oriented programming put fuzzy design into the context of modern software engineering. A full source code example of a fuzzy kernel and a Windows-based design tool called FuzzyLab are all you need to get a good foundation in the principles and practical uses of fuzzy logic."

Tom Williams, Computer Design Magazine

"A worthwhile and comprehensive look at the theory and application of fuzzy logic."

R. Colin Johnson, Advanced Technology Editor,
EE Times

Fuzzy Logic

for Real World Design

Ted Heske
Jill Neporent Heske

Annabooks

San Diego

Fuzzy Logic for Real World Design
by
Ted Heske
Jill Neporent Heske

PUBLISHED BY

Annabooks
11838 Bernardo Plaza Court
San Diego, CA 92128-2414
USA

619-673-0870

Copyright © Ted Heske and Jill Neporent Heske 1996

All rights reserved. No part of the contents of this book may be reproduced or transmitted in any form or by any means without the prior written consent of the publisher, except for the inclusion of brief quotations in a review.

Printed in the United States of America

ISBN 0-929392-24-8

First Printing January 1996

Information provided in this publication is derived from various sources, standards, and analyses. Any errors or omissions shall not imply any liability for direct or indirect consequences arising from the use of this information. The publisher, authors, and reviewers make no warranty for the correctness or for the use of this information, and assume no liability for direct or indirect damages of any kind arising from technical interpretation or technical explanations in this book, for typographical or printing errors, or for any subsequent changes.

The publisher and authors reserve the right to make changes in this publication without notice and without incurring any liability.

All trademarks mentioned in this book are the property of their respective owners. Annabooks has attempted to properly capitalize and punctuate trademarks, but cannot guarantee that it has done so properly in every case.

About the Authors

Ted and Jill Heske live and consult in Atlanta, GA. Ted has BS degrees in EE and Applied Physics and an MS in Physics. He holds several US patents. Jill has a BA in Foreign Studies, an MS in CIS, and speaks several foreign languages. Together the authors have 20 years of experience in the computer industry and specialize in designing and implementing hardware and software intelligent system solutions. Their combined practical experience ranges from ASIC and board design to object-oriented technologies and end-user applications development.

Dedication

To Nana Sadie

Acknowledgements

Our very special thanks go to Adam Dixon, Dave Selig, and Larry Turvy for their consistent support and valuable contributions during the entire writing of this manuscript. We would also like to thank Mario Lopez, Robert Koss, and Rick Rudowicz for their reviews of selected chapters; Greg Viol, David Gray, and Duane Shippy for reviewing the O-O design and poring over source code; Craig Franklin, Karen Wilson, and Greg Criswell for participating in the usability study of **FuzzyLab**; and Shapeware Corporation for providing us with VISIO for Chapters 8 and 9.

Thanks also to the staff and editors at Annabooks for helping to produce the final product. We are especially grateful to our editor, John Choisser, for his wholehearted enthusiasm for this project.

We would like to extend thanks to some people who did not participate directly in the production of this book, but have educated, inspired, encouraged, and otherwise helped us along the way: Arlene and Lew Neporent; Brian Burke and Joe Scherrer of Germantown Academy; and Norman Badler, University of Pennsylvania. Thanks also to Steve Marsh and Motorola for awakening interest in fuzzy logic so successfully in industry, and to David Brubaker for his review and constructive feedback.

And finally, we wish to express special appreciation to Lotfi Zadeh for being an inspiring example of intelligence, hard work, and persistence.

Contents

1. **Introduction** — 1
   - The Nature of Fuzzy Logic — 2
   - History of Fuzzy Logic — 3
   - Why Use Fuzzy Logic? — 9
   - Fuzzy Logic Is Already Here — 24
   - Fuzzy Logic Is Here To Stay — 27
   - In The Following Chapters.... — 27

2. **Terminology** — 29
   - Membership Functions and Fuzzy Sets — 29
   - Experiment with the Mind — 29
   - The Human Link — 31
   - The Meaning and Elements of a Membership Function — 32
   - Summary — 68

3. **Fuzzy Rules and Operations** — 69
   - The Meaning and Elements of a Fuzzy Rule — 70
   - Combining Antecedents — 79
   - Inference — 89
   - Rulebase Evaluation — 93
   - Summary — 103

4. **Fuzzy Engineering** — 105
   - When to Use Fuzzy Logic — 105
   - How to Use Fuzzy Logic — 107
   - Summary — 139

5. **Fuzzy Engineering II** — 141
   - Pendulum Revisited — 142
   - Fuzzy-PID Control — 159
   - Decision Support System — 179
   - Summary — 185

6. **Embedded Systems Considerations** — 187
   - Goals and Constraints — 190
   - Solutions — 197

7.	**Implementation Choices**	**213**
	Software	214
	Hardware	237
	Hybrid	242
8.	**Object Oriented Design of a Fuzzy Application**	**243**
	The Static Design	244
	The Dynamic Design	255
	Reusing Our Design	262
	Summary	266
9.	**Object Oriented Implementation of a Fuzzy Application**	**267**
	Getting the Most from this Chapter	268
	TEMPCNTL.EXE	268
	Mapping Classes from the OO Design to C++	301
	Construction of Objects	304
	Application Control Flow	312
	How to Build Your Own Fuzzy Application Using Our Code	318
	Enhancing Our Code	320
	Summary	321
10.	**FuzzyLab Primer**	**323**
	Installation	323
	Getting Started	323
	What's on the Screen	325
	Running the Simulation	329
	Editing the Pendulum	330
	Editing the Rulebase	332
	Strategies for Exploring FuzzyLab	335
	How the Pendulum Controller Works	338
	Appendix A: Resources	**341**
	Bibliography	342
	Fuzzy Logic Development Tools	345
	Vendor Reference	352
	On-Line Resources	354

Appendix B: Cross Compiling the Fuzzy Kernel   359
    Target: Motorola 68HC11   359
    Target: Motorola 68HC16   384
    Target: Intel 80C196   390

Glossary   405

Index   417

Foreword

The past two years have witnessed a rapid growth in the number of books dealing with fuzzy logic and its applications. Along with the growth came a change in direction. Earlier books tended to be focused on foundations and basic theory. The change in direction is manifesting itself in much stronger emphasis on applications and much greater stress on implementation, software tools, and embedded systems concepts. *Fuzzy Logic for Real World Design*, by Ted and Jill Heske epitomizes this trend and speaks the language of a design-oriented engineer who is very much at home in the world of toolboxes, graphical user interfaces, object-oriented languages, and software kernels. To those who are in this world, *Fuzzy Logic for Real World Design* has a great deal to offer. With high expository skill and liberal use of graphics, the authors communicate to the reader their extensive experience in the design of fuzzy systems. The use of the words "Real World Design" in the title of the book is fully justified by its contents.

There are a few points - relating mostly to semantics - that I should like to comment on.

There are many misconceptions about fuzzy logic that stem from the fact that the term "fuzzy logic" has two different meanings. In its narrow sense, fuzzy logic is a logical system which is aimed at a formalization of modes of reasoning which are approximate rather than exact. In this sense, fuzzy logic is an extension of multivalued logical systems, but its agenda is quite different both in spirit and in substance.

In its wide sense - which is in predominant use today - fuzzy logic, or FL for short, is almost synonymous with the theory of fuzzy sets, that is, classes of objects with unsharp boundaries. Today, most of the practical applications of fuzzy logic involve FL rather than fuzzy logic in its narrow sense.

What is important to recognize is that in most applications of FL what is used is a largely self-contained subset of FL which centers on the manipulation of fuzzy if-then rules. The subset in question is in effect a calculus of fuzzy rules, or CFR for short. A key role in CFR is played by what may be referred to as the Fuzzy Dependency and Command Language (FDCL). Although the term FDCL is not employed by the authors, the underlying concept plays a pivotal role in their exposition - a role which reflects the fact that the use of fuzzy rules lies at the center of human reasoning.

Another important point that needs to be recognized is that, in most of the applications of fuzzy logic, the point of departure is a human solution. In this perspective, a fuzzy logic solution may be viewed as an articulation of a human solution in the language of fuzzy if-then rules. Thus, much of the calculus of fuzzy rules is concerned with the execution of instructions expressed in this language. By focusing their exposition on CFR rather than on other parts of FL, the authors make it easier for a design-oriented engineer to proceed from the basics of FL to its real-world applications.

The authors' approach clarifies the distinction between FL and earlier attempts to find an accommodation with the pervasive imprecision of the real world. In the first place, fuzziness differs from vagueness in that fuzziness relates to the unsharpness of class boundaries whereas vagueness relates to insufficient specificity. Thus, vagueness is a property of propositions rather than predicates. In this sense, "I will be back in a few minutes" is fuzzy but not vague, while "I'll be back sometime" is both fuzzy and vague.

In the final analysis, the main contribution of fuzzy logic is a methodology for computing, reasoning, and programming with words - a methodology based on the concept of a linguistic variable and the calculus of fuzzy rules. By their conscious decision to focus on this methodology and their extensive use of modern software tools, the authors have produced a reader-friendly book that is packed with up-to-date information about computer-oriented design of real world fuzzy-logic-based systems. *Fuzzy Logic for Real World Design* is in the vanguard of books conceived in this spirit.

Ted and Jill Heske have written a book that is a must reading for anyone who has a serious interest in applying fuzzy logic to real world design. Their book is an important contributon to making fuzzy logic an effective tool for the conception, design, and deployment of systems which are smart, robust, and easy to use.

Lotfi A. Zadeh

Berkeley, CA

November 20, 1995

Foreword

If I get too warm, I take my jacket off. If the radio is slightly too loud, I turn the volume down a little bit. These and other rules define the common sense by which I operate. My son has different rules. For him, if the radio is slightly too loud, he cranks the volume up a lot. Regardless of your rule set, the rules are fuzzy by nature. Everything with which you may interface is a candidate to employ fuzzy logic as well.

I first heard of fuzzy logic in the mid-seventies. Having recently "conquered" Boolean logic, I presumed that fuzzy logic was anti-logic and hoped the fad would pass. It did.

It was much later that I encountered "the elevator" that started my odyssey with fuzzy logic. I had just arrived in Japan. My first business appointment was scheduled for that same afternoon. The plane arrived an hour late. Being late to a business appointment in Japan is an absolute breech of protocol. I ran into my hotel with the faint hope that I could make my business appointment on time. The hotel was busy. I pressed the button for "the elevator" and after a short delay, the door opened. I pressed the button for the ninth floor. The door closed, then immediately opened again. Having a low tolerance for stupid elevators (especially when I'm in a hurry), I mumbled something and reached to press the button for the ninth floor again. Then I noticed it. *I was on the ninth floor.* Somehow, the elevator transferred me nine floors without perceptible movement, and very quickly at that. Fascinated, I pushed more buttons. Down to three, up to twelve.. the movement was discernible with concentration but was nonetheless amazing. Perhaps because it could whisk me to my floor so fast, it did not stop at intermediate floors to pick up additional passengers. My contacts in Japan told me that the elevator used fuzzy logic and that the vendor claimed that the system time savings was 30% better than traditional elevators. I was quite late for my appointment.

On that same trip, I asked my Japanese customers if they used fuzzy logic. Most did. The ones that didn't had been asked by their marketing departments to do so. At that time, the rest of the world had no clue what was going on in Japan with regard to fuzzy logic. Many US customers thought that precise input values to a fuzzy system would give an unpredictable output (not true).

It was obvious that the vast majority of fuzzy logic applications could be implemented in software on standard microcontrollers. Having significant marketshare in the microcontroller arena, we (Motorola) decided to introduce fuzzy logic in the form of an educational promotion, complete with design contest. In preparing the educational program, we discovered that most material available was academically oriented rather than engineering focused. Microcontroller-oriented software appeared to be nonexistent. The terms used to describe many aspects of fuzzy logic were inconsistent. Development of Motorola's educational program for engineers was truly a pioneering effort.

Ted Heske was one of a small group of winners in the United States fuzzy logic design competition. Drs. Lotfi Zadeh, Earl Cox and others presented concepts beyond the scope of the original education program during a week-long seminar presented to the winners in Hawaii.

Two years after the introduction of the educational kit, we conducted a marketing survey in which over 50% of the respondents indicated that their company planned to use fuzzy logic in their products.

While giving numerous seminars on fuzzy logic, it became apparent that engineers easily grasped the concepts of fuzzy logic, but had difficulty applying those concepts to their application at hand. an important feature of this book is that it teaches fuzzy logic concepts in a way that helps the engineer bridge the gulf of understanding between conventional and fuzzy implementation. Only after fuzzy logic concpets are comprehended within a familiar context can engineers effectively add fuzzy logic to their bag of tools.

Steve Marsh

Director, Strategic Operations

Motorola

Preface

In the past, applying fuzzy logic to solve a problem required an expert in fuzzy logic. This was due primarily to the fact that the relevant information was scattered. An individual was compelled to comb through research papers, academic journals, and conference proceedings to learn the fundamentals. After understanding the fundamentals, though, applying the technology was still an effort. There existed even less information relevant to the practice of fuzzy engineering. The need for a practical, example-based guide to designing, constructing, and fine-tuning fuzzy systems provided our motivation for writing *Fuzzy Logic for Real World Design*.

This book is intended to serve the needs of the curious and practical-minded. It is targeted at those who have real-world problems to solve and are willing to look a little further than the traditional problem solving approaches. In addition to the fundamentals of fuzzy logic, we have collected, under one cover, the tools, tips, techniques, and references necessary to successfully field a fuzzy logic based solution. We do not know of any other work on fuzzy logic that provides as complete an approach to applying this effective and versatile technology to everyday engineering problems.

by Ted Heske and Jill Neporent Heske

Chapter 1:
Introduction

Lotfi Zadeh, a U.C. Berkeley professor, introduced the idea of fuzzy sets to the world in 1965 in a paper simply titled *Fuzzy Sets*[1]. Even though the application of fuzzy set theory has been developed to comercial fruition elsewhere in the world, its adoption in the U.S. has been ironically slow. One reason often cited for the slow adoption of this technology has been that it suffers from a bad name. People associate the term *fuzzy* with sloppiness and vagueness. However, the mathematical foundations of fuzzy logic are far from sloppy. Professor Zadeh knowingly christened his new field with a provocative name. That name has provoked both great curiosity and some scorn. In the words of Zadeh " it's very difficult to find a person who is hostile and well-informed". The goal of this book is to inform.

The purpose of this chapter is to introduce various aspects of fuzzy logic and the issues which pertain to its use in solving real world problems. In order to understand the role that fuzzy logic plays in modern problem solving we must start with the basics. Next, we'll pay our respects to the founders of fuzzy logic by briefly recapping the history and evolution of fuzzy logic. We then explore its intrinsic properties in order to understand why fuzzy logic is the first artificial intelligence technology responsible for an entire generation of successful consumer products. Finally, we will examine the diversity of current applications of fuzzy logic.

---

[1]Fuzzy Sets. *Information and Control* 8 (1965): pp. 338-353.

The Nature of Fuzzy Logic

For too long we have constrained our problem-solving approaches by aligning our thinking with the most powerful technological advance of our generation -- the binary-driven digital computer. Fuzzy logic frees us from binary constraints by embracing and exploiting the notion that things are not just true (1) and untrue (0). It admits a continuous range of truth from 0 to 1, and acknowledges that things can be true and untrue to varying degrees at the same time. In contrast to its name, fuzzy logic is built upon solid mathematical foundations that are no less rigorous than the foundations of calculus or physics. When applied to problem solving, fuzzy logic principles empower our own natural capacities for classification and approximate reasoning by providing a framework to represent degrees of truth in a computer-compatible way.

Today's fuzzy logic diverges from bivalued logic in two important aspects. The first, and most obvious divergence is the continuum of truth values permitted. Fuzzy logic is a continuously valued elaboration of the bivalued theme established by Aristotle, Boole, and Frege (as we will discuss in the history below). Truth in the fuzzy logic framework comes in degrees.

For example, if we ask someone whether it's hot out and they answer, "Well, it is and it isn't," we usually get the point. "Very warm" can mean the same thing as "sort of hot." Does this mean it can be warm and hot at the same time? The answer is yes, it can be some *degree* of both. When it is warm to a high degree (very) it is also hot to a low degree (sort of). Our definitions of hot and warm overlap, and we use qualifiers like "very" and "sort of" to describe varying degrees of hotness or warmness.

Bivalued logic is a special case of fuzzy logic in the sense that if the truth values of a fuzzy system are restricted sufficiently we arrive at a bivalued logic system again. This is the equivalent of saying that it must be either hot or not hot, allowing no room for varying degrees of these two values.

The second aspect of fuzzy logic that diverges completely from the classical logic is that truth itself can exist as a fuzzy quantity. Fuzzy truth values might be expressed in statements like "somewhat likely" or "sort-of true". Truth is not automatically assumed to be a crisp value. By allowing truth to be fuzzy, we are able to more completely model human concepts of truth in the real world.

Together the two aspects of fuzzy logic just outlined serve as a basis for automating common-sense reasoning. Fuzzy logic takes the latest steps in the manipulation and representation of human knowledge, which started two thousands years ago with Aristotle.

History of Fuzzy Logic

No matter which historian of logic one reads, every history of logic seems to start with Aristotle (384-322 B.C.). Aristotle, the student of Plato, thought of Logic as the science of knowing. Aristotle's most germane, and certainly his most quoted contributions to the foundations of fuzzy logic are two axioms named the Law of Contradiction and the Law of the Excluded Middle. The Law of Contradiction says that a thing cannot both belong to a class and not belong to a class at the same time. This is like saying that it cannot be both *raining* and *not raining* at the same time. The Law of the Excluded Middle says that a thing must either belong to a class or not belong to a class. In other words, it must be *raining* or *not raining*. Together these two axioms leave no room for such concepts as *sort-of raining* or *slightly raining*.

Aristotle's two axioms paint the panorama of logical argument with only two colors, black and white. In the Aristotelian palette white represents complete truth and black represents complete falsehood. Together, the Law of Contradiction and the Law of the Excluded Middle forbid the continuum of truth values that provides the backbone of fuzzy logic. Only with the more expressive palette of grays, or degrees of truth, is it possible to more accurately capture the profundity and subtlety of human reasoning.

Both of Aristotle's laws seem reasonable. Aristotle felt these laws reflected an intuitive understanding that was clearly in synchrony with common sense. Across the two millennia since he walked the earth, Aristotle's axioms of logic have served well. To most succeeding scholars Aristotle held the position of originator of the study of logic and writer of the axioms upon which the whole subject rested.

The logic of Aristotle attempted to reduce all rational argument to a deductive sequence of syllogisms, whose prototype appears as follows:

All men are mortal
Socrates is a man, therefore
Socrates is mortal.

By means of such syllogisms, he investigated whether certain propositions could be proven true using other propositions which were already known to be true. Eventually, the most impactive use of Aristotle's logic came to be the proof of mathematical propositions.

The analysis of mathematical truth proceeded under the strict guidance of the syllogistic approach for two thousand years. In 1847, George Boole formally established logic as a branch of mathematics by publishing Mathematical Analysis of Logic. Boole developed an algebraic framework for the analysis of the Aristotelian syllogism. That framework is known today as Boolean Algebra. While the algebra of Boole clearly revealed the inadequacies of syllogistic reasoning, it provided the starting point for various fields in mathematics and computer science.

Mathematical logic may be regarded as one of the defining successes of modern mathematics. Armed with the new algebra of logical proof, and guided by the overall philosophy of reducing all mathematics to formal logic, mathematicians made great strides in the hundred years following Boole's work. The contributions of Frege, Cantor, Hilbert, Whitehead and Russell, Godel, Tarski, Robinson, and others provided a rigorous logical justification of most, if not all, of mathematics. The logic that these great minds developed not only served to illuminate the fundamental validity of mathematics, but it also claimed to be a formalization of rational thought. It was assumed that human reasoning could be broken down into subunits which could then be manipulated according to the same logical formalisms that applied to mathematics. It is the application of their logic to the representation and manipulation of knowledge that most interests us.

The work of Gottlob Frege, first order predicate calculus, created a mathematical language for the precise description of propositions, and truth value assignments. In addition, Frege's first order predicate calculus created the tools necessary for automated reasoning: a language for expressions, a theory for assumptions related to the meaning of expression, and rigorous mathematical methods for inferring new true expressions. Frege's calculus plays a fundamental role in the development of formal or classical artificial intelligence (AI).

By the time the digital computer was invented, the formal mathematical and logical foundations for the study of artificial intelligence already existed. The goal of automated reasoning now seemed achievable. By the end of the 1940s the creation of computer-based artificial intelligence seemed possible. The computer possessed enough memory and processing power to implement and test formal reasoning methodologies. Virtually all early AI efforts were automated versions of Frege's first order predicate logic. Frege's logic has been an enduring component of AI. Even today, a substantial community of AI researchers believe that AI should be based on first order predicate logic.

Traditional AI approaches have celebrated some successes, particularly in the realm of the so-called expert system. DENDRAL[2] inferred the molecular structure of organic compounds from their chemical formulae. MYCIN[3] diagnoses bacterial blood infections, spinal meningitis, and even prescribes treatments. PROSPECTOR[4] infers likely location and type of metal ore deposits. Yet, even a cursory look at the typical expert system quickly reveals that truth is assumed to be one of only two values, either completely true or completely false. This aspect of expert systems may not sound alarming. However, when we reflect upon the demonstrated inability of such systems to capture

---

[2]Lindsay et al, 1980. *Applications of artificial intelligence for organic chemistry: the DENDRAL project.* New York: McGraw-Hill.

[3]B.G. Buchanan and E.H. Shortliff, eds. 1984. *Rule-Based Expert Systems: The MYCIN Experiments of the Stanford Heuristic Programming Project.* Reading, MA: Addison-Wesley.

[4]Duda et al. *Model design in the PROSPECTOR consultant system for mineral exploration.* In Michie (1979).

what is best described as common sense reasoning, their bivalued truth system looks inadequate.

Vagueness plays a central role in common-sense knowledge and reasoning. One can find endless examples of vagueness in language. Our human languages have been evolving for tens of thousands of years. Yet, we hear vague statements like Jon is a good listener, or please wait a few moments, or I'll take about a pound of that sliced turkey. What do "good listener", "few moments", and "about a pound" really mean? Of course, these words express vague (or fuzzy if you like) concepts. Human languages have evolved in a direction that provides us with very efficient means of expressing vague notions. (This is likely due to our perceptions of the world being fundamentally fuzzy.) So vagueness in language may be by design and not at all a shortcoming.

A number of philosophers of the late nineteenth and early twentieth centuries probed at the ubiquitous vagueness they perceived. Charles Peirce proposed that any concept described by a continuum, such as weight, height, time, emotion, beauty, etc., would be subject to vagueness. He asserted that such continuums govern knowledge, and that " vagueness is no more to be done away with in the world than friction in mechanics"[5]. In On Liberty (1859) John Stuart Mill observed "There is no such thing as absolute certainty, but there is assurance sufficient for the purposes of human life."[6] In his musings on the impact of human perception on the study of natural phenomena Albert Einstein wrote "physical concepts are free creations of the human mind, and are not, however it may seem, uniquely determined by the external world."[7] However, recognizing the vast vagueness around us was just the first step toward a comprehensive theory and consistent treatment of it.

---

[5]D.McNeil and P.Freiburger, *Fuzzy Logic*, 1993: New York, Simon & Schuster.

[6]J. S. Mill, *On Liberty*, 1859. Excerpted from *Bartlett's Familiar Quotations*, (Little, Brown and Company, 1980), Microsoft Bookshelf 1992, s.v. "John Stuart Mill: There is no such thing as ..."

[7]Albert Einstein, *Evolution of Physics*, 1938. Excerpted from *Bartlett's Familiar Quotations*, (Little, Brown and Company, 1980), Microsoft Bookshelf 1992, s.v. "Albert Einstein: Physical concepts are free creations of ..."

Chapter 1: Introduction

Breaking the bivalued truth barrier in a formal mathematical way started at the beginning of this century with work on multivalued logic. In 1920 Jan Lukasiewicz published a paper describing a multivalued logic that used only three truth values. In addition to the standard *true* and *false*, he added the value 1/2 representing *possible*. For the first time, partial truth and partial contradiction were permitted. With that small addition, Lukasiewicz opened the door to infinite possibilities. Although he himself realized that three truth values could be expanded to greater numbers of levels, we have limited evidence that he pursued it further.

The expansion of Lukasiewicz's multivalued logic generalizes to a continuum of truth values. Max Black recognized the continuum of truth values possible from such an expansion in a 1937 article.[8] In his paper Black introduced the vague set, which we now call the fuzzy set. He mapped the continuum of the vague onto the formal mathematical framework of logic. While the truth values of Black's vague sets took the form of probabilities, the continuously valued logic that he formalized finally was able to model the fuzziness inherent in common-sense reasoning. Although in its own day the theory of vague sets fell on unresponsive ears, Max Black survived to see his vague sets revived and praised as the newly named fuzzy logic.

The next, and certainly most significant, steps in the development of fuzzy logic were taken in 1965 in a paper called *Fuzzy Sets*. Lotfi Zadeh of the University of California at Berkeley took those steps by coining the name fuzzy logic and defining the mathematical notions of inclusion, union, intersection, complement, relation, convexity, etc. for such sets. In this boldly original work, Zadeh recognized that "... the classes of objects encountered in the real physical world do not have precisely defined criteria of membership." Furthermore, he wrote "...the fact remains that such imprecisely defined classes play an important role in human thinking ...."

During the next decade Zadeh continued laying down the foundations of fuzzy logic with a variety of works: *Probability Measures*

---

[8]M. Black, "Vagueness: An Exercise in Logical Analysis," *Philosophy of Science*, 4, 1937, pp.427-455.

of *Fuzzy Events*[9], *Decision-Making in a Fuzzy Environment*[10], *Similarity Relations and Fuzzy Orderings*[11], *Outline of a New Approach to the Definition of Complex Systems and Decision Processes*[12], *A Fuzzy Algorithmic Approach to the Definition of Complex or Imprecise Concepts*[13], and *Fuzzy Sets as a Basis for a Theory of Possibility*[14].

Zadeh then turned his attention to applying his newly minted fuzzy framework to modeling and automating modes of human reasoning. APPROXIMATE REASONING is the term used by Zadeh to describe the human ability to process imprecise, incomplete, and possibly unreliable information while reaching concrete conclusions. Zadeh followed his earlier works with *The Concept of a Linguistic Variable and Its Application to Approximate Reasoning Parts 1, 2, and 3*[15],, *A Theory of Approximate Reasoning*[16], *The Role of Fuzzy Logic in the Management of Uncertainty in Expert Systems*[17], *A Theory of Commonsense Knowledge*[18], and *Syllogistic Reasoning in Fuzzy Logic and its Application to Usuality and Reasoning with Dispositions.*[19] Of course, Zadeh produced much more work than is cited here. These works are merely representative of the genesis and progression of Zadeh's contributions.

---

[9]L.A. Zadeh, appeared in *Journal Mathematical Analysis and Application,* 10, 1968, pp. 421-427.

[10]R.E. Bellman and L.A. Zadeh, appeared in *Management Science,* 17:4, 1970, pp. 141-164.

[11]L.A. Zadeh, appeared in *Information Sciences,* 3, 1971, pp.177-200.

[12]L.A. Zadeh, appeared in *IEEE Transactions on Systems, Man, and Cybernetics,* SMC-3, 1973, pp. 28-44.

[13]L.A. Zadeh, appeared in *International Journal of Man-Machine Studies,* 8, 1976, pp. 249-291.

[14]L.A. Zadeh, appeared in *Fuzzy Sets and Systems,* 1, 1978, pp. 3-28.

[15]L.A. Zadeh, parts 1 and 2 appeared in *Information Sciences,* 8, 1975, pp. 199-249. part 3 appeared in *Information Sciences,* 9, 1976, pp. 43-80.

[16]L.A. Zadeh, appeared in *Machine Intelligence,* Vol. 9, 1979, pp.149-194.

[17]L.A. Zadeh, appeared in *Fuzzy Sets and Systems,* 11, 1983, pp. 199-227.

[18]L.A. Zadeh, appeared in H.J Skala et al. Eds., *Aspects of Vagueness,* 1984, pp. 257-296.

[19]L.A. Zadeh, appeared in *IEEE Transactions on Systems, Man and Cybernetics,* SMC-15, 1985, pp. 754-763.

Today Professor Zadeh is referred to as the father of fuzzy logic for a number of reasons. The first reason is that he provided the world with a comprehensive mathematical and logical foundation for fuzzy logic. The second reason relates more to Zadeh's perseverance in the face of the reluctance and occasional hostility with which the academic community received his ideas. In this regard the quote from Niccolò Machiavelli (1469-1527) well describes the situation Zadeh faced: "There is nothing more difficult to take in hand, more perilous to conduct, or more uncertain in its success, than to take the lead in the introduction of a new order of things." Originally perceived of as having minimal importance, Zadeh's fuzzy logic now impacts an ever widening range of applications. The third reason for the paternalistic description of Zadeh is the expansive body of work he contributed to the domains of fuzzy logic and related fields including approximate reasoning, common sense knowledge, and meaning representation in natural language, to name but a few. The final reason to refer to professor Zadeh as the father of fuzzy logic is his well-known generosity to colleagues and novices alike. For all of these reasons, Lotfi Zadeh is a scholar and man whom history will remember.

Why Use Fuzzy Logic?

Instead of justifying the adoption of fuzzy logic by simply stating that Japanese engineers have designed hundreds of products with it, or relating NASA's use of fuzzy logic on the space shuttle, we'll explore the reasons that fuzzy logic fundamentally improves problem solving. In this section we discuss some compelling motivations for the use of fuzzy logic in problem solving, which derive from the fundamental properties and advantages inherent to the logic of fuzzy sets. Motivations fall into two categories: philosophical, and practical. The advantages in the former category stem from theoretical foundations, while application-specific concerns underscore the members of the latter category.

Philosophical Motivations for Fuzzy Logic

Inadequacy of Boolean Models of the World

In the Boolean view of the world, as depicted in Figure 1.1(a), there are only two truth values. 0 signifies complete non truth, and 1 signifies complete truth. Throughout the rest of this book we will be referring to sets and membership in those sets. It is natural then to recast the Boolean truth values in terms of set membership. A truth value of 0 corresponds to complete non membership in a set, and a truth value of 1 corresponds to complete membership in that set.

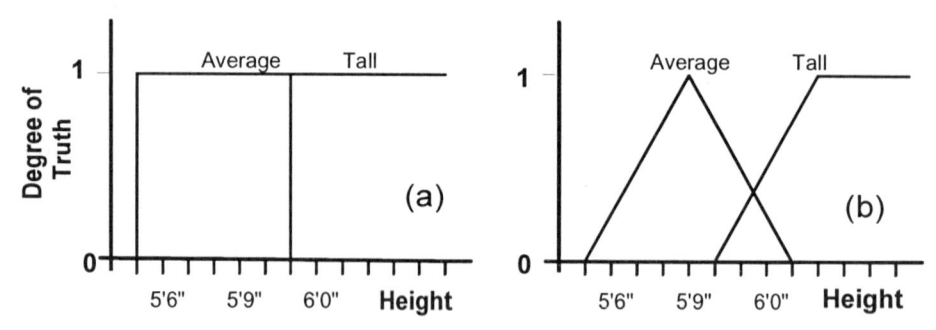

Figure 1.1: (a) Boolean classification of Average and Tall height, and (b) fuzzy logic classes.

Figure 1.1(a) helps to illustrate the one-to-one correspondence between truth value and set membership. All heights that fall in the *Average* set are complete members of that set. Heights of 5'6", 5'8", and 5'10" are all members of the *Average* set, each with a degree of membership equal to 1. All heights that fall outside of that range are complete non members of the *Average* set. You'll notice this Boolean representation of height has some strange properties. Consider two people, one who is 5' 10.99" and another who is 5"11.01". Because of the way that the Boolean classifications are drawn, one is considered *Average*, while the other is considered *Tall*. Adding just .02" inches to one height is enough to switch the classification from *Average* to *Tall*. The transition from *Average* to *Tall* is instantaneous and unequivocal.

Chapter 1: Introduction

The slope of that transition is mathematically infinite. We can ask the question: Is this a good model for the human concept of height? Certainly the validity of the mathematics underlying Boolean representation is not in question. What is in question, however, is the appropriateness of applying Boolean rules of membership to intuitive human classifications like *Average* height.

The weakness of the Boolean representation of height is the fact that all Boolean sets have crisp edges. Those crisp edges cause instantaneous transitions from complete membership to complete non membership. Does this mirror common sense or mock it? Does subtracting .02 inches from a *Tall* person make them *Average* ? We inherently know that if a *Tall* person shrinks by .02 inches they are still *Tall*, though a little less *Tall* than before. Subtract a little more height, and they become a little less *Tall*. Subtract a lot of height and they will be *Tall* to a much smaller degree. We inherently understand that there exist degrees of TALLNESS. Using a Boolean representation reduces all of the possible degrees of TALLNESS to just two values: 0 and 1. Without degrees of TALLNESS, only instantaneous transitions between classes are allowed in the Boolean system. Yet we must have gradual transitions between *Average* and *Tall* to accurately model our notions of TALLNESS. So the Boolean representation is inadequate to model the concept of height.

Fuzzy logic provides a good solution to this problem. In Figure 1.1(b), the *Average* and *Tall* classifications have been drawn as fuzzy sets. The fuzzy set *Tall* covers a range of crisp values that have non-zero degrees of membership in the *Tall* set. A height near the center of the *Tall* fuzzy set has a higher degree of membership in that set than does a height near the edge of the set. The shape of the fuzzy sets permits gradual transitions between *Average* and *Tall*. As height increases above 5'8", degree of membership in *Average* decreases and degree of membership in *Tall* increases. The fuzzy sets accurately model what happens if we add .02 inches to a 5'10.99" person. The degree of membership in the *Tall* set increases slightly, which means that the person is a little taller. Also, the degree of membership in the *Average* set decreases slightly, which means that person is a little less *Average*.

So, the fuzzy set representation is much more in line with our understanding of the concept *height*.

Height, as far as concepts go, is not terribly unique. There is nothing special about height that makes it more difficult to model (*Short, Average, Tall*, etc.) than other concepts. Height is one of a vast number of concepts and properties which we use a collection of NATURAL CLASSIFICATIONS to describe. Natural classifications pervade our speech and thoughts as we describe the world around us. These classifications are so deeply ingrained in us that they seem innate rather than learned. Of course, we have words to express our classifications.

Classifications -
- Short, Tall,
- Small, Medium, Large,
- Heavy, Light,
- Bright, Dim,
- Hot, Cold,
- Fat, Skinny,
- Wet, Dry,
- Wide, Narrow,
- Beautiful, Ugly,
- Frequently, Seldom,
- Many, Several, Few,
- Likely, Not Likely, etc.

And we use modifiers to shade those classifications in ever finer degrees.

Qualifiers -
- Very, Extremely,
- Nearly, More or Less, Somewhat, Almost,
- Rather, Fairly, About, Around,
- Moderately, Close to, Roughly

Most natural classifications and qualifiers, like those listed above, have unsharp, or fuzzy boundaries. Every day, as we safely negotiate through the physical world and communicate facts and ideas to each other, we make constant use of these natural classifications and the relationships between them The lack of precision in our language has seemed an annoying drawback to some. But in reality, the lack of precision in our language, the fuzziness of it, exactly mirrors the way that we perceive the world.

Opposing Needs for Precision and Complex Problem Solving

The idea of using less precision to achieve a solution is unsettling. It saws against the grain of engineering and scientific principles that have guided us into the age of computers. The words of William Thomson, Lord Kelvin, sum up the prevailing thinking:

> *"When you can measure what you are speaking about, and express it in numbers, you know something about it; but when you cannot measure it, when you cannot express it in numbers, your knowledge is of a meager and unsatisfactory kind: it may be the beginning of knowledge, but you have scarcely, in your thoughts, advanced to the stage of science."*
> Popular Lectures and Addresses [1891-1894][20]

This quotation has inspired the search for scientific truth for the last 100 years. It points out that the need for exactness and measurement have been pervasive through out the scientific community for as long as there have been scientists. The corollary to Lord Kelvin's words is: the more precisely you can measure what you are speaking about, the more satisfactory is your knowledge of it.

Precision, however, carries a price. The price that is paid for precision rises sharply as the complexity of a particular problem increases. The electronics engineer knows from experience that constructing an analog-to-digital data acquisition system having 8-bits

---

[20]*Bartlett's Familiar Quotations*, (Little, Brown and Company, 1980), Microsoft Bookshelf 1992, s.v. "William Thomson, Lord Kelvin: When you can measure what you ..."

of resolution requires much less design effort and cost than constructing one having 16-bits of resolution. Further still, a 24-bit resolution system is all but impossible to build. The price paid for precision is intractability. Intractability in this context means that as required precision of a solution increases, our ability to solve that problem diminishes. In any computer system, higher precision requires more calculations and faster computation. It also requires more memory, and probably consumes more power. In some cases we can justify the additional resources needed to support higher precision in our systems. However, if our goal is to make computers smarter, increasing the precision of measurements or increasing the resolution of the internal numeric representation leads to a dead end far short of the goal.

Raising the complexity of a problem places precision and our ability to solve that problem at odds. The reason for this was expressed succinctly by L.A. Zadeh as the Principle of Incompatibility:

> "..the essence of this principle is that as complexity of a system increases, our ability to make precise and yet significant statements about its behaviour diminishes until a threshold is reached beyond which precision and significance (or relevance) become almost mutually exclusive characteristics."[21]

Raise the complexity of a problem enough and you will not find a precise solution. An example will drive this point home. Consider parallel parking a car. Suppose that you were given a set of specifications for parking your car that included the following requirements: distance from car in front, distance from curb, distance from car in back, and parallel alignment to the curb. At first you are given a wide range of values for each of the requirements and you find you are able to park easily. However, as the allowable range of values shrinks (i.e. greater precision is required) you will find parking more difficult, perhaps requiring several tries to get it right. Beyond some

---

[21] L.A. Zadeh, "Outline of a New Approach to the Analysis of Complex Systems and Decision Process", *IEEE Trasactions of Systems, Man, and Cybernetics*, SMC-3 (1973), pg. 28.

point, increases in the precision of your parking specifications will prevent you from completing the task at all.

With all of its precision and computational speed, there does not yet exist a computer system that can parallel park a car in an arbitrary street environment. And yet, we successfully parallel park our cars, even in the tightest of places. We don't use a ruler or protractor, so we don't work with very precise measurements. We adjust the car's position a little this way or little that way based on our judgments of the distances and angles involved. *So the reasoning that we use must not require high precision.* Daily, we solve hundreds of problems whose complexity is far beyond the reach of any computer. We possess a reasoning capability that relies on approximations, and yet produces results profoundly more intelligent than even the fastest super computer. Our lack of reliance on precision does not hamper our capabilities. It enhances them. By exploiting the tolerance for imprecision in solutions to complex problems we enable much deeper cognitive power. We possess the ability to reason with incomplete, vague, and ambiguous data to reach concrete decisions. This ability can be called approximate reasoning.

Fuzzy logic enables computers to think more like we do by endowing them with approximate reasoning capabilities. Fuzzy logic provides the algorithmic and philosophical framework necessary to exploit the tolerance for imprecision and arrive at answers to more complex problems.

Every day, engineers engage in developing solutions to problems. They provide solutions whose cost and performance, taken together, solve their particular problem well enough. Saying that a solution is *good enough* is both vague and profound. Assessing what constitutes *good enough* is a judgment call. Finding the *best* solution to a complex problem is seldom more than an intellectual exercise. Certainly, that search can produce valuable insights, but rarely is justifiable as an end in itself. The *best* solution often takes too much money and/or time to find and/or implement. And, for a sufficiently complex problem, finding the *best* solution is impossible.

The traveling salesman problem (TSP) is one that neatly points out the value of finding a solution that is *good enough*. It also illustrates the

Fuzzy Logic for Real World Design

opposing natures of precision and tractability. The TSP is stated as follows: A traveling salesman is given a list of cities to visit and must pick the single route which minimizes the cumulative distance traveled to reach all cities on the list. Deterministic solutions to the TSP, i.e., solutions that look at all possible routes to determine the shortest path, are a computational nightmare. The number of calculations required of a deterministic solution to the TSP exponentially increases with the number of cities. (Actually, the number of distinct routes involved is a factorial function.)

Number of Cities	Number of Distinct Routes
1	1
2	2
5	120
10	3,628,800
20	2.43290 E 18
50	3.04141 E 64
100	9.33262 E 157
200	7.88658 E 374

Table 1.1: Number of cities versus unique routes for the Traveling Salesman Problem.

The numbers in Table 1.1 are staggering. A computer capable of 100 million floating point operations per second would require 55 years to evaluate all routes between just 20 cities. Clearly, we cannot wait around for the best solution to be ground out. What we can do however, is sacrifice some precision in order to obtain a solution that is good enough. The New York Times reported just such a solution (3-12-91) to the TSP:

Number of Cities	Accuracy of Result	Time to Compute
100,000	1%	2 days
100,000	0.75%	7 months
1,000,000	3.5%	3.5 hours

Table 1.2: Approximate solutions to an otherwise impossibly complex problem.

So by sacrificing a small amount of accuracy in the final result we arrive at an answer to an otherwise effectively unsolvable problem. The important lesson here is that obtaining a good solution is always better than no solution at all.

Evolution in the Language used for Problem Definition and Solution

Fuzzy logic offers a different way of specifying solutions. Computer programming is the process of translating a solution that exists in the mind into language that the computer can understand. For most computers, that language consists of ones and zeros. Luckily, we no longer have to write our programs using just ones and zeros. Interpreters and compilers now translate higher level programming languages like BASIC and C++ into ones and zeros for us. The computer science community has tirelessly developed better and better programming languages. As a result, the level at which we express solutions to problems has risen from the ones and zeros that were first used. One goal of an advanced computer languages is to make it easier for us to describe the solution to the computer. Nevertheless, we still spend much of our time and energy translating our thoughts into semantically precise and syntactically rigid language that the compiler or interpreter understands. With fuzzy logic, the computer handles much more of the translation work.

As a simple example consider the problem of balancing an inverted pendulum, like the one shown in Figure 1.2.

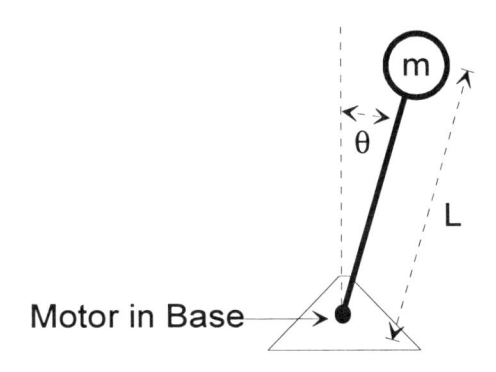

Figure 1.2: Inverted pendulum apparatus

Fuzzy logic allows the expression of the solution in the form of FUZZY RULES as follows:

IF theta is *Near_Zero* AND dtheta is *Near_Zero* THEN motor torque is *Near_Zero*

IF theta is *Small_Positive* AND dtheta is *Small_Positive* THEN motor torque is *Small_Negative*

IF theta is *Large_Positive* AND dtheta is *Large_Positive* THEN motor torque is Very *Large_Negative*

etc.

This simple example assumes two input variables, theta and dtheta, and one output variable, motor torque. Through the definition of fuzzy sets, such terms as *Near_Zero*, *Small_Positive*, and *Large_Positive*, have meanings that are easily manipulated and calculated by computer. Most importantly, these are terms that have clear meanings to us. Figure 1.3 depicts one possible definition of the fuzzy sets of the theta variable.

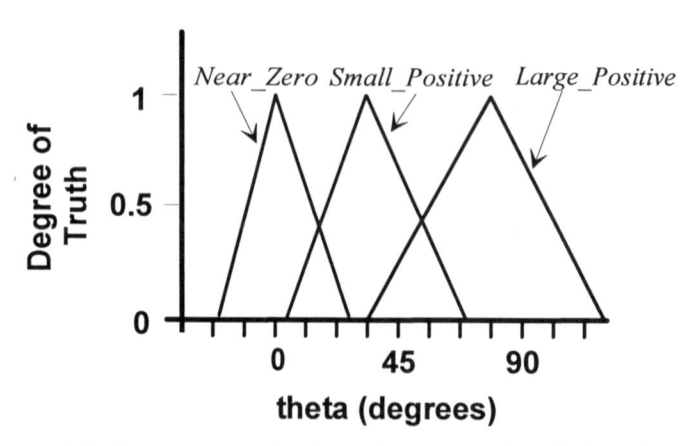

Figure 1.3: Fuzzy sets defined as Near_Zero, Small_Positive, and Large_Positive.

The three rules listed above look like free form natural language and provide an intuitive approach for system definition. Although the inverted pendulum is easily modeled mathematically, and adequate non-fuzzy solutions exist, the ease of system definition with fuzzy logic is undeniable.

Our own language should be the natural choice when we come to problem definition and problem solution. Fuzzy logic allows the computer to take a step in our direction by allowing us to specify solutions in terms of the natural classifications and qualifiers that we understand.

Practical Motivations for Fuzzy Logic

Model-free Approach does not require Mathematical Equations

Certainly, the linguistic IF-THEN rules of fuzzy logic are uniformly easier to write than mathematical equations. However, being numerically model-free yields another important advantage in handling complex systems. The model-free nature of fuzzy systems means specifically that descriptions of the system of interest use the linguistic constructs of IF-THEN rules rather than mathematical equations.

The advantages of a model-free approach appear quickly as problem complexity increases. While it may be possible to write a complete mathematical description for the kinds of simplified systems that one might see in a physics text, few real world systems are that simple. Consider the following example of a pole and cart balancing task.

The pole and cart balancing problem receives much coverage in adaptive control texts. It is inherently unstable, and is considered easy to model. The apparatus, shown in Figure 1.4, consists of a cart that freely moves along one axis, and a pole mounted to the moving cart. The pole freely rotates in one plane from a pivot attached to the cart. The control task is twofold. First, the pole must be balanced atop the platform. Second, the $x(t)$ displacement must be held as close to zero as possible. The only means of balancing the pole is to drive the cart to the left or right. This situation is a simplified variant of balancing a pen on the tip of your finger.

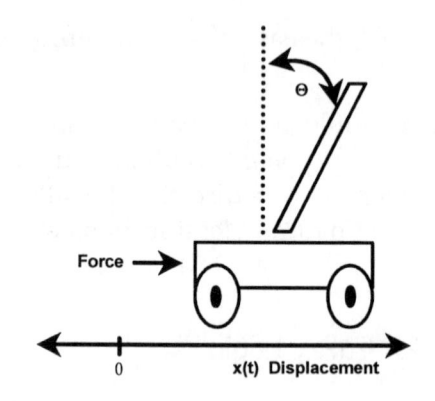

Figure 1.4: The pole and cart apparatus.

What do the mathematics of the system look like? The equations of motion for this system include the angular acceleration of the pole $\alpha(t)$ at time t:

$$\alpha(t) = \frac{g(m_c + m_p)\sin\theta - \cos\theta\left[F(t) + \lambda m_p \omega^2 \sin\theta\right]}{\frac{4}{3}\lambda(m_c + m_p) - \lambda m_p \cos^2\theta}$$

and the acceleration of the cart $a(t)$ at time t:

$$a(t) = \frac{F(t) + m_p \lambda\left[\omega^2 \sin\theta - \alpha(t)\cos\theta\right]}{m_c + m_p}$$

where the variables used are defined as:

$\theta(t) \equiv$ pole angle with respect to vertical at time t
$\omega(t) \equiv \dot{\theta}(t) \equiv$ angular speed
$x(t) \equiv$ cart position
$v(t) \equiv \dot{x}(t) \equiv$ cart speed
$m_c \equiv$ mass of cart
$m_p \equiv$ mass of pole
$\lambda \equiv$ length of pole
$g \equiv$ gravitational constant (9.82 $meters/sec^2$)
$F(t) \equiv$ force applied by control system

The important point here is that this idealized physical system is considered to be *simple*. Adding real world terms such as friction, non-linear forcing functions etc. complicate the equations of this *simple* system beyond practical usefulness.

Most useful systems have far more complicated mathematics than the example above. When complex systems are required to operate across a wide range of parameter values, and real world constraints are factored into the mix, exact mathematical models may be impossible. Model-free solutions, like linguistically defined systems based on fuzzy logic, do not require any mathematical models.

No math means that fuzzy logic does not hit the brick wall of unobtainable mathematics when a problem becomes complex. This in turn means that fuzzy logic is capable of extending the reach of control systems to address systems that were formerly beyond the scope of the classical math-centric methods of control and modeling.

Inherent Capability to Handle and Exploit Non-Linearities

In order to understand the usefulness of the non-linear capabilities of fuzzy logic, we first need an understanding of the difference between linear and non-linear systems. Unfortunately, a rigorous treatment of these differences hinges upon detailed mathematics well beyond the scope of this introduction. We must, instead, aim for a qualitative appreciation of the differences.

The study and control of all kinds of systems is called *SYSTEM THEORY*. The real world application of system theory dominates the efforts of most engineering disciplines. In addition to the more obvious examples like electronic and mechanical systems, system theory applies with equal validity to socio-economic, business, finance, thermal, fluidic, biological, and chemical systems. The unifying theme of systems theory that allows the various systems to be analyzed and modeled in similar ways is the abstraction depicted in Figure 1.5.

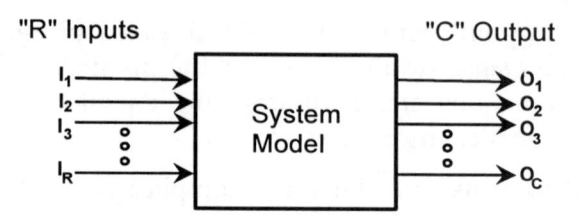

Figure 1.5: The Input-System Model-Output concept central to system theory.

The block diagram of Figure 1.5 points out that, regardless of the type of system, certain quantities serve as inputs, other quantities serve as outputs, and the inputs relate to the outputs via the system model. Such a representation then makes it possible to consider the effects of varying the system inputs or the system structure on the system outputs. Once a problem is cast into the form suggested by Figure 1.5, the underlying mathematical framework of system theory takes over. The entire spectrum of problems and systems can be treated in the same manner because the math remains the same. This powerful approach allows any system capable of being modeled like Figure 1.5 to be dealt with in a uniform manner dictated by a common set of mathematics.

Now we are ready for a working definition of non-linearity. Nonlinear terms involve algebraic or other more complicated functions of the system variables. For example, in a system with two variables x and y, expressions such as x^3 or $\cos(x)$ or $3xy$ would be nonlinear terms.

In the framework of system theory, our ability to model and control such systems depends only upon the mathematical techniques we use. Any inherent or self imposed limitations in the mathematics will affect our solution. One predominate self-imposed limitation is summarized by the following quote, which comes from a college level text on system theory:

"In this text ... the models considered are said to be linear, which eliminates many systems from consideration, and requires *rather crude approximations of the relationships among the quantities describing others."*[22]

Notice that the crux of the restriction is the linear nature of the models considered. That voluntary restriction eliminates an important area: non-linear systems. Why restrict oneself to a subset of problems? The authors provide the usual, but not altogether acceptable answer to that question by stating:

"... linear systems form a very important subset of systems because of the simplicity of their models, the ease with which solutions are obtained, and the generality of the solutions."[23]

Linear systems have been intensively studied precisely because the math is easier. For the same reason, tools and techniques for the analysis and design of linear systems predominate. It is fair to say that most engineers have nothing but a glancing acquaintance with non-linear techniques. Typically, when a system is known to be non-linear, messy linearization techniques are employed which yield an approximation to the original system. The resulting approximation can then be manipulated with linear techniques. The lesson here is that when the only tool you have is a hammer (linear systems methods) everything looks like a nail (linear system).

Non-linear effects in systems increase in importance as the complexity of those systems increases. And as the magnitude of non-linear effects increases, linearizations become increasingly inappropriate. Non-linearity is neither desirable nor undesirable: it exists everywhere in nature and must be dealt with.

One way to handle nonlinear systems is with nonlinear techniques like fuzzy logic based systems. Fuzzy logic is well suited to this task. Fuzzy systems have been proven general enough to perform any nonlinear control actions and to model any non-linear system.[24]

---

[22]G.V. Lago and L.M. Benningfield, *Circuit and System Theory*, 1979, New York, John Wiley & Sons. pg.3.

[23]Ibid., pg. 3.

[24]Li-Xin Wang, *Adaptive Fuzzy Systems and Control*, 1994, Prentice-Hall. pp. 210-220.

Another proven theorem states that any continuous nonlinear function can be approximated as exactly as needed with a finite set of fuzzy variables, values, and rules.[25] While it may be true that working with nonlinear systems can be tricky, there exist well-founded techniques that simplify such efforts.

Speed of Development and Ease of Implementation

Fuzzy logic is easy to understand because it emulates human decision making. Its lack of explicit mathematical structure makes it more accessible to non-math experts. Given equivalent performance across several approaches, an engineer will choose the approach that yields easier development and maintenance. Fuzzy logic provides both due to its reliance on linguistic variables and intuitive IF/THEN rules.

In addition to low development and maintenance costs, low implementation cost is one reason for the success of commercial fuzzy applications. Fuzzy logic lends itself to efficient implementation on even the smallest general purpose microcontrollers. Thus it can be added to most consumer and industrial electronics with zero to little additional equipment cost.

Fuzzy logic must be regarded as a powerful tool added to the tool belt of researchers, scientists, and especially engineers. Fuzzy logic travels in the opposite direction of traditional AI, which has a large following of academics but a minuscule following of engineers. Fuzzy logic currently enjoys a ground swell of interest among the active practitioners of the engineering community. Due to its ease of understanding, ease of maintenance, and low implementation cost, fuzzy logic augments the existing state of the art in dramatic new ways.

Fuzzy Logic is Already Here

Fuzzy logic may seem like the "new kid on the block" of the artificial intelligence world, but it has already produced an entire generation of successful consumer products. Here is just a sampling of some of the successful products on the market. You might be surprised at how

---

[25] Bart Kosko, *Neural Networks and Fuzzy Systems*, 1992, Prentice-Hall.

many of them you already have in your home and office. Some of them even have the words "fuzzy logic" imprinted on them.

PRODUCT	COMPANY	FUZZY LOGIC ROLE
Anti-lock brakes	Nissan	Controls brakes in hazardous cases based on car speed, wheel speed and acceleration.
Copy Machine	Canon	Adjusts drum voltage based on picture density, temperature and humidity.
Dishwasher	Matsushita	Adjusts cleaning cycle and rinse-and-wash strategies based on the number of dishes, and type and amount of food encrusted on the dishes.
Health Management	Omron	Tracks and evaluates employees' health and fitness.
Palmtop Computer	Sony	Recognizes handwritten Kanji characters.
Kiln Control	Mitsubishi Chemical	Mixes cement.
Golf Diagnostic System	Maruman Golf	Selects golf club based on golfer's physique and swing.
Stock Trading	Yamaichi	Manages portfolio of Japanese stocks based on macroeconomic and microeconomic data.
Sendai Subway System	Japan	Controls acceleration, deceleration, and ride-smoothness.
Refrigerator/Freezer	Sharp, Whirlpool	Sets defrosting and cooling times based on usage. A neural network learns usage habits and tunes fuzzy rules accordingly.

Table 1.3: Early examples of fuzzy intelligent products[26].

Notice the veritable honor role of Japanese companies listed in Table 1.3. Astute observers debate the reasons why Japanese corporations pursued practical uses of fuzzy logic much earlier than their counterparts in other countries. Most arguments center on the philosophical alignment between fuzzy logic and the traditional religious/

---

[26] Bart Kosko, *Fuzzy Thinking* (1993) pp.184-187.

metaphysical beliefs held by many Asian cultures[27]. Rather than viewing the world as an absolute, black and white environment, so the argument goes, they embrace a view which celebrates the shades of gray which are inherent in nature and pervade intellectual thought. This may be true. However, a technology does not flourish based only on mystical beliefs. The Japanese recognized significant potential in the application of fuzzy logic. In addition to possibly coincidental philosophical leanings, the huge Japanese market share of worldwide electronic consumer products brought preparation and opportunity together. When the powerful new technology of fuzzy logic met Japanese marketing muscle it produced the first generation of intelligent electronic goods.

In the United States, significant work is under way to incorporate the intelligence of fuzzy logic into products and systems. Table 1.4 lists a few of the most recent developments offered by American engineers and scientists.

PRODUCT	COMPANY	FUZZY LOGIC ROLE
Uninterruptible Power Supply	Controlled Power	Widens the range of power fluctuation before switching to battery power
Single Target Tracker	Sensor Data Integration	Independently tracks azimuth and elevation movements using error, change in error, and previous outputs.
Health Care Fraud Detection	Metus Systems	Screens insurance claims for fraudulent charges and practices.
Laser Beam Alignment	NASA Lewis Research Center	Aligns optical elements in three axes by mimicking trained human operators.
Automatic Automobile Transmission	Saturn Division of General Motors	Down-shifts transmission like expert driver using throttle position, grade, speed, and brake application time.
Computer Disk Drive	Seagate Technology	Spindle motor controlled by fuzzy PI model for minimum spin-up time .

Table 1.4: Recent U.S. additions to the burgeoning pool of fuzzy intelligent products.

---

[27]Daniel McNeill et al., *Fuzzy Logic*, pp. 128 -136.

Fuzzy Logic is Here to Stay

Fuzzy logic is not, as some detractors have asserted, a tool used to justify sloppy science or lazy engineering. It is philosophically aligned with the way that we think and reason, so it is well-suited to the task of translating our classifications and approximate reasoning capability into enhanced machine intelligence. Fuzzy logic is a tool that allows us to trade in an amount of our previously cherished precision to gain a deeper level of machine reasoning. If your objective is to compute the first 2000 digits of PI, then fuzzy logic will disappoint you. However, if your goal is faster development of more robust, smarter computer systems then you best not overlook fuzzy logic because your competition certainly will not.

In the Following Chapters.........

This book was written to specifically address the needs of the individual who, possessing little to intermediate knowledge of fuzzy logic, wishes to learn about and apply fuzzy logic to problem solving. We endeavored to collect under one cover the information necessary to take fuzzy logic from initial concepts to final implementation. We avoid the extensive mathematical treatments found in less readable texts in favor of a more applications-oriented understanding of the concepts and constructs.

Accordingly, the material in this book is divided into roughly three areas. In order of appearance, we cover the basic concepts, design methodologies, and finally, a variety of implementations. Chapters Two and Three present the basic framework of fuzzy logic concepts. Chapter Four introduces a structured methodology to the design of fuzzy models. Chapter Five illustrates that methodology with a variety of easily followed examples. Chapters Six, Seven, Eight, and Nine complete the picture by presenting design choices and several specific implementations.

This book also contains at least two elements that you won't find anywhere else. Chapter Ten, provided for the immediate use of practitioners, is a sourcebook listing information about commercially available development tools and a study guide listing complementary educational materials. The companion disk contains all of the ASSEMBLER, C, and C++ source code that appears in the book. And finally, the companion disk contains **FuzzyLab**, an interactive educational software system developed for the PC Windows environment. **FuzzyLab** features comprehensive on-line help, and allows the extensive interactivity required to truly teach the fundamentals of fuzzy control.

Chapter 2:
Terminology

This chapter will introduce and illustrate the fuzzy logic terminology used in this book. A variety of fuzzy logic terminology is in common use today. That variety can lead to confusion for the beginner who strives first to understand the concepts. Here we aim for clarity and consistency. Since we have the opportunity to choose from equivalent terminology, we present those terms whose names best convey the essence of the item. Where appropriate, commonly accepted synonyms of the primary terms will follow in the text describing those terms. The reader can refer to the Glossary if any questions remain.

Membership Functions and Fuzzy Sets

Fuzzy logic is first and foremost a logic of classes that have unsharp boundaries. In order to set the stage for further exploration, a qualitative understanding of the terms *class* and *unsharp boundary* is necessary.

Experiment with the Mind

The ability to classify is a unique human trait. We manage the flood of data that our senses pick up from the world by summarizing and classifying. Being able to describe an item with one word, like fruit, vegetable, art, crime, etc., that more or less summarizes its properties is invaluable to human thought. The process of classification amounts to grouping items according to their properties and characteristics. For instance, items in the class Fruit share certain properties like sweetness, edibility, fleshiness, etc.

The following experiment will take you just a few minutes to complete, but illustrates your understanding of classes and their fuzziness with great impact. The task is to rate the words in each list, on a scale of one to seven, as to how well each word typifies its category. For instance, if the category is Fruit, rate Orange a "one" if you think it is a great example of Fruit, or rate it a "seven" if you think it is a poor example.

Fruit	**Vegetable**	**Art**
Orange	Broccoli	Ballet
Avocado	Cucumber	Doodles
Kiwi	Avocado	Sculpture
Raspberry	Onion	Graffiti
Banana	Tomato	Impressionist Painting
Tomato	Potato	House Painting

Eleanor Rosch, a UC Berkeley psychologist, pioneered this type of experiment in 1973. Her aim was to prove the fuzziness of natural classes like Fruit and Vegetable. She took a group of over one hundred college students and presented them with six words in each of eight classes: Fruit, Science, Sport, Bird, Vehicle, Crime, Disease, and Vegetable. The students had the same task of rating each word in its class, on a scale from one to seven, as to how well that word typifies its class.[28]

Rosch noted several interesting outcomes of her experiments. Students immediately understood the directions for the experiment, and averaged only three seconds per choice. Furthermore, the ratings agreed very closely from student to student. From these three factors, Rosch concluded that the experiment assessed every day mental processes. In other words, every day mental processes involve judging

---

[28]Eleanor Rosch, "On the Internal Structure of Perceptual and Semantic Categories," in Timothy Moore, ed., *Cognitive Development and the Acquisition of Language*, pp. 111-114.

a degree of fit between items and their classifications. In fuzzy logic, this degree of fit is called a *DEGREE OF MEMBERSHIP*. The degree of membership is a measure of how well the current item fits into a classification.

As you try this experiment on your colleagues, observe how your own individual perceptions overlap theirs. Just as Rosch found, you'll probably find that the strongest agreement among your colleagues comes from the best examples of each class. The weaker ratings indicate borderline members with greater fuzziness and hence more uncertainty in classification. As you repeat this experiment, keep in mind that human classifications are not limited to just Fruit and Vegetable. The results of this experiment apply to all areas involving human judgment, reasoning, and classification, including science, medicine, engineering, and many other disiplines.

The Human Link

Human-derived classifications tend to be general rather than precise. Using only a few different terms to describe the entirety of foodstuffs leaves certain ambiguities. Fuzziness exists inescapably at the boundaries of our classifications. Consider the classification of a tomato, which you and your experimental subjects rated in the preceding paragraphs. Is it fruit or vegetable? On one hand, high vitamin C content, and juicy flesh suggest that tomatoes are fruit. On the other hand, tomatoes lack sweetness, and you rarely see anyone pull a ripe tomato from their lunch box and eat it like an apple. So, maybe it is a vegetable. Is this uncertainty a problem?

Just because we can't stuff tomatoes into one single class doesn't mean that our classifications have failed. The reason that we can't use a single class to describe tomatoes is that the boundary between fruit and vegetable is unsharp. A food item can simultaneously have fruit and vegetable properties. The difficulty in finding a single class for tomatoes derives from the fact that Fruit is a fuzzy set and Vegetable is also a fuzzy set. As a consequence of unsharp boundaries between classes, an item, like the tomato, can be a member of more than one class to various degrees. Depending on whom you ask, the tomato is both a fruit and a vegetable to some degree. An orange is a member of the fruit class to a higher degree than a tomato. Broccoli is a member of the vegetable class to a higher degree than a tomato. Oranges and broccoli happen to fall farther away from the fuzzy boundaries of their respective classes than tomatoes do, so they more clearly fall into a single class.

Fuzzy logic is uniquely equipped to handle human classifications and their inherently fuzzy boundaries. A simple construct, called the membership function, is the keystone that supports the rest of the fuzzy logic apparatus. We now look at the features and interpretation of membership functions.

The Meaning and Elements of a Membership Function

It is an easy matter to look at someone and declare them *Tall*. To make that declaration in no way suggests that we have measured their height, rather it reflects our innate ability to classify. Based on our own experience, we easily classify people as *Short, Average, and Tall*. Our classifications enable a very compact description of tallness. With only a word or two others quickly understand the meaning and extent of our descriptions of tallness. We do not need a tape measure to tell us that the person we are observing is *Tall*. In fact, using a tape measure to determine if a person is *Tall* confuses the issue. Certainly, the measurements can be made accurately, and tell us precise information about the subject. However, precise information is not necessarily meaningful information. If those measurements lack a fundamental tie to our notion of tallness, very little meaning can be gleaned from the measurements alone. The intention of the fuzzy construct called the MEMBERSHIP FUNCTION, or equivalently called a FUZZY SET, is to provide that fundamental tie between precise measurements and general classifications.

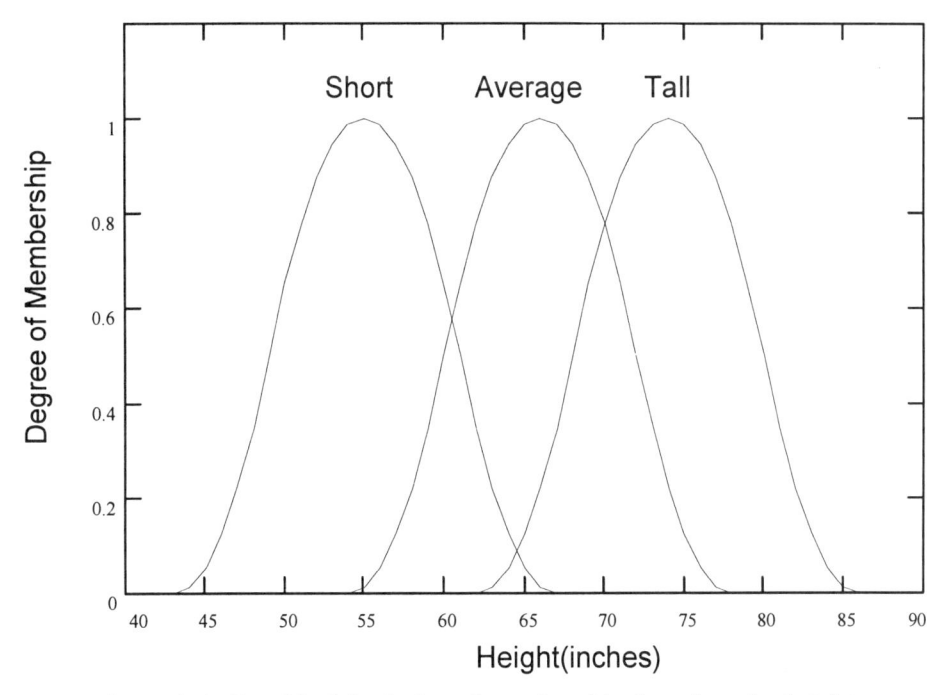

Figure 2.1: Graphical depiction of membership functions for height.

Viewed graphically, as in Figure 2.1, a membership function is two-dimensional. The horizontal axis contains the range of precise measurements that might be used to define height. The vertical axis indicates the *DEGREE OF MEMBERSHIP*, and is called the *TRUTH VALUE* axis. The curves in Figure 2.1 map precise height measurements to degrees of membership in the fuzzy sets *Short*, *Average*, and *Tall*. Thus the shape of the membership functions assign, for the fuzzy sets *Short*, *Average*, and *Tall*, a degree of membership to each possible precise measurement of height. Put another way, the degree of membership measures the compatibility between a crisp value like 5'9", and a classification like *Average*.

At this point, we'll look at a bit of mathematical notation that will function as an effective shorthand when dealing with membership functions and degrees of membership. The first and most important piece of notation is the mathematical expression for degree of membership. Generically, the notation $\mu_A(x)$ expresses the degree to which the crisp value x is a member of the fuzzy set A.

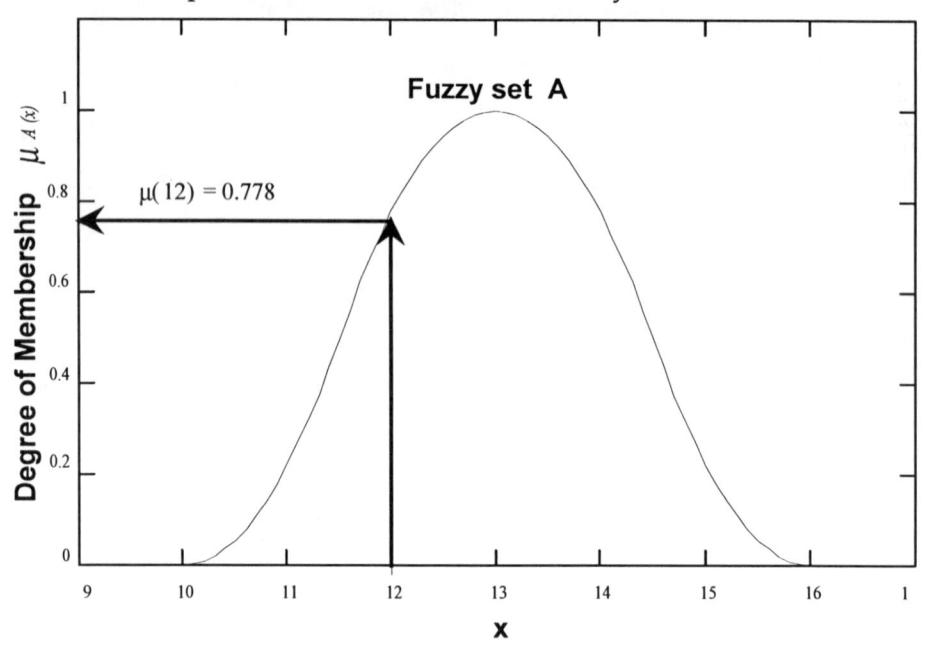

Figure 2.2: Graphical interpretation of membership degree in fuzzy set A.

Figure 2.2 shows the graphical interpretation of $\mu_A(x)$. Only two things are required to determine the degree of membership: a membership function, and a crisp input. A *CRISP INPUT* is simply a non-fuzzy quantity, which can be represented as a discrete, real number. In the figure, the crisp input is 12.0, and $\mu_A(12) = 0.778$. As a practical illustration $\mu_{Short}(x)$ is the degree of membership of the variable x in the fuzzy set *Short*. In this instance, the variable x takes on values (feet, meters, etc.) consistent with expressions of height. In practice, x is usually a continuous variable like temperature, pressure, humidity, etc. However, the methods of fuzzy logic remain valid for discrete variables as well.

Chapter 2: Terminology

For the purposes of this section, graphical determinations of degrees of membership suffice to illustrate the concept. However, as we will see in following chapters, any computer implementation of fuzzy logic must be mathematically oriented. Fuzzy logic scores points on this account, as the calculations necessary to render a membership function and associated degrees of membership are straightforward and computationally simple.

Although the rendering of degrees of membership from a range of input values is its prime characteristic, the membership function has a number of other important properties. The range of heights for which the fuzzy set *Average* has non-zero degree of membership is called the *SCOPE*, or the *MEMBERSHIP FUNCTION DOMAIN*. In Figure 2.3, the scope of *Short* is the range 43 to 67 inches. In the same figure, the scope of *Tall* is the range 62 to 87 inches.

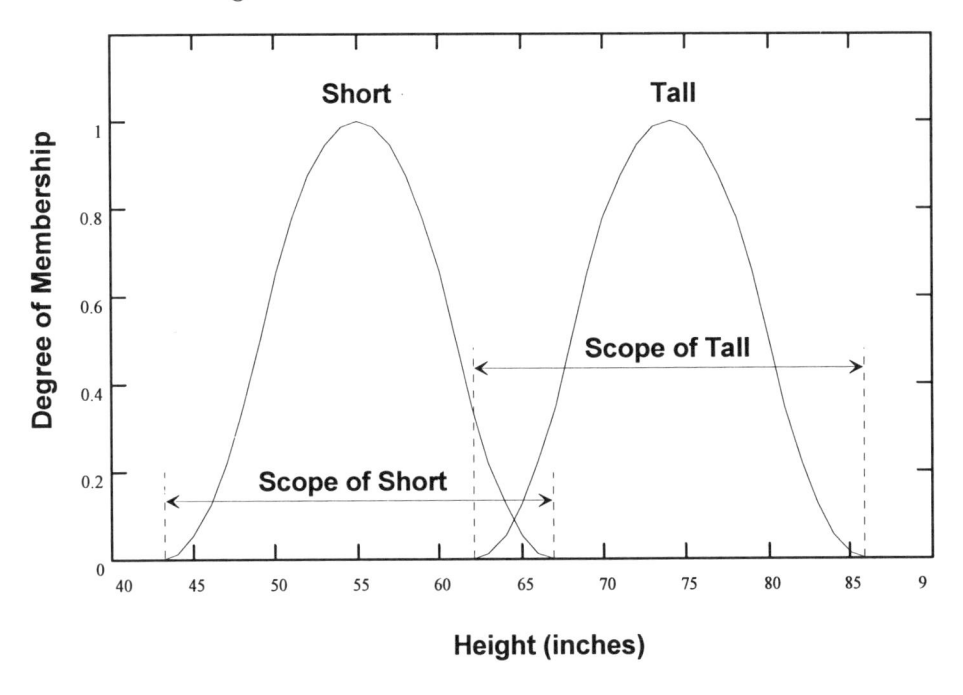

Figure 2.3: Scope of membership functions Short *and* Tall.

35

The membership function name is commonly referred to as a LINGUISTIC LABEL. Linguistic labels, like *Short* and *Tall*, reflect the common sense, intuitive labeling of fuzzy sets that is the hallmark of fuzzy logic. Since fuzzy logic was conceived of as a tool to capture human reasoning, correctly naming membership functions requires the natural language that we use to describe humanistic concepts. As in the case of the labels *Short* and *Tall*, many linguistic labels are domain specific. In other words, it makes sense to use the labels *Short, Average*, and *Tall* to describe height. Using the same labels to describe temperature would be confusing and inappropriate. It makes far more sense to use the linguistic labels *Cool, Warm*, and *Hot* to describe temperature.

Even when domain-specific labels like *Cool* or *Short* don't exist, more generic linguistic labels convey significant meaning. For instance, the labels *Small, Medium*, and *Large* apply to a vast number of possible variables. When positive and negative values must be expressed, labels like *Positive Small* or *Negative Large* easily suffice.

Fuzzy Variables

As a generic term, FUZZY VARIABLE describes a variable that is used in a fuzzy system. A fuzzy variable will be defined by a set of membership functions. In the context of a FUZZY SYSTEM, a fuzzy variable will be either an input to the system or an output from the system. Whether used as input or output variables, fuzzy variables share a variety of properties that we will now examine.

Typically, overlapping membership functions define a fuzzy variable across its UNIVERSE OF DISCOURSE. The universe of discourse for a fuzzy variable is its range of interest. At one end of the universe of discourse is the minimum value of interest. At the other end of the universe of discourse is the maximum value of interest. As shown in Figure 2.4, a minimum of 40 degrees and a maximum of 120 degrees define the universe of discourse for the fuzzy variable "temperature" used in a residential furnace controller.

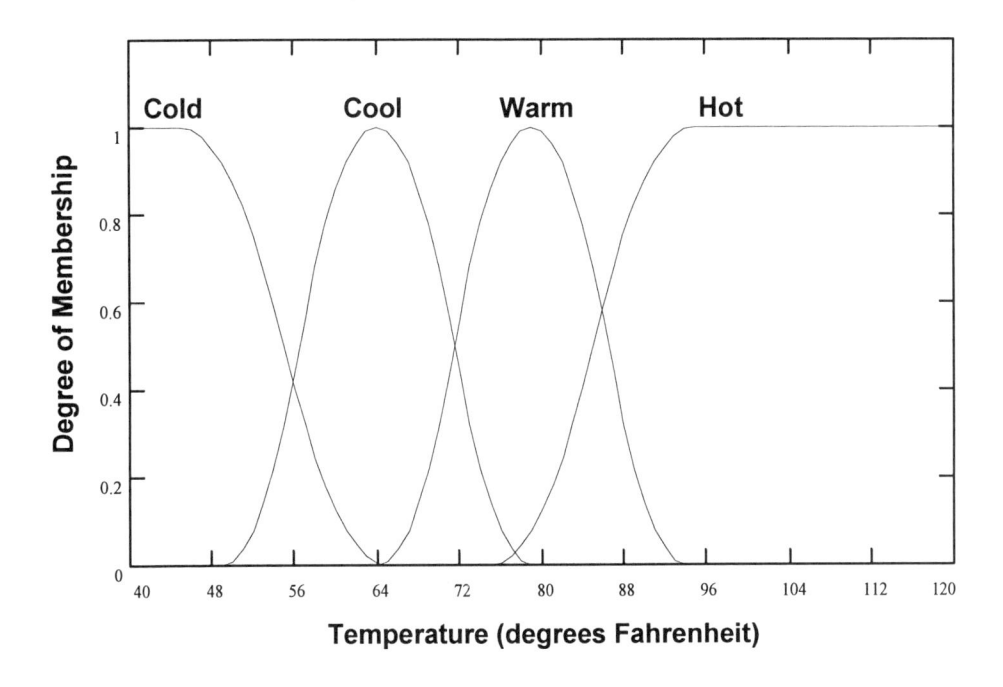

Figure 2.4: Universe of discourse for Temperature corresponding to indoor temperature.

The universe of discourse for a fuzzy variable is problem dependent. While Figure 2.4 defines temperature in the context of an indoor temperature, Figure 2.5 defines temperature in the context of a cement kiln[29]. Clearly, changing the problem changes the context for the fuzzy variable, hence altering its universe of discourse.

---

[29]Generally regarded as the first commercial application of fuzzy logic, the Smidth & Co. cement kiln in Denmark was successfully operated by a fuzzy controller in June of 1978.

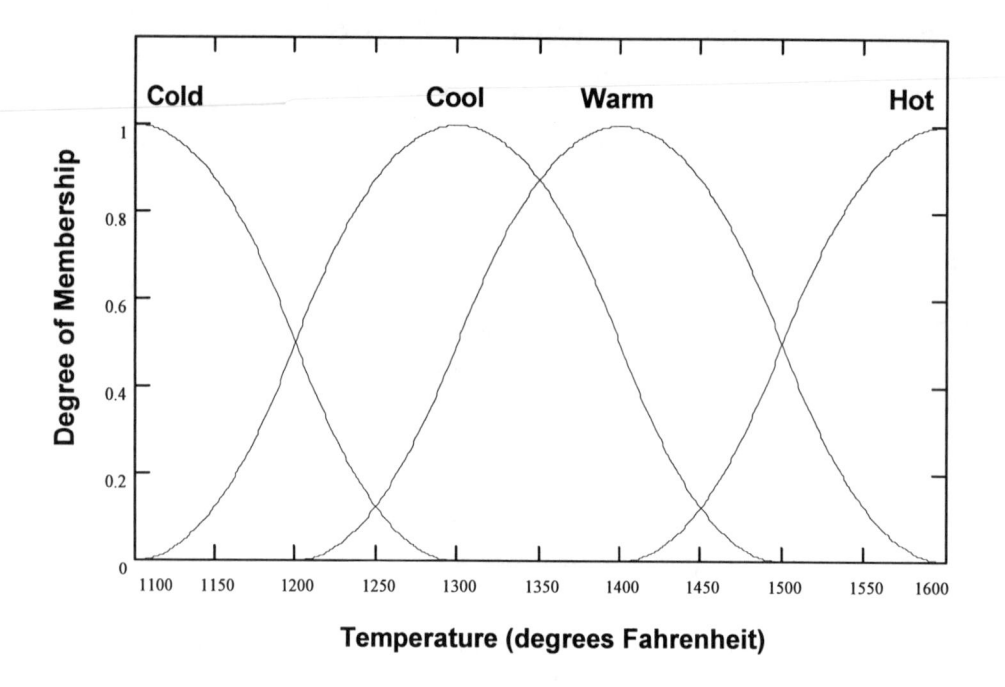

Figure 2.5: Temperature membership functions for cement kiln operation.

Fuzzy variables have also been called LINGUISTIC VARIABLES. As L.A. Zadeh first defined them, linguistic variables are variables whose values are the linguistic labels of fuzzy sets.[30] For instance, Age is a linguistic variable if its values are young, middle aged, old, etc. Linguistic variables are the stuff of human reasoning. Can you spot the linguistic variables in the following statements?

From an investment text: "Liquidity and safety usually go hand in hand, but for a higher return you generally have to sacrifice some liquidity or safety, or both."[31] (Answer: liquidity and safety).

---

[30]Lotfi A. Zadeh, "Outline of a New Approach to the Analysis of Complex Systems and Decision Processes," *IEEE Transactions on Systems, Man, and Cybernetics*, SMC-3 (1973), pg. 28.
[31]Janet Bamford et al., *Complete Guide to Managing Your Money*, p. 314.

From an electronics engineering text: "In selecting an A/D converter, choices must be made that trade off speed, resolution, accuracy, and noise immunity."[32] (Answer: speed, resolution, accuracy, noise immunity).

And, from a macroeconomics text: "It is harder for a family to reduce its expenditures from a high level than for a family to refrain from making high expenditures in the first place."[33] (Answer: high expenditures and expense level).

From a computer programming text: "If there is a single dominant theme in this book, it is that practical methods of numerical computation can be simultaneously efficient, clever, and clear."[34] (Answer: efficiency, cleverness, clarity).

The statements above use fuzzy variables of two different types. The first type, and the easiest to understand, are the fuzzy variables like speed, resolution, or high expenditures, that can usually be measured accurately. Imposing membership functions on the first type of fuzzy variable amounts to dividing the easily specified universe of discourse into overlapping membership functions. The second type of fuzzy variable, of which cleverness and clarity are examples, cannot be measured accurately. Although we understand the distinction between, *very clever*, *somewhat clever*, and *not very clever*, a much finer grained distinction is not possible. Fuzzy variables of this second type can be considered as pure linguistic variables because their membership functions can only be described as linguistic labels like *somewhat clever*, *not very clever*, etc.

Each of the statements above expresses some human expertise or human knowledge that is not very computer compatible in the typical Boolean sense. And yet, to make machines smarter we must find a way to encode human expertise in a computer compatible way. The fuzzy logic framework enables the representation and manipulation of

---

[32]Richard J. Higgins, *Electronics with Digital and Analog Intergrated Circuits*, p. 273.
[33]J.C.Poindexter, *Macroeconomics*, p. 140.
[34]William H. Press et. al., *Numerical Recipies in C* , p. xii.

human expertise through the use of linguistic, fuzzy variables and their associated membership functions.

Logical Operations on Membership Functions

Fuzzy logic, just as any logic, needs a set of well defined operators to manipulate complex statements and arrive at conclusions. Fuzzy logical operators operate on the fuzzy sets of a linguistic variable in a way that is analogous to Boolean operations on Boolean variables. The first set of operators, NOT, AND, and OR, will look familiar and are similar to the Boolean operators of the same names.

NOT (Complement)

NOT is the complement operator. Figure 2.6 shows the graphical result of complementing the membership function *Cool*. The membership function NOT *Cool* is described by the equation $\mu_{NOTCool}(x) \equiv 1 - \mu_{Cool}(x)$. Complementation of fuzzy sets also has the property that $\mu_{NOTNOTCool}(x) = 1 - \mu_{NOTCool}(x) = \mu_{Cool}(x)$.

Using the NOT operator in fuzzy logic brings about a very interesting result which is illustrated by the shaded area in Figure 2.7. The shaded area is the region where $\mu_{Cool}(x) \neq 0$ and $\mu_{NOTCool}(x) \neq 0$.

Thus, it can be *Cool* and NOT *Cool*, to varying degrees, simultaneously. In a Boolean system no temperature can be both *Cool* and NOT *Cool* at the same time. This property of Boolean systems is called the Law of Non-Contradiction.[35] Here we mark the first glaring departure that fuzzy logic makes from the established road of Boolean logic, but it will not be the last.

---

[35] Aristotle's search for the fundamental principles of logic produced two axioms. The Law of Non-Contradiction states that an item cannot be both A and NOT-A. The Law of the Excluded Middle states that an item must be either A or NOT-A.

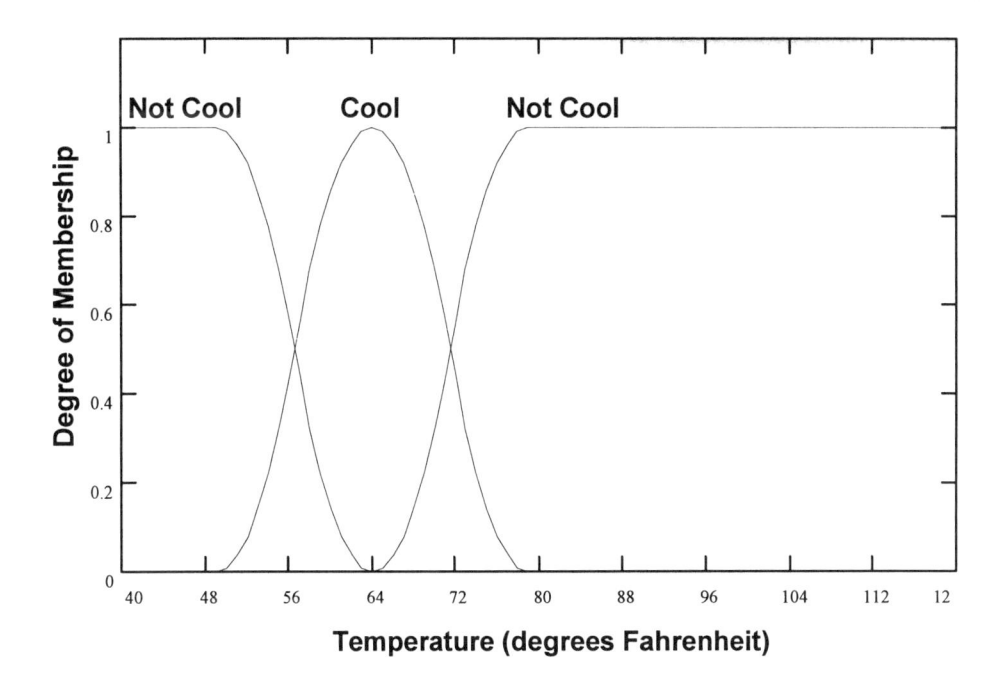

Figure 2.6: Effect of the NOT operator on fuzzy set Cool.

AND (Intersection)

AND is the intersection operator. The AND operator requires at least two arguments. With fuzzy logic, those arguments are membership functions. In this book, the AND operator will be used in two different contexts. One usage will be the ANDing of fuzzy expressions of two different fuzzy variables, like temperature and humidity. We will cover that particular usage in the sections concerning fuzzy propositions, fuzzy rules, and fuzzy rule evaluation. For now, the AND operator is explained in the context of two membership functions of the same fuzzy variable. In that context Figure 2.8 depicts two temperature membership functions *Warm* and *Cool*, and the fuzzy set {*Warm* AND *Cool*} is shaded where the two overlap.

Fuzzy Logic for Real World Design

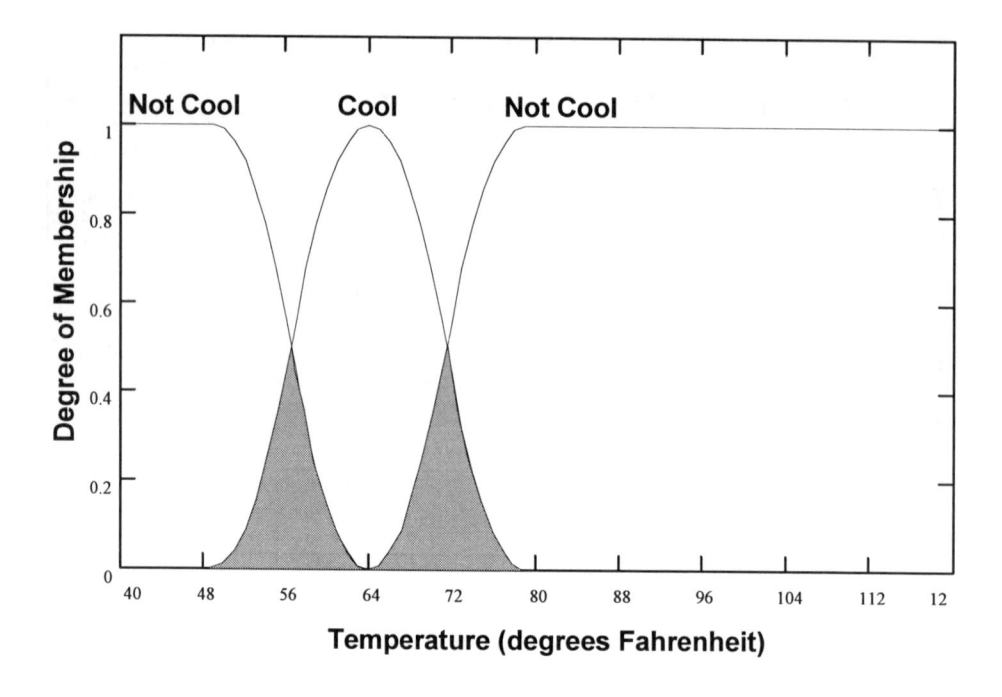

Figure 2.7: Overlap between Cool *AND NOT* Cool.

Using the intersection operator ∩, the mathematical notation for the intersection of the Warm and Cool fuzzy sets is: $\mu_{Warm \cap Cool}(x) \equiv Min[\mu_{Warm}(x), \mu_{Cool}(x)]$. This equation describes the intersection of *Warm* and *Cool* as the pointwise minimum of the two fuzzy sets. The shaded region of Figure 2.8 shows this graphically. As a shorthand to the previous notation, we will use the ∧ symbol to denote the pointwise minimum between two membership functions. So rewriting the previous equation gives: $\mu_{Warm \cap Cool}(x) \equiv \mu_{Warm}(x) \wedge \mu_{Cool}(x)$. Notice that in this context the result of ANDing two fuzzy sets yields another fuzzy set.

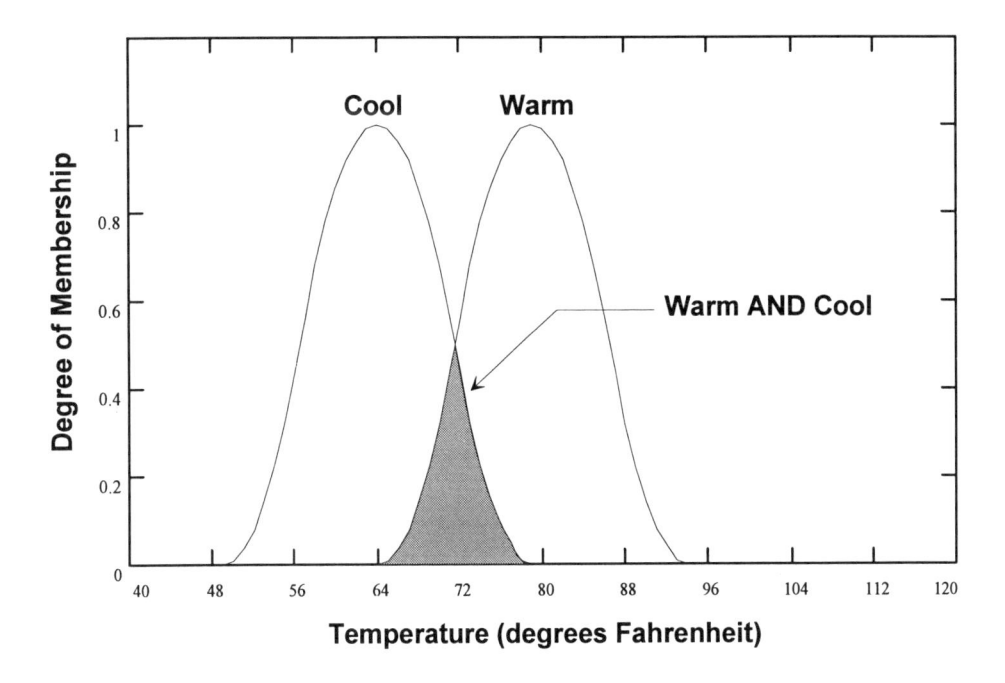

Figure 2.8: Fuzzy set produced by Warm *AND* Cool.

OR (Union)

OR is the union operator. The OR operator also requires at least two arguments. Like the AND operator, the OR operator is used in two different contexts. One usage, which deals with fuzzy rules, will be covered in the sections concerning fuzzy rules and fuzzy rule evaluation. The other context, which we cover here, concerns the ORing of two membership functions of the same fuzzy variable.

Fuzzy Logic for Real World Design

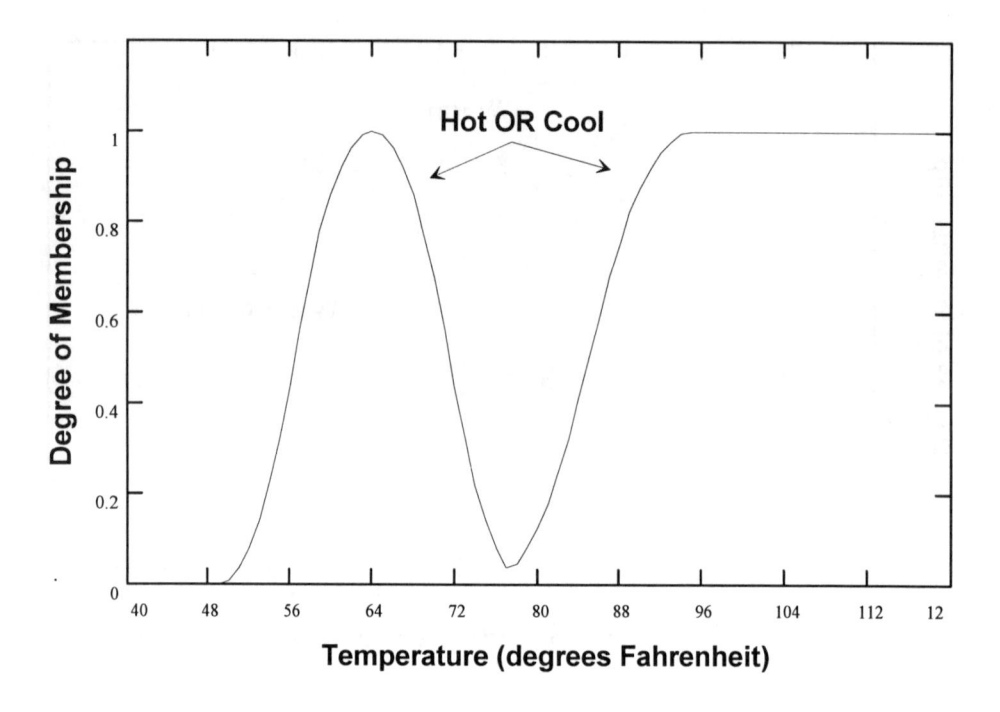

Figure 2.9: Fuzzy set produced by Hot OR Cool.

Using the union operator ∪, the mathematical notation for the union of the *Hot* and *Cool* fuzzy sets depicted in Figure 2.9 is: $\mu_{Hot \cup Cool}(x) \equiv Max[\mu_{Hot}(x), \mu_{Cool}(x)]$. The shaded region of Figure 2.9 shows the union of *Hot* and *Cool* graphically. As a shorthand we will use the ∨ symbol to denote the pointwise maximum between two membership functions. Rewriting the previous equation gives: $\mu_{Hot \cup Cool}(x) \equiv \mu_{Hot}(x) \vee \mu_{Cool}(x)$. Notice that the OR operator results in another fuzzy set.

NORM (Normalization)

Although the NORM operator does not share the same logical context as the NOT, AND, and OR operators, this is an appropriate place to define its action. The need for a normalization operator derives from the continuous truth values possible in fuzzy logic. If the maximum

Chapter 2: Terminology

degree of membership of a fuzzy set is less than one, NORM will rescale that fuzzy set so that the highest element has a membership grade equal to one. The highest membership grade for fuzzy set A will be denoted $\overline{\mu}_A(x)$. Figure 2.10 shows the effect of the normalization operation: $NORM(\mu_A(x)) = \mu_A(x) / \overline{\mu}_A(x)$.

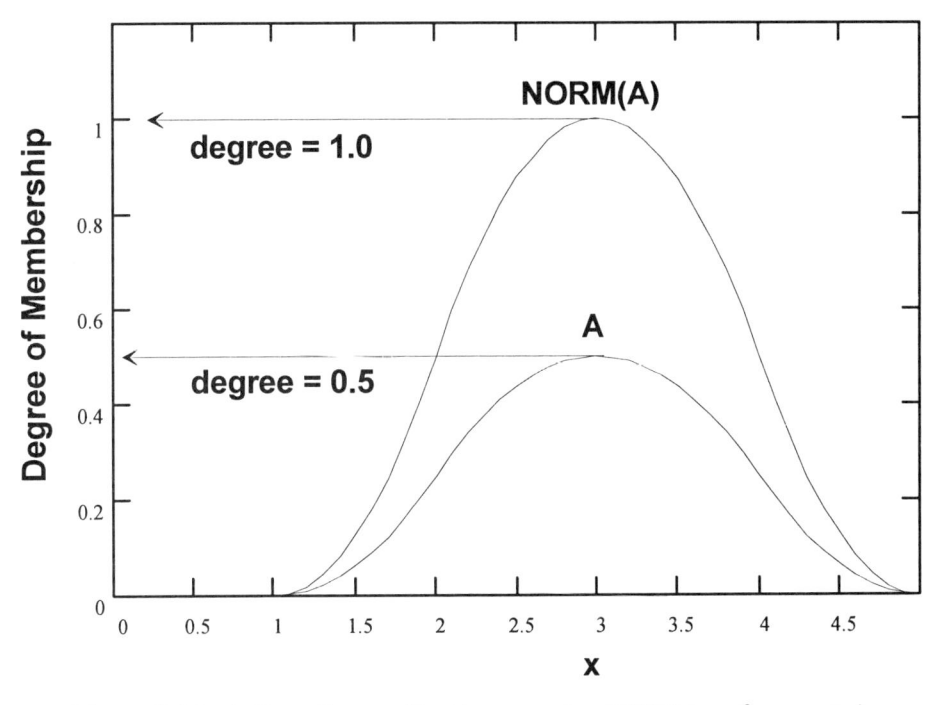

Figure 2.10: Action of normalization operator NORM on fuzzy set A.

Alpha-cuts

A membership function *ALPHA-CUT* establishes the minimum degree of membership that a crisp value must obtain before it is considered a member of that fuzzy set. If we establish an alpha-cut of α for fuzzy set A, any crisp value with degree of membership less than or equal to α is not considered a member of fuzzy set A.

$$\mu_A(x,\alpha) = \mu_A(x), \mu_A(x) \geq \alpha$$
$$= 0, \mu_A(x) < \alpha$$
[36].

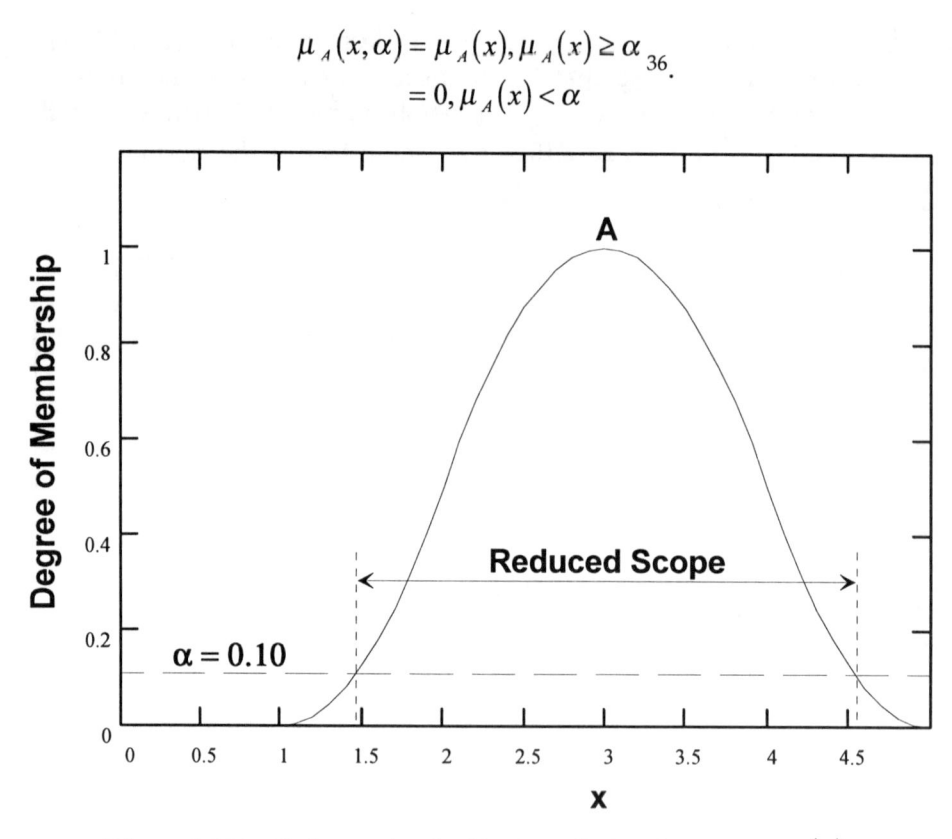

Figure 2.10b: Alpha-cut level of 0.10 applied to fuzzy set $\mu_A(x)$.

As shown in Figure 2.10b, an alpha-cut reduces the scope of the original fuzzy set. Since high level alpha-cuts cause a fuzzy system to tend toward a Boolean system of reduced scope, only low level alpha-cuts are used in practice.

Alpha-cuts are particularly useful in dealing with membership functions that have long 'tails'. Membership functions with long tails tend to yield low degrees of membership across a wide scope, which may not contribute significantly to the fuzzy system. In such a case,

---

[36] The equation shown is that of the strong alpha cut. The weak alpha cut replaces the >= with >.

alpha-cuts reduce the computational burden on the system without sacrificing much fidelity.

Alpha-cuts are also effective at dealing with noisy input data by eliminating low membership values which may be caused by random fluctuations. We will revisit the alpha-cut concept in the next chapter when we look at rule level alpha-cuts.

Fuzzy Analogs to Boolean Algebra

The NOT, AND, and OR operators allow us to manipulate and combine fuzzy sets in a familiar Boolean way. The Boolean familiarity does not end there, however. Fuzzy logic operators share other similarities with their limited Boolean counterparts. The reader is encouraged to prove the following properties by sketching membership functions formed by executing the operations specified in each statement.

Dominance

The four dominance properties are the most intuitive of the set of properties that we will consider in this section. The properties are:

$$\mu_{Cool}(x) \cup 1 = 1,$$
$$\mu_{Cool}(x) \cup 0 = \mu_{Cool}(x),$$
$$\mu_{Cool}(x) \cap 1 = \mu_{Cool}(x), \text{ and}$$
$$\mu_{Cool}(x) \cap 0 = 0.$$

In the preceding equalities, the symbol 1 denotes a membership function with $\mu_1(x) = 1, x \in X$. Also, the symbol 0 denotes a membership function with $\mu_0(x) = 0, x \in X$.

Associativity

The associativity property holds true for both AND and OR. Associativity means that $Hot \cap (Warm \cap Cool) = (Hot \cap Warm) \cap Cool$, and $Hot \cup (Warm \cup Cool) = (Hot \cup Warm) \cup Cool$. In words, the fuzzy set formed by the statement *Hot* AND *Warm* AND *Cool*, is the same regardless of which AND operation is executed first. Also, the fuzzy set formed by the statement *Hot* OR *Warm* OR *Cool* is the same regardless of which OR operation is performed first.

Commutativity

The commutative property states that $Hot \cap Cool = Cool \cap Hot$, and $Hot \cup Cool = Cool \cup Hot$. The fuzzy sets resulting from the AND and OR operations are formed, respectively, by a pointwise Minimum, and Maximum function (reference the prior section). Minimum and Maximum functions give the same result regardless of the order of their arguments, so the fuzzy AND and fuzzy OR are commutative.

Distributivity.

The distributive property also holds true for the fuzzy AND and OR. Distributivity means that
$Hot \cap (Warm \cup Cool) = (Hot \cap Warm) \cup (Hot \cap Cool)$, and
$Hot \cup (Warm \cap Cool) = (Hot \cup Warm) \cap (Hot \cup Cool)$.
Explained in words, the first equation says that the fuzzy set formed by the statement *Hot* AND (*Warm* OR *Cool*) is equivalent to the fuzzy set formed by the statement (*Hot* AND *Warm*) OR (*Hot* AND *Cool*). In words, the second equation says that *Hot* OR (*Warm* AND *Cool*) is equivalent to (*Hot* OR *Warm*) AND (*Hot* OR *Cool*).

De Morgan's Theorem

The familiar Boolean De Morgan laws also have valid fuzzy analogs. The first law states $(Hot \cup Warm)' = Hot' \cap Warm'$. The second law states $(Hot \cap Warm)' = Hot' \cup Warm'$.

The similarity of fuzzy logical operations to Boolean logical operations ends here. Thus far, fuzzy sets and the operators used to manipulate them have been described in a way that is meant to draw upon the reader's previous experience with Boolean systems. However, exploring the full power of fuzzy logic requires a departure from this approach. Fuzzy logic possesses a much richer set of properties and operations than Boolean metaphors can express. Henceforth we dispose of the Boolean comparisons and dive deeper in the exploration of fuzzy logic.

Semantic Operations on Membership Functions

As we saw in the previous sections, linguistic labels describe the values of linguistic variables. In one example, the linguistic labels *Cold, Cool, Warm,* and *Hot* were used to denote the values of the linguistic variable *Temperature*. Each linguistic label in turn was given a concrete meaning by defining its membership function over a certain range of temperatures. Once membership functions were given a well-defined mathematical form, the development of fuzzy logic as a problem solving tool began with an examination of the operators NOT, AND, and OR. While the NOT, AND, and OR operators are sufficient for Boolean representations, they barely scratch the surface of possibilities for fuzzy logic.

The linguistic nature of fuzzy variables and their membership functions suggests extensions to the mechanics of fuzzy logic which are rooted in our language itself. Specifically, fuzzy logic mathematically defines the use of *HEDGES* like *very* or *somewhat*, and *USUALITY* and *FREQUENCY MODIFIERS* like *usually* or *often*. Hedges and modifiers act upon base membership functions like *Cool* to produce different but related membership functions labeled *very Cool, usually Warm, etc.* Hedging and modifier operators are called semantic operators because they alter the meaning of a base fuzzy set to generate a new fuzzy set. Hedges, modifiers, and other semantic operators give fuzzy logic an unprecedented ability to accurately and easily translate human expertise into the computational realm of the computer.

Hedges

The linguistic hedge is the most frequently occurring semantic operator in everyday speech. The hedge is also the operator which most often finds a place in fuzzy systems designed for real world problem solving. Consider the following partial list of frequently occurring linguistic hedges:

- *very, extremely,*
- *nearly, almost, close to*
- *fairly, about, more or less, somewhat, around,*
- *moderately, rather, roughly*
- *slightly, hardly, etc.*

In our everyday language, using different linguistic hedges enables the verbal expression of very fine shades of meaning. In fuzzy logic, hedging operators permit the encoding and manipulation of the fine shades of meaning inherent to our language. Notice that a hedge does not stand on its own. A base fuzzy set, like *Tall*, must be acted upon by a hedging operator, like *very*, to produce a different but related fuzzy set *very Tall*. As we will see, fuzzy hedges behave in a well-defined way that is mathematically straightforward.

Very

Very is one of a set of CONCENTRATION operators. A CONCENTRATION operator reduces the degree of membership of elements of a fuzzy set that have a relatively low degree of membership. In addition, concentration operators affect elements with high degrees of membership minimally. Applying the concentration operator produces a subset of the original fuzzy set.

The word "very" is defined as "in a high degree."[37] So "very green" means green to a high degree. Figure 2.11 shows *very green* in relation to the root fuzzy set *green*. That figure makes plain the constricting action of the hedge *very* on the root fuzzy set *green*.

As another example of the *very* hedge, Figure 2.12 shows *very Tall* in relation to *Tall*.

The membership function *very Tall* is a subset of the membership function *Tall* formed by the following hedging operation: $\mu_{veryTall}(x) = (\mu_{Tall}(x))^2$.[38] For simplicity we will use the notation $\mu^2_{Tall}(x) \equiv (\mu_{Tall}(x))^2$. Once a fuzzy set is produced by a hedging operator, it behaves the same as any other fuzzy set. Figure 2.12 also illustrates the fuzzy set formed by the expression NOT *very Tall*, which is: $\mu_{veryTall'}(x) = 1 - \mu_{veryTall}(x) = 1 - \mu^2_{Tall}(x)$.

---

[37]*The American Heritage Dictionary of the English Langauge*, p.1425.
[38]Zadeh, "A Fuzzy-Set Theoretic Interpretation of Linguistic Hedges," *Journal of Cybernetics*, 2, pp 4-34.

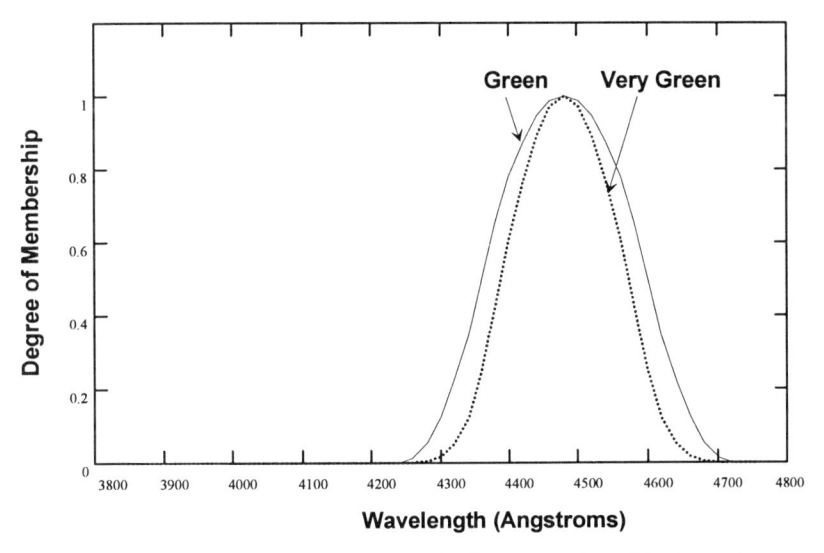

Figure 2.11: The effect of hedge very *of the base membership function* green.

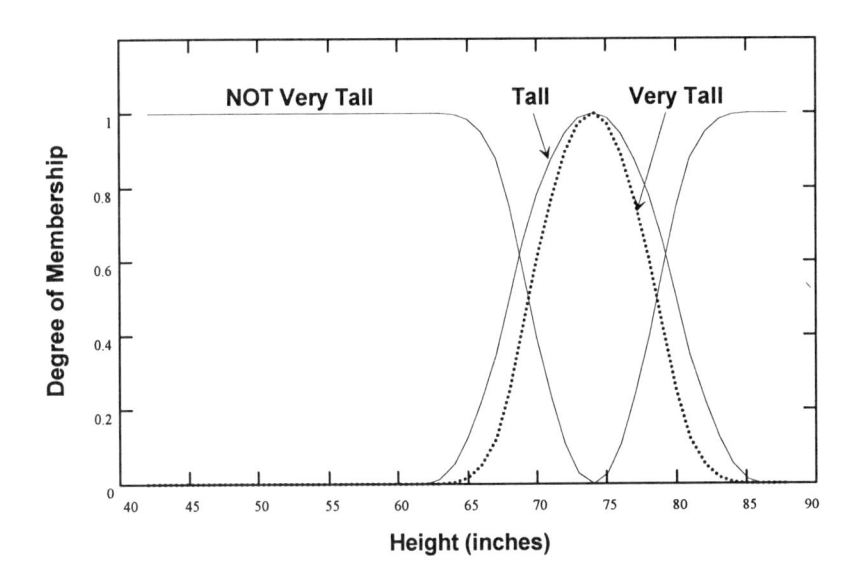

Figure 2.12: The effect of hedge very *on base membership function* Tall.

In evaluating a string containing multiple operators, like NOT *very Tall*, it must be noted that operators generally do not commute. That is, the order in which the operators are applied to a membership function affects the result. To illustrate the ordering dependence of operators, compare NOT *very Tall* with *very* NOT *Tall*. In the previous paragraph we established that the mathematical expression for NOT *very Tall* is $\mu_{NOTveryTall}(x) = 1 - \mu^2_{Tall}(x)$. However the mathematical expression for *very* NOT *Tall* is $\mu_{veryNOTTall}(x) = (1 - \mu_{Tall}(x))^2$. Clearly, $1 - \mu^2_{Tall}(x) \neq 1 - 2\mu_{Tall}(x) + \mu^2_{Tall}(x)$, for all $\mu_{Tall}(x)$. So, in a practical sense, the *very* and NOT operators do not commute.

Somewhat

Somewhat is an example of a *DILATION* operator. *DILATION* operators increase the degree of membership of elements that have relatively low degrees of membership. A dilation operator minimally affects elements of the fuzzy set that have relatively high degrees of membership. Applying the dilation operator to a fuzzy set produces a superset of the original.

As originally proposed by Zadeh, *somewhat* was defined as the inverse of *very*.[39] While *very* tightens the membership requirement in a fuzzy set, *somewhat* loosens the membership requirement. Figure 2.13 shows the effect of the linguistic hedge *somewhat* on the fuzzy set *Warm*. In that figure, *somewhat* has the definition $\mu_{somewhatWarm}(x) = (\mu_{Warm}(x))^{\frac{1}{2}} = \mu^{.5}_{Warm}(x)$. The operator acts on the base fuzzy set, *Warm*, to produce a superset, *somewhat Warm*.

The operators *somewhat* and *very* commute. This means that *somewhat very Warm* is equivalent to *very somewhat warm*. The two operators are also inverses. *Very* reverses the effects of *somewhat*, and vice-versa. We can see the commutative and inverse properties mathematically given by the definitions already established for both operators.

---

[39]Ibid.

Chapter 2: Terminology

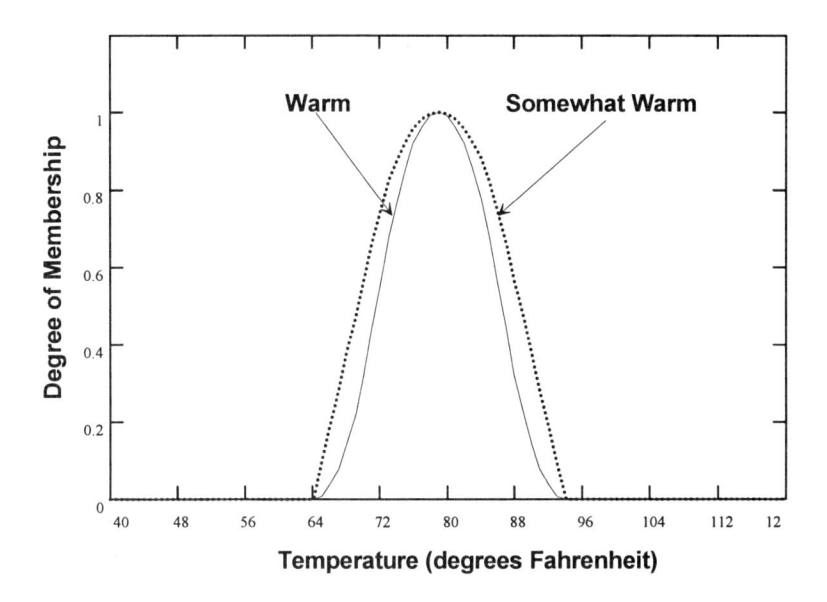

Figure 2.13: Effect of hedge somewhat *on base membership function* Warm.

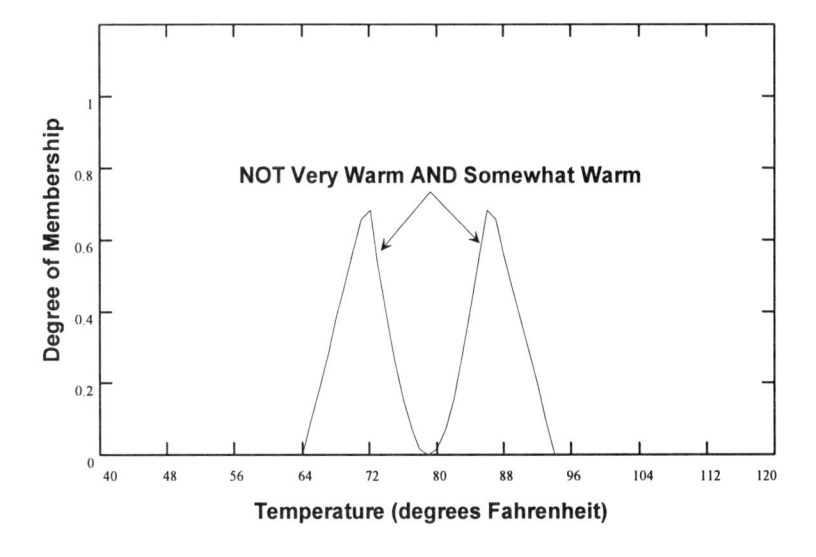

Figure 2.14: Fuzzy set produced by compound statement: NOT very Warm *AND somewhat* Warm.

53

Fuzzy Logic for Real World Design

$$\mu_{somewhatveryWarm}(x) = \left(\mu^2_{Warm}(x)\right)^{.5} = \mu_{Warm}(x)$$

$$\mu_{verysomewhatWarm}(x) = \left(\mu^{.5}_{Warm}(x)\right)^{2} = \mu_{Warm}(x)$$

Figure 2.14 shows many of the operators we have observed thus far by depicting the compound statement NOT *very* Warm AND *somewhat* Warm. As an exercise, the reader can work out the mathematical expression equivalent to the curve in Figure 2.14.
(Answer: $Min\left[1-\mu^2_{Warm}(x), \mu^{.5}_{Warm}(x)\right]$).

There is nothing sacred about the mathematical definitions of *very* or *somewhat* presented here. Our definitions simply reflect the most common usage of these two hedges. Readers should feel free to define new mathematical operators to model the human sense of words upon which the hedges are based.

Slightly

To this point, we have defined two types of operators. The first type, exemplified by the hedges *very* and *somewhat*, transform single membership functions. The second type combines, like AND and OR, multiple membership functions. We will describe the hedge *slightly* as a combination of both types of operator.

In our language the term *slightly* connotes that something is true to a small degree. The operator *slightly* emphasizes elements of a fuzzy set with relatively low degrees of membership. *Slightly* also de-emphasizes elements of a fuzzy set that have high degrees of membership. One of many possible definitions of *slightly* is

$\mu_{slightlyCold}(x) = NORM\left[\mu_{Cold}(x) \cap \mu_{NOTveryCold}(x)\right]$. Simplifying mathematically yields: $\mu_{slightlyCold}(x) = NORM\left[Min\left(\mu_{Cold}(x), 1-\mu^2_{Cold}(x)\right)\right]$.

Chapter 2: Terminology

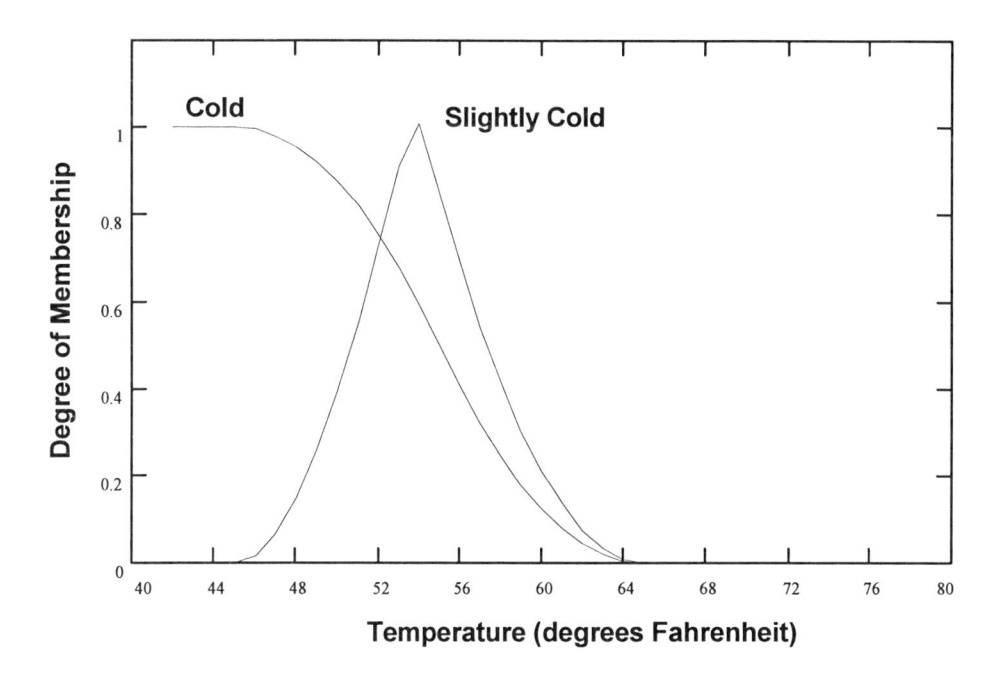

Figure 2.15: Effect of the hedge slightly *on the membership function* Cold.

From the example depicted in Figure 2.15, 56 degrees is good example of a *slightly Cold* temperature. Using the same figure, 44 degrees is a good example of a *Cold* temperature.

Sort of

The final linguistic hedge we will look at is *sort of*. Like *slightly*, *sort of* is one of a family of hedges that reduce the degree of elements having a high degree of membership, and increase the degree of elements with a low degree of membership. One possible definition of this operator is $\mu_{sortofHot}(x) = NORM[\mu_{NOTveryveryHot}(x) \cap \mu_{somewhatHot}(x)]$. Simplifying mathematically yields: $\mu_{sortofHot}(x) = NORM[Min[(1-\mu_{Hot}^4(x)), \mu_{Hot}^{.5}(x)]]$.

55

Fuzzy Logic for Real World Design

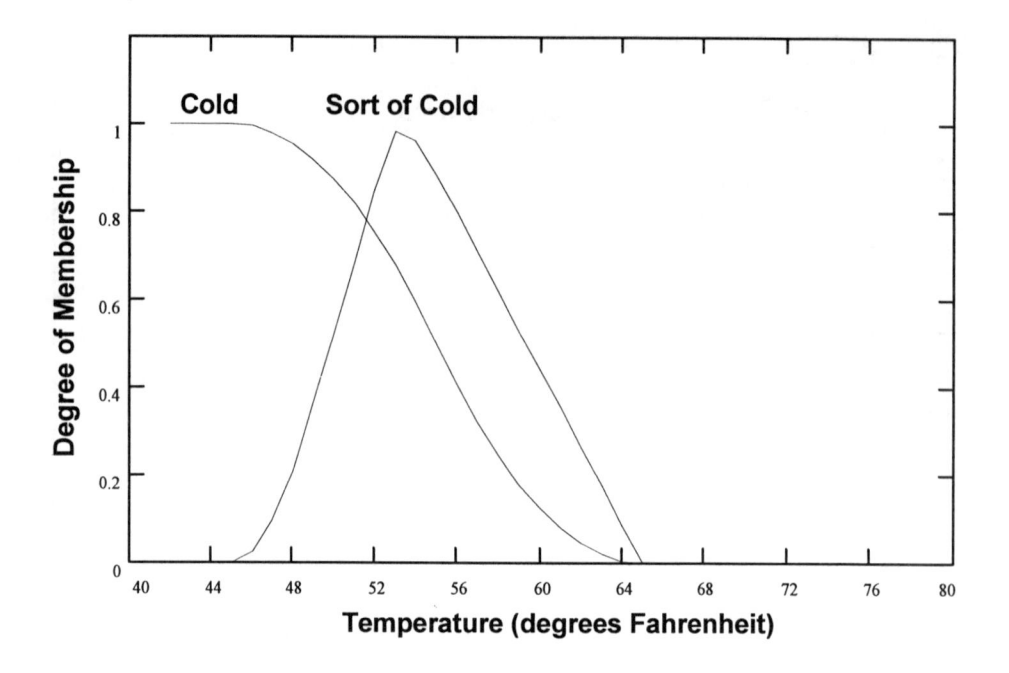

Figure 2.16: Effect of hedge sort of *on membership function* Cold.

There is nothing sacred about the mathematical definitions of *slightly* and *sort of* presented here. As an exercise, the reader should consider making up alternate definitions for *slightly*, *sort of,* and other linguistic hedges listed at the beginning of this chapter. Keep in mind that the objective in defining any fuzzy logic operator is to match the effect of the operator with the semantic content of the word itself.

Membership Function Shapes and Their Interpretations

At this point it probably appears that membership functions for fuzzy variables take on only a few different general shapes. However, this impression is due to the fact that the examples used thus far have consisted of only one family of curves. While that one family of curves sufficed to illustrate the mechanics of hedges, logical operators, etc., we

Chapter 2: Terminology

are now ready for more detail on the actual construction of membership functions.

The fuzzy methodology permits as many membership function shapes as there are colors in the rainbow. The robustness of the fuzzy methodology permits the use of a variety of shapes without appreciably affecting the overall meaning of the underlying fuzzy set. The following sections detail the common approaches to defining membership function shapes.

S, Z, and PI curves

S, Z, and PI curves are so named because of the shapes they trace on a graph. They provide smooth membership functions that resemble bell curves. This family of curves is parameterized so that one need only specify the midpoint and width to obtain the desired membership function. The parameterized expressions for each type of curve follow with graphs illustrating each type. This family of curves allows the smoothest possible transitions as the fuzzy system output changes.

S-Curve

The shape of the S-Curve is produced by a quadratic equation. While there is a historical precedent set by Zadeh for the definition of the S-Curve[40], its functional form should not be considered as immutable. Membership functions shapes can be improved upon at the users discretion. The three parameter expression for the Zadeh S-Curve is:

$$S(x,a,b,c) = 0; x \leq a$$

$$= 2\left(\frac{x-a}{c-a}\right)^2 ; a \leq x \leq b$$

$$= 1 - 2\left(\frac{x-c}{c-a}\right)^2 ; b \leq x \leq c$$

$$= 1; x \geq c$$

---

[40] Zadeh, "A Fuzzy-Algorithmic Approach to the Definition of Complex or Imprecise Concepts," *International Journal of Man-Machine Studies, 8,* pp 249-291.

x is the independent variable and a,b,c are parameters describing the shape of the curve. The Zadeh S-Curve has a first derivative given by the equations:

$$\frac{d}{dx}S(x,a,b,c) = 0; x \leq a$$

$$= 4\frac{x-a}{(c-a)^2}; a \leq x \leq b$$

$$= -4\frac{x-c}{(c-a)^2}; b \leq x \leq c$$

$$= 0; x \geq c$$

Figure 2.17 shows the general shape of the Zadeh S-Curve, and its first order derivative.

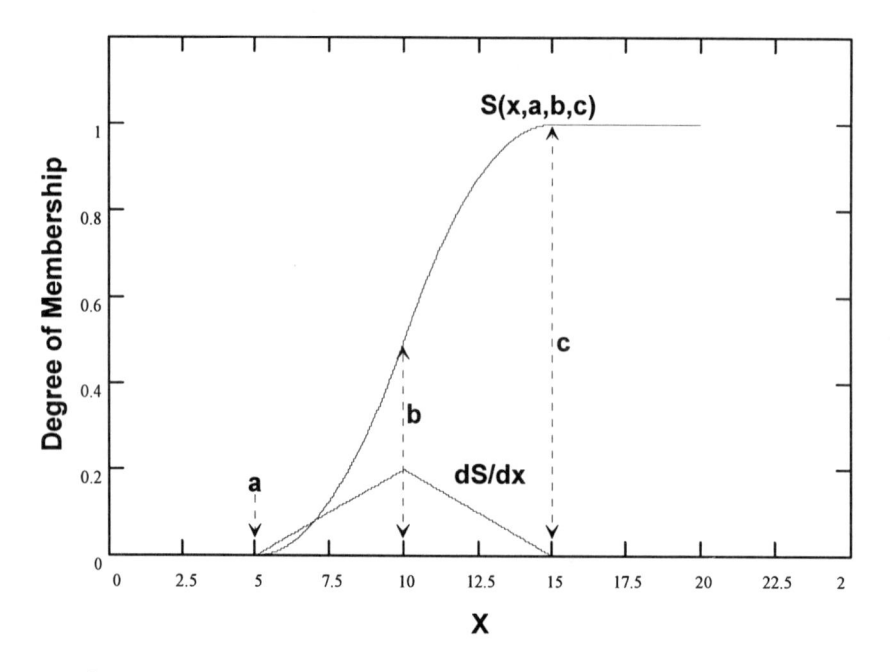

Figure 2.17: Graph of Zadeh S-Curve defined by $S(x,5,10,15)$ and its first derivative.

In some cases it may be more convenient to use a two parameter formulation of the S-Curve. The following equations define a commonly used alternate S-Curve definition:

$$S(x,a,b) = 0; x < a-b$$
$$= \frac{(x-(a-b))^2}{2b^2}; a-b \leq x \leq a$$
$$= 1 - \frac{((a+b)-x)^2}{2b^2}; a < x \leq a+b$$
$$= 1; x > a+b$$

This S-curve function takes two parameters, midpoint and width. The continuous first derivative of this function is:

$$\frac{d}{dx}S(x,a,b) = 0; x < a-b$$
$$= \frac{x-(a-b)}{b^2}; a-b \leq x \leq a$$
$$= \frac{(a+b)-x}{b^2}; a < x \leq a+b$$
$$= 0; x > a+b$$

Both the function and its derivative are graphed in Figure 2.18. The midpoint is given by parameter a, and width is given by parameter b.

Fuzzy Logic for Real World Design

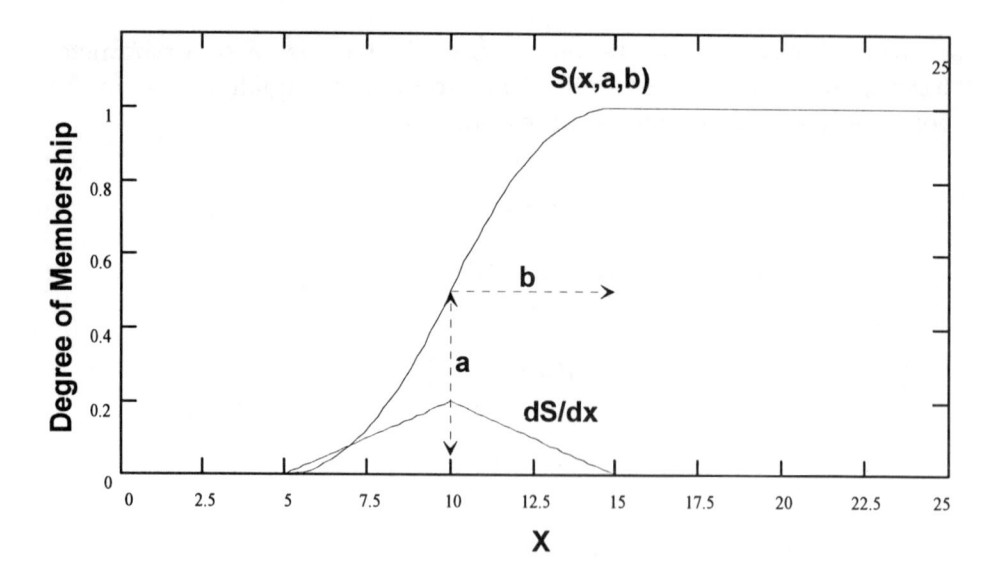

Figure 2.18: The two parameter S-Curve and its first derivative. S(x,10,5) is shown.

Z-Curve

The Z-Curve is defined using the definition already established for the two parameter S-Curve. The expression for the parameterized Z-Curve is: $Z(x,a,b) = 1 - S(x,a,b)$. Stated another way:

$$Z(x,a,b) = 1; x < a - b$$

$$= 1 - \frac{(x-(a-b))^2}{2b^2}; a - b \leq x \leq a$$

$$= \frac{((a+b)-x)^2}{2b^2}; a < x \leq a + b$$

$$= 0; x > a + b$$

Again, the independent variable is x, the midpoint is given by parameter a, and width is given by parameter b. The improved Z-Curve defined by Z(x,10,5) and its first derivative are graphed in Figure 2.19.

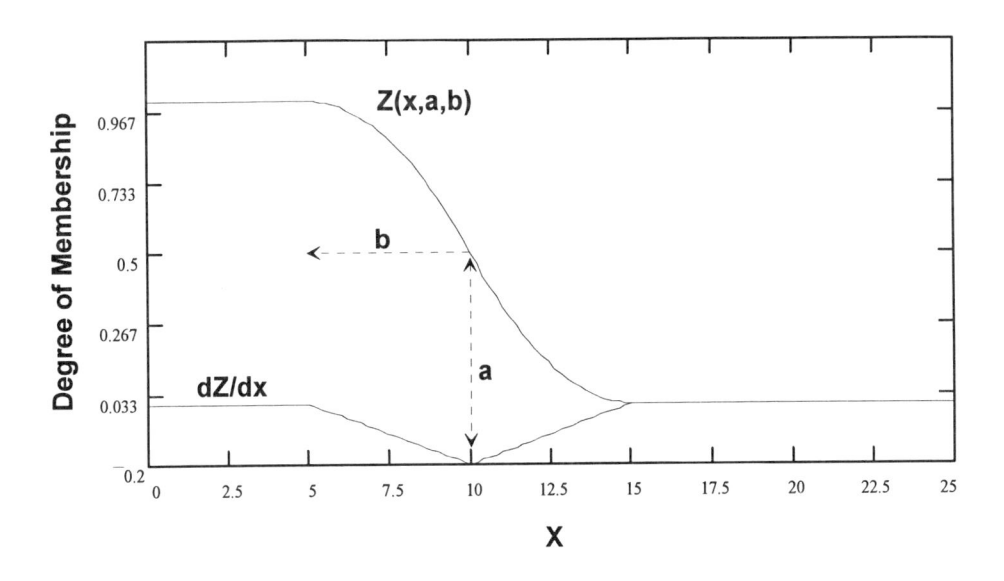

Figure 2.19: Z-Curve and its first derivative. Z(x,10,5) is shown.

Pi-Curve

The Pi-Curve is generally bell shaped and is formed by placing S- and Z- curves back-to-back. The expression is:

$$Pi(x,a,b) = S(x, a - \frac{b}{2}, \frac{b}{2}); x \leq a$$

$$= Z(x, a, a + \frac{b}{2}, \frac{b}{2}); x \geq a$$

Fuzzy Logic for Real World Design

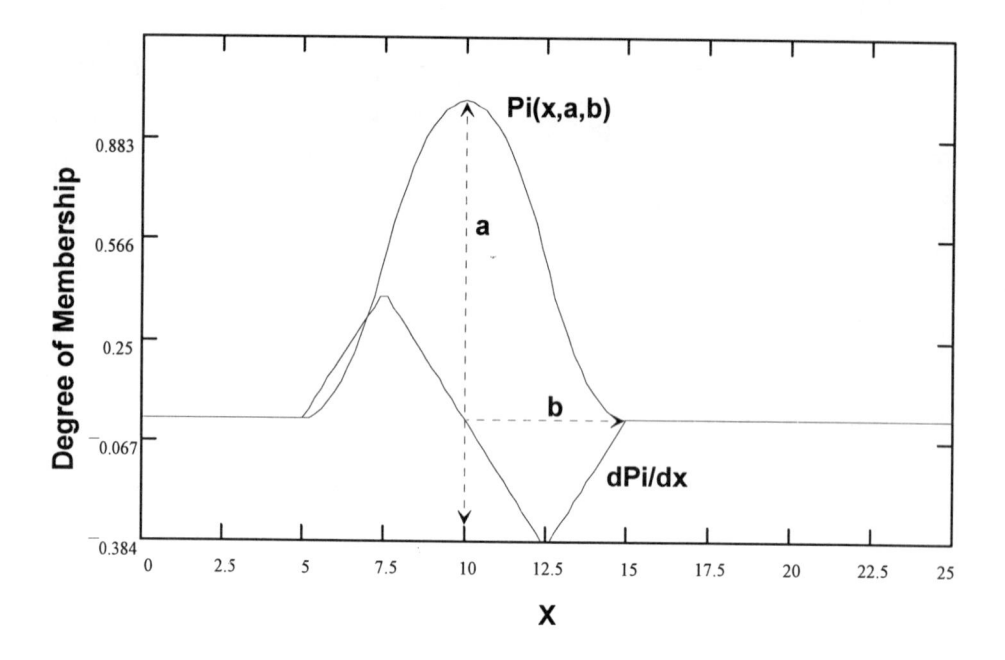

Figure 2.20: Pi-Curve and its first derivative. Pi(x,10,5) is shown.

Gaussians

Otherwise known as the bell curves, the gaussian family of curves describe the so-called normal distribution arising from the study of uniform random variables and seen throughout the field of statistics. Two parameters describe the gaussian family, μ, and σ. One parameter locates the center of the curve, μ, and the second parameter, σ, fixes the width.[41] The degree of membership is given by: $G(x,\mu,\sigma) = e^{\frac{-(x-\mu)^2}{\sigma^2}}$. Figure 2.20a shows G(x,10,5). The drawback to the gaussian family that results in infrequent use in fuzzy representations is the infinite tail on each side of center. Mathematically, no finite value of x results in a zero degree of membership. Also, since the gaussian family only describes

---

[41]In the mathematical literature σ^2 is the variance of the distribution, and μ is the mean.

symmetrical curves, its shapes do not easily generalize to all applications of fuzzy logic.

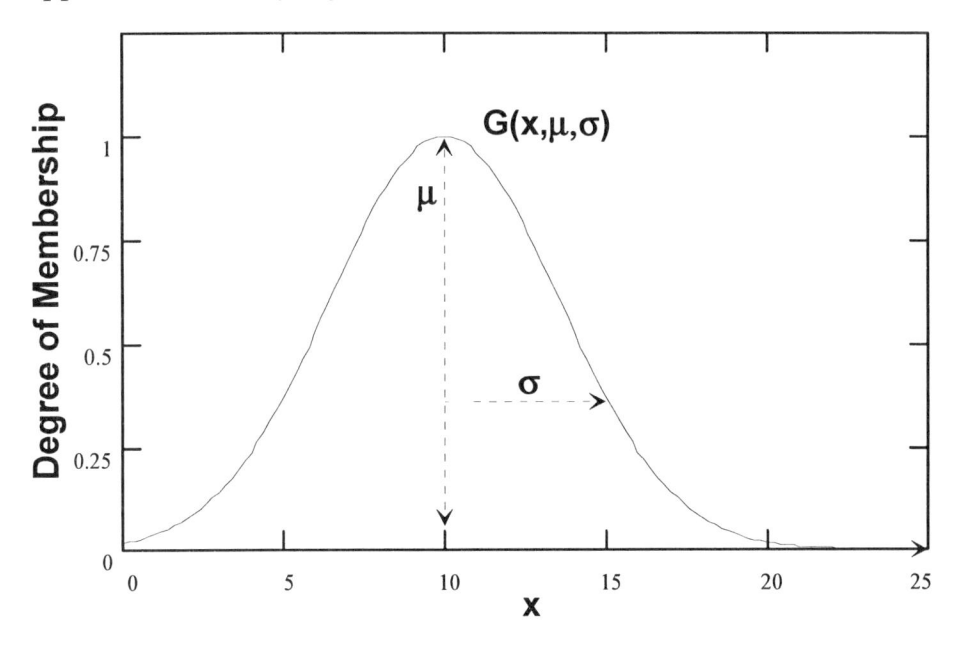

Figure 2.20a: Gaussian Curve. G(x,10,5) is shown.

Trapezoids and Triangles

The memory requirements and execution time restrictions in embedded systems don't often allow the use of S-, Z-, and Pi-Curves. Often, the smoothness of output that can be obtained using the S-Curve family is not necessary to acceptably solve the problem at hand. The number of calculations required to render a complete set of membership functions can be reduced by using piecewise linear functions instead of the smoothly varying but quadratic S-, Z-, and Pi-Curves. The use of trapezoidal and triangular membership functions reduces the computational overhead of the fuzzy system further still. However, two drawbacks exist in all of the piecewise linear membership function representations.

The following graphs will clearly demonstrate that linear functions do not blend smoothly at the extremes of membership. Discontinuities exist in the slopes of all piecewise linear membership functions at truth values of 1 and 0. Any discontinuity in the membership function slope means that the first derivative of that function is discontinuous. Discontinuity of the membership function's slope is not enough to prevent the successful application of fuzzy logic; however, it does prevent certain automated methods aimed at automatically tuning fuzzy system parameters.[42] In addition, discontinuous functions result in discontinuous responses, which may not be desired.

In practice, trapezoidal and triangular membership functions have proved sufficient for the vast majority of embedded control applications. Their ease of use, coupled with their miserly use of memory and CPU bandwidth, make them the primary choice of fuzzy system designers.

Trapezoids

The trapezoid can be used as a rough approximation to the previously defined Pi-Curve. The advantages of using the piecewise linear function are two-fold. Firstly, the trapezoid requires fewer IF-THEN comparisons. Eliminating IF-THEN comparisons speeds up computer calculation by reducing program branching. Program branching is notorious in its negative performance impact on a CPU. Secondly, the trapezoid eliminates the costly quadratic calculations required by the Pi-Curve. One possible four parameter definition for the trapezoid is:

$$\begin{aligned} Trap(x,a,b,c,d) &= 0; x \leq a \\ &= 1 - \frac{b-x}{b-a}; a < x \leq b \\ &= 1; b < x \leq c \\ &= \frac{d-x}{d-c}; c < x \leq d \\ &= 0; x > d \end{aligned}$$

---

[42]Brian W. Grant et.al., "The Use of Neural Networks for Automation of Fuzzy Knowledge Base Creation," in *Proceedings of Fuzzy Logic '93*.

Chapter 2: Terminology

Here again, x is the independent variable, and a,b,c,d are the vertices of the trapezoid. Figure 2.21 shows the graph of the trapezoid membership function.

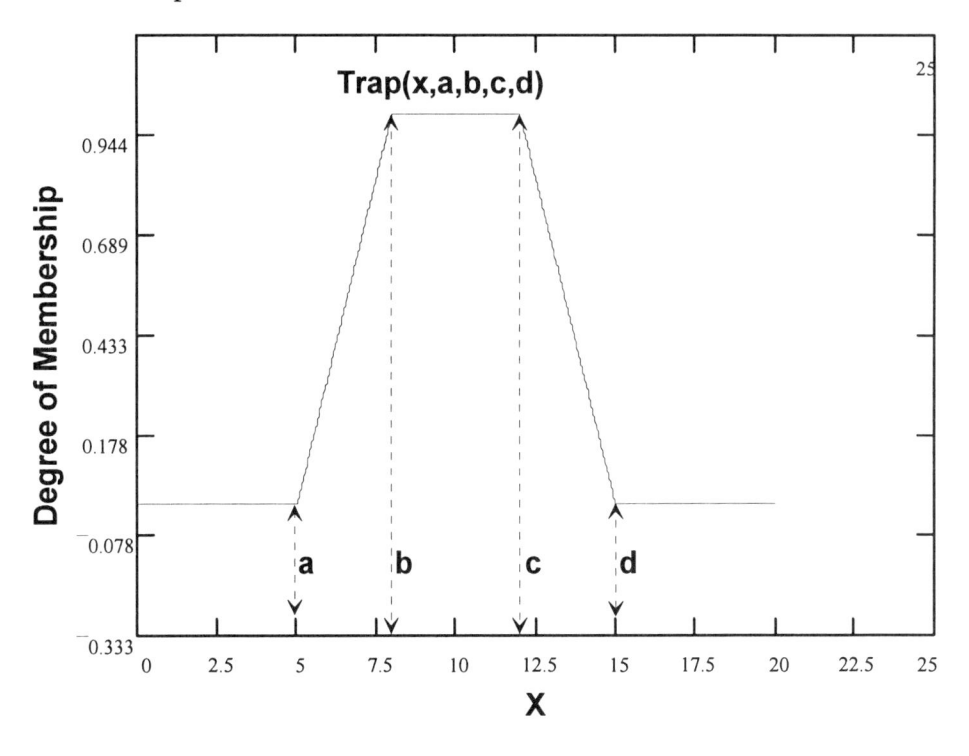

Figure 2.21: Trapezoidal membership function Trap(x,5,8,12,15).

Triangles

Triangular membership functions are even simpler than their trapezoidal cousins. In fact, triangular membership functions are a special case of trapezoidal functions where $b = c$. We will use the definition:

65

Fuzzy Logic for Real World Design

$$Tri(x,e,f,g) = 0; x \le e$$
$$= 1 - \frac{f-x}{f-e}; e < x \le f$$
$$= \frac{g-x}{g-f}; f < x \le g$$
$$= 0; x > g$$

Parameters e, f, g are the vertices of the triangle. Figure 2.22 shows the graph of the triangular membership function.

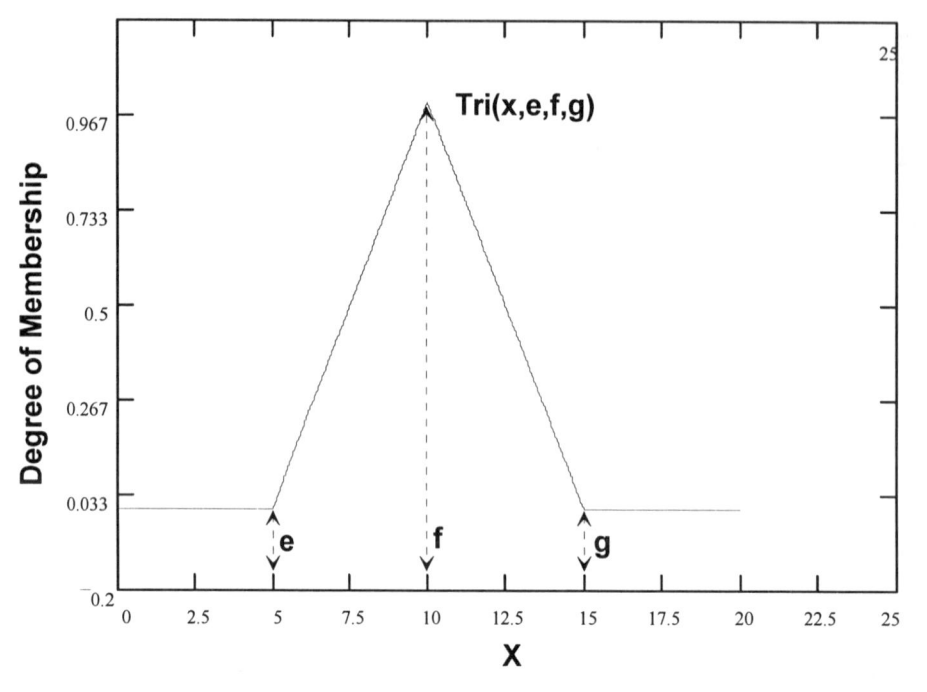

Figure 2.22: Triangular membership function $Tri(x,5,10,15)$.

Singletons

The fuzzy singleton represents a departure from all of the previous membership function definitions. While all of the previous functions are suitable for both fuzzy input and output variables, the fuzzy singleton is only appropriate for output variables. As depicted by its graph in Figure 2.23, the so-called fuzzy singleton is not fuzzy at all. Non-zero at only one point, the singleton greatly simplifies calculations needed to produce fuzzy outputs. The fuzzy singleton is reminiscent of the ubiquitous Dirac delta function used in science and engineering. The fuzzy singleton takes the form:

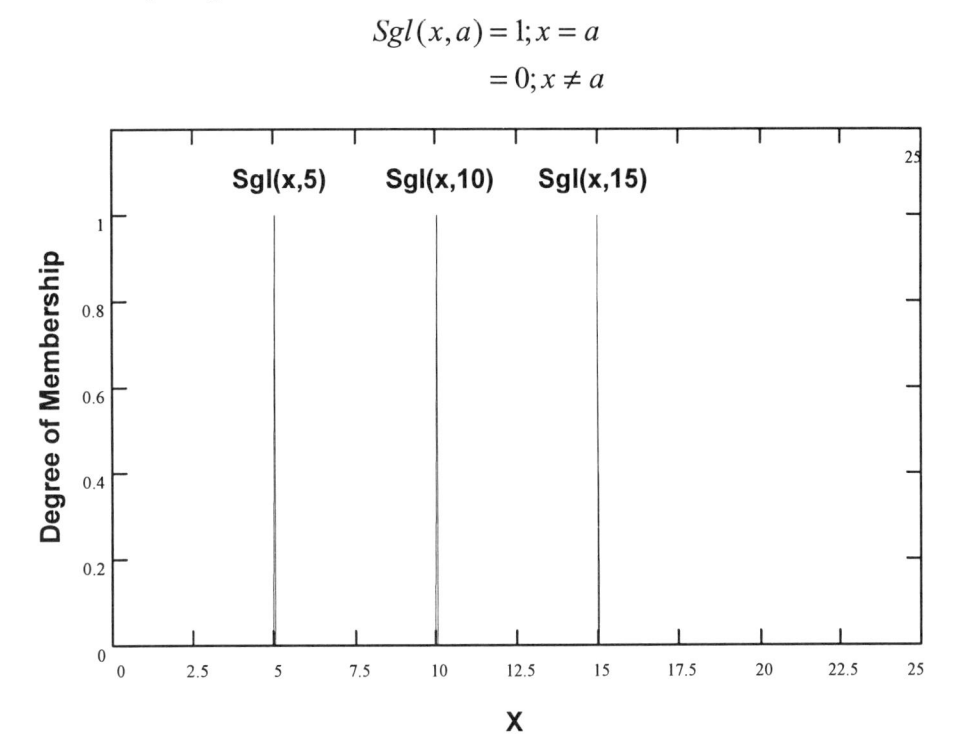

$$Sgl(x,a) = 1; x = a$$
$$= 0; x \neq a$$

Figure 2.23: Three fuzzy singletons, Sgl(x,5), Sgl(x,10), and Sgl(x,15).

Summary

In this chapter, we started our exploration of fuzzy logic by looking at its first key construct, the fuzzy set. While the formal math behind fuzzy logic is no less rigorous than calculus or geometry, we strove for a less mathematical appreciation of the fuzzy set. In seeking a more personal understanding of fuzzy logic, we drew upon our every day experience with fuzzy classes and concepts.

We defined much of the jargon encountered when discussing fuzzy sets and introduced a consistent notation that will be used throughout the balance of the book. After covering some of the basic mathematics that Boolean and fuzzy sets share, we introduced fuzzy-specific operations.

In exploring the interaction and combination of fuzzy sets, we liberaly illustrated basic operations such as AND, OR, and NOT. We discovered that fuzzy logic allows semantic hedges like *very* and *somewhat* in a mathematically relevant and consistent way. We also reviewed a variety of membership function shapes and their mathematical specifications. Most of all we became aware that our discovery of the richness and utility of fuzzy logic had only just begun.

In the next chapter we will accelerate the presentation of material as we outline the second key construct of fuzzy logic, namely fuzzy IF/THEN rules.

Chapter 3:
Fuzzy Rules and Operations

With the help of membership functions, fuzzy rules encode the intelligence of a fuzzy system. Because they are written using the linguistic variables discussed in the previous chapter, fuzzy rules read like free-form natural language. Fuzzy rules capture the essence of knowledge about the system by expressing how system inputs relate to system outputs. Most importantly, rule-writing captures the knowledge of the human in a form native to that expert. The ability to express rules in our own natural language speeds the process of engineering a solution to a given problem.

Using the IF-THEN syntax of fuzzy rules eliminates the intermediate and time-consuming step of translating our knowledge into mathematical equations, Boolean decision trees, or computer software. Few of us realize that when we set about solving a problem, we are actually solving two problems. The second problem is to translate our notions of the solution to the first problem to the tools we have available. Fuzzy rules are written in our language, thus making the job of translating our knowledge much quicker.

Without perceiving it, we often constrain our problem solving approaches with the foreknowledge of the tools and techniques which we anticipate using. In other words, if we intend to use a digital computer, we may look for solutions which are easily expressed on a digital computer. Or, in programming, we may jump ahead in our thinking to evaluate whether a particular data structure is easily programmed in our favorite computer language. The data structure that we choose will likely be the one which is easiest to construct, debug, and maintain with the tools and techniques at hand. Fuzzy logic does not inherently constrain our problem solving approaches because fuzzy rules adequately represent human knowledge. So, instead of translating our unique human reasoning to some other form (Boolean decision trees for instance), we can express our knowledge more naturally and completely.

In the following sections, we build upon our knowledge of membership functions by introducing fuzzy propositions, fuzzy rules, and the math needed to manipulate them. Combining fuzzy propositions will form fuzzy rules. Finally, combining multiple rules and their consequent actions will fully express the underlying knowledge in the fuzzy system.

The Meaning and Elements of a Fuzzy Rule

Fuzzy Propositions

Fuzzy logic rules are built by combining *FUZZY PROPOSITIONS*. Examples of fuzzy propositions are: *Temperature is Hot, Humidity is Low, Angle is Acute, Risk is Minimal*, etc. Each fuzzy proposition makes a statement about the possible value of the underlying linguistic variable. In the proposition, *Temperature is Hot*, where Temperature is the linguistic variable, *Hot* is one possible value for Temperature. The usefulness of a fuzzy proposition derives from the fact that we can measure the truth of the proposition. With a well-defined *Hot* membership function, and a crisp temperature, we easily compute the degree of truth of the proposition.

FUZZIFICATION is the commonly used term to describe the process of testing a crisp input against a fuzzy proposition. Figure 3.1

graphically depicts how the fuzzification process works. Notice from the figure that when a fuzzy variable is partitioned into overlapping membership functions, multiple fuzzy propositions may have non-zero degrees of truth for a given crisp input. Using Figure 3.1 with the crisp temperature of 78 degrees Fahrenheit, the proposition *Temperature is Hot* is true to degree 0.42, and the proposition *Temperature is Warm* is true to degree 0.33. The proposition *Temperature is Cool* is true to degree 0.00 because the crisp value of 78 degrees is outside the scope of fuzzy set *Cool*.

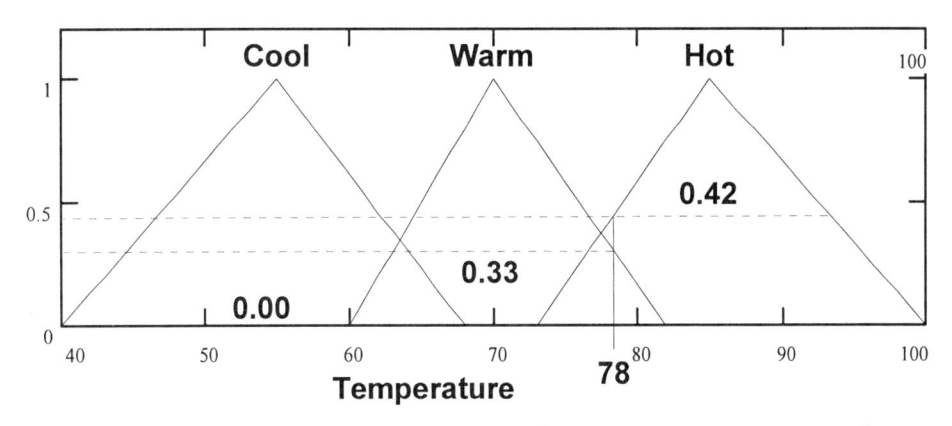

Figure 3.1: Fuzzification of 78 degrees into fuzzy sets Cool, Warm, *and* Hot.

Antecedents and Consequents

Two distinct usages of fuzzy propositions exist in fuzzy systems. An *ANTECEDENT* is a fuzzy proposition based on an input fuzzy variable. A *CONSEQUENT* is a fuzzy proposition based on an output fuzzy variable. These two usages combine to give us fuzzy rules.

A fuzzy rule relates antecedents to consequents[43]. Stated another way, a fuzzy rule maps an input fuzzy proposition to an output fuzzy proposition. The general form of a fuzzy rule is :

---

[43] A fuzzy rule does not always require an antecedent. Another valid fuzzy rule construct is called an *ASSERTION*. It consists of only a consequent, and acts to

> IF antecedent_1 AND antecedent_2 AND ... THEN consequent_1 AND consequent_2 AND ...

The IF-side of the rule is composed of one or more antecedents, while the THEN-side is composed of one or more consequents. In examining this concept, we'll first look at the simplest case, a rule composed of one antecedent and one consequent.

The structure of the rule is interpreted in a way that is analogous to the familiar BOOLEAN IF-THEN statement. The IF-side of the rule specifies the conditions on the fuzzy system inputs that must be true in order for the actions specified on the THEN side of the rule to take effect.

Figure 3.2 illustrates the rule *IF Temperature is Warm THEN Fan_Speed is Low*. The figure is shaded such that the darker the region, the stronger the rule. Notice that the mapping produced by this rule is not crisp. The scope of the membership functions *Warm* and *Low* define the region where this rule has applicability. As Temperature moves away from maximal warmness (at 70 degrees), the rule has less and less validity. The rule's transition from valid to invalid is, characteristically, a smooth process with no sharp breaks.

Because membership functions permit smoothly varying degrees of truth, the previous rule also expresses the following concept: *IF Temperature is* approximately *Warm THEN Fan_Speed is* approximately *Low*. This concept is central to the issue of approximate reasoning. Approximate reasoning describes the ability to reason in qualitative, imprecise terms. Deducing conclusions from a of collection imprecise premises is the hallmark of approximate reasoning in humans.[44] The ability to represent approximate relations like the one above provides evidence that fuzzy logic was conceived with the requirements of approximate reasoning in mind.

---

bias the output of the fuzzy system. This can be considered a rule with an antecedent that is TRUE across the entire input space.

[44]Zadeh, "A Theory of Approximate Reasoning", Machine Intelligence, 9.

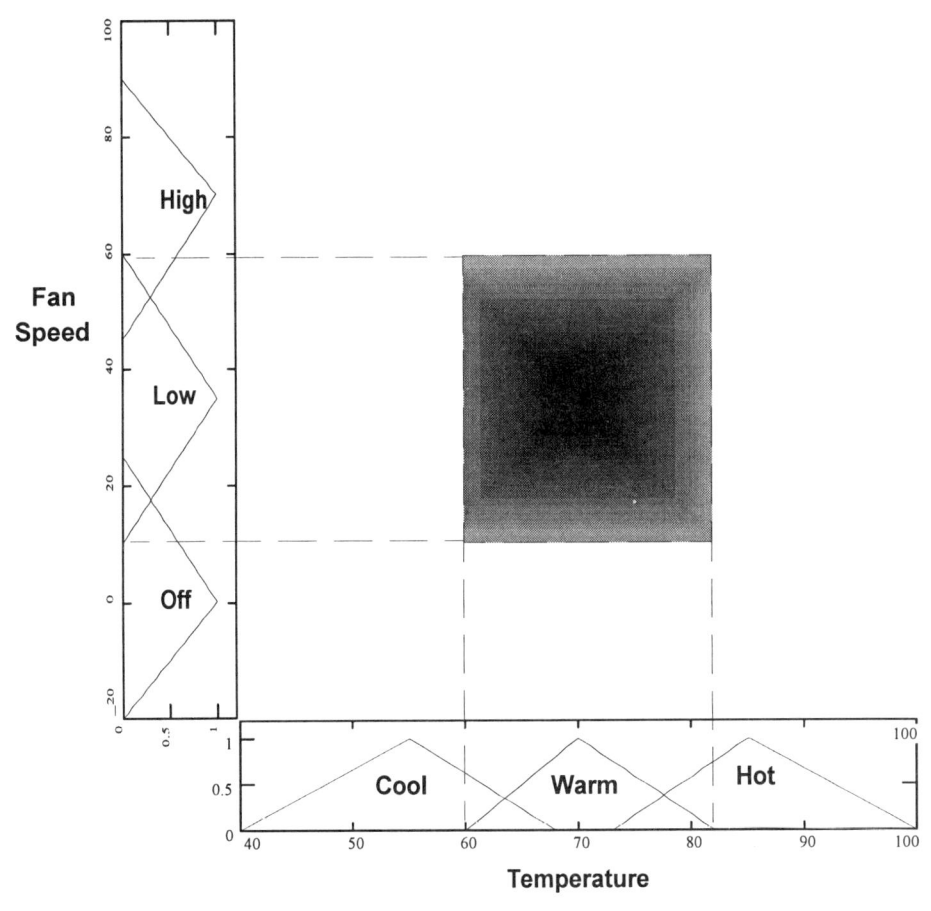

Figure 3.2: Fuzzy mapping produced by rule IF Temperature is Warm THEN Fan_Speed is Low.

Rules and Fuzzy Relations

Building upon the idea of a single rule yielding a fuzzy relation between membership functions we now turn our attention to multiple rules. Multiple fuzzy rules can be viewed as single rules operating in parallel. So a set of fuzzy rules yields a set of fuzzy relations between fuzzy inputs and fuzzy outputs. This idea is illustrated in Figure 3.3. The figure shows the mapping established by the three rules:

Fuzzy Logic for Real World Design

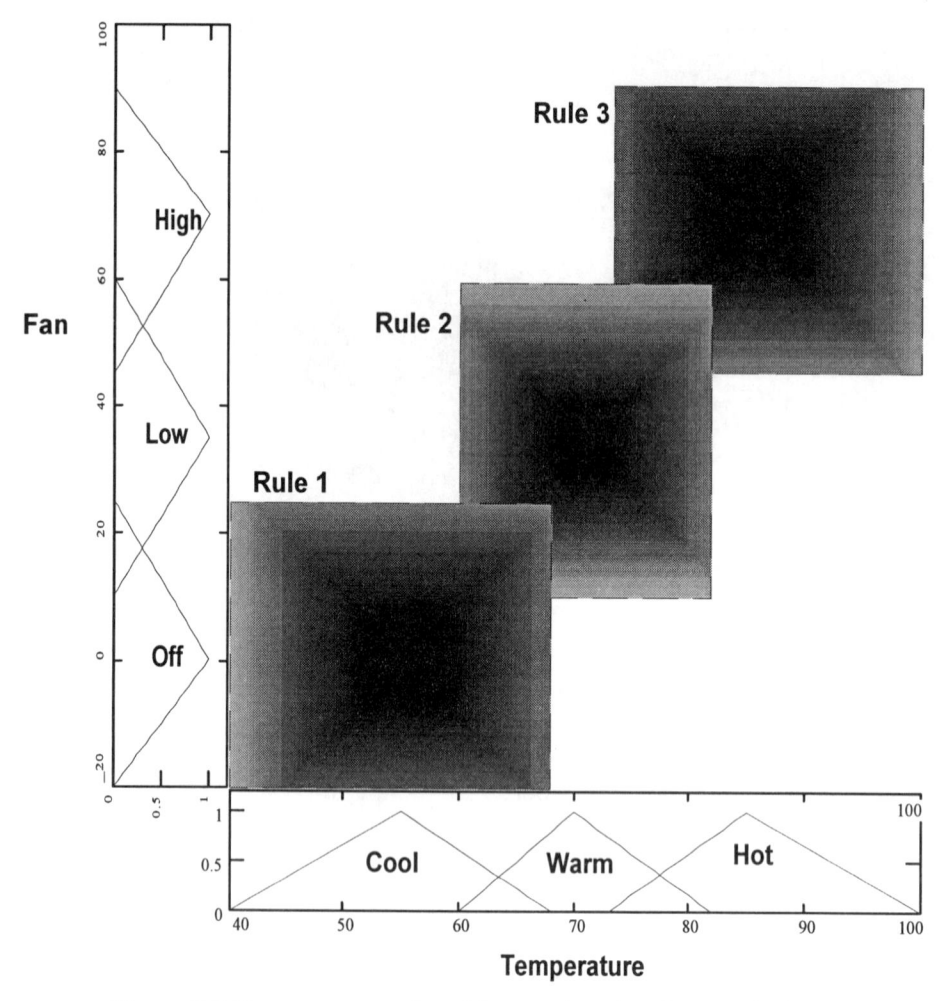

Figure 3.3 Fuzzy map produced by overlapping fuzzy relations.

Rule_1: IF Temperature is Cool THEN Fan_Speed is Off,

Rule_2: IF Temperature is Warm THEN Fan_Speed is Low,

Rule_3: IF Temperature is Hot THEN Fan_Speed is High

Figure 3.3 shows three overlapping fuzzy relations, one for each rule. The overlapping membership functions *Cool, Warm, Hot,* and *Off, Low, High* cause the overlap of the fuzzy relations produced by each rule. In a qualitative sense, a fuzzy rule set yields its strongest conclusions at the points of maximum degree of membership of the individual membership functions in each rule. In the case of the rule set which generated Figure 3.3, Rule_1 is strongest where *Cool* and *Off* are true to degree 1.0. Rule_2 is strongest where *Warm* and *Low* are true to degree 1.0. Likewise, Rule_3 is strongest where *Hot* and *High* are true to degree 1.0.

Figure 3.3 suggests an important consideration in the design of fuzzy systems. Greater overlap of membership functions generates a denser fuzzy map. As membership function overlap is increased, the fuzzy relations produced by each rule are squeezed closer together. In the real world, greater overlap of fuzzy sets results in smoother and finer control, and hence is an important design consideration. We will explore this aspect, and many other design considerations in Chapter 4.

Fuzzy Function Approximation

So far, the primary advantage of using fuzzy logic that we've highlighted is the ability to represent fuzzy classes and reason in the face of imprecision and fuzziness. Approximate reasoning embodied by the simple three rule fan controller above is a clear example of this use of fuzzy sets (*Hot*, *Warm*, etc.). However, the usefulness of fuzzy modeling is not limited to just knowledge and classes which are fuzzy by their nature.

Fuzzy logic easily models mathematical functions, both linear and non-linear[45]. Modeling of mathematical functions makes use of the interpolative properties of a fuzzy system. The interpolative properties of a fuzzy system arise out of the idea that if we know that a particular statement like: *IF X is A THEN Y is B*, is true, then the statement: *IF X is around A THEN Y is around B*, is true to some degree. Using a fuzzy model to approximate a function like $y = x^2$, might look like Figure 3.4. The accuracy of the approximation depends on the number of

---

[45]With an appropriately constructed fuzzy model, any real continuous function can be approximated to any accuracy desired. For a proof and examples, see L.X.Wang and J.M.Mendel, "Generating Fuzzy Rules from Numerical Data, with Applications," *USC-SIPI Report #169*.

membership functions used for both x and y, the shape of those membership functions, and the rules expressing the relationship between those membership functions.

Approximating a mathematical function with a fuzzy model is qualitatively similar to the well-known mathematical tools of Taylor power series expansion and Fourier expansion. Both the Taylor power series expansion and the Fourier expansion specify an infinite series of terms that add up to an exact match to the original function. Of course, in a practical application of either the Taylor or Fourier methods, we cannot wait around to calculate an infinite series of terms. Ultimately a decision must be made when to stop calculating terms in the series. So the practical application of the Taylor and Fourier methods can also be viewed as an approximation.

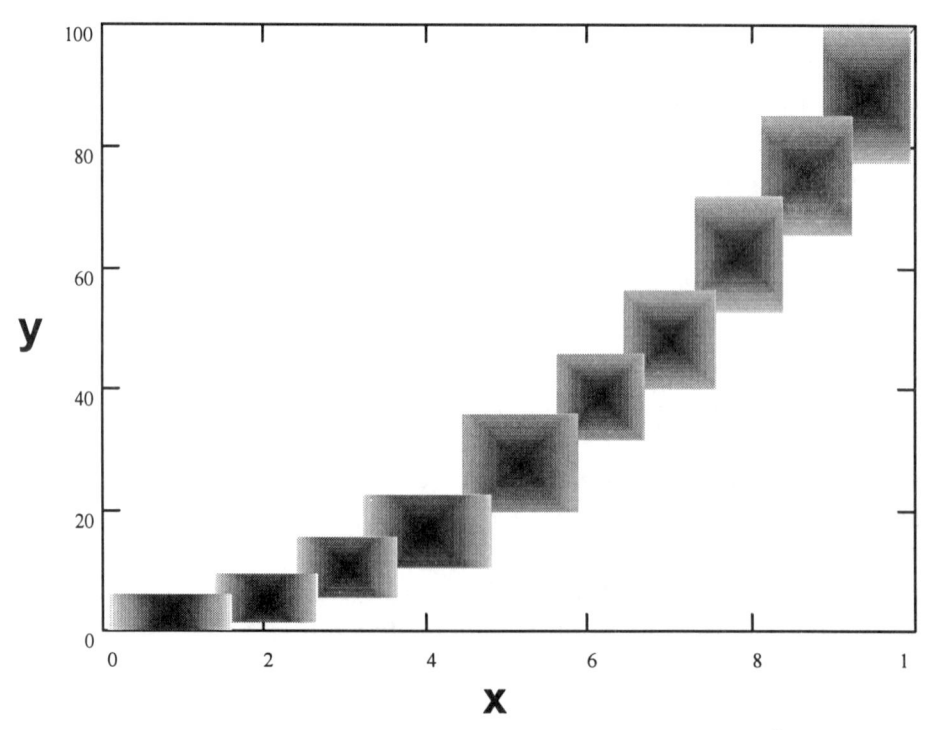

Figure 3.4 Fuzzy approximation to the function $y = x^2$.

In spite of the argument presented in the last paragraph that claims a similarity between fuzzy modeling and Taylor and Fourier methods, not all function approximations were created equal. The advantage that the Taylor and Fourier methods have over a fuzzy model is the ease with which a function can be calculated to any accuracy desired. By simply adding more terms to the Taylor series or the Fourier expansion we increase the precision of the resultant function.

Increasing the accuracy of a fuzzy model requires any or all of the following: adding more membership functions, changing existing rules, adding rules, and/or changing the shape of membership functions. Imposing any of those changes on a fuzzy model causes a ripple effect. No component of the fuzzy system acts in isolation. For instance, changing the shape of a membership function will affect all rules that include that membership function. Adding or deleting membership functions results in rule changes, additions, or deletions. Improving the precision of a fuzzy model is not accomplished by simply adding more terms.

Overall, there are still plenty of excellent uses for fuzzy function modeling. Two basic questions should guide the selection of fuzzy model versus exact mathematical calculation. First, how much time is available to calculate an answer? Second, how accurate must the result be? In cases where a mathematical expression is costly (in terms of execution time), a fuzzy model can be the speedier alternative. Generally, as a mathematical equation becomes more complex, a fuzzy model becomes increasingly expedient. In a live control system environment, like robotic control which typically involves extensive matrix multiplication, execution time overrides other concerns. Usually, the accuracy penalty paid for using the fuzzy model is a secondary consideration in such control system environments.

As an example of fuzzy function modeling consider the function: $y = e^{\sin(x)}$. One way to evaluate this function is to expand it out into the equivalent Taylor power series:

$$e^z = 1 + z + \frac{z^2}{2!} + \frac{z^3}{3!} + ...$$

$$\sin(x) = x - \frac{x^3}{3!} + \frac{x^5}{5!} - \frac{x^7}{7!} + ...$$

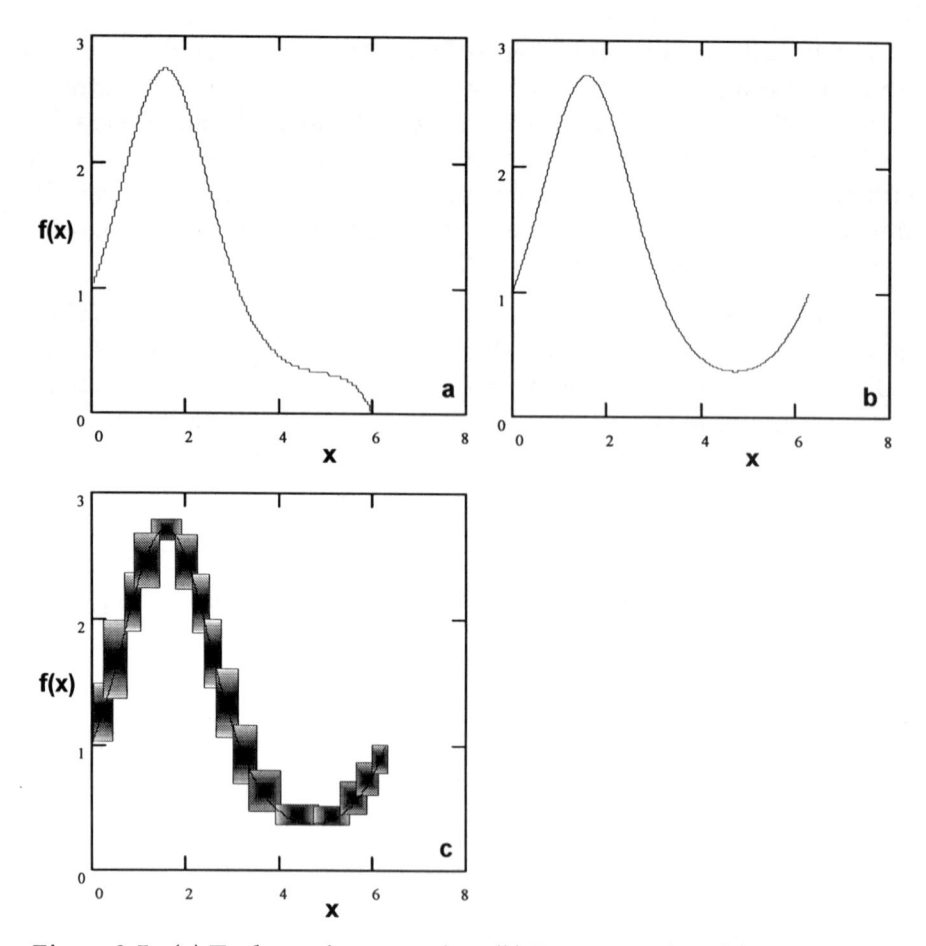

Figure 3.5: (a) Taylor series expansion, (b) Exact equation, (c) Fuzzy model alternative.

and calculate enough terms to satisfy our precision requirements. First of all, the two series' composing the function approximation converge slowly. Figure 3.5a was generated using the two finite approximations:

$$e^z = 1 + z + \frac{z^2}{2!} + \frac{z^3}{3!} + \frac{z^4}{4!} + \frac{z^5}{5!}$$

$$\sin x = x - \frac{x^3}{3!} + \frac{x^5}{5!} - \frac{x^7}{7!} + \frac{x^9}{9!} - \frac{x^{11}}{11!}$$

Compared to the exact function, shown in Figure 3.5b, and in spite the of large number of terms included, this approximation is quite poor. The fuzzy model alternative, which executes in far less time, is suggested in Figure 3.5c.

Up to this point we have looked at fuzzy propositions and fuzzy rules only in a qualitative sense. In order to illustrate the concept of a fuzzy mapping we have avoided the mathematics that generate crisp actions from a set of fuzzy rules. In addition, most practical fuzzy models are composed of more than one input variable, and involve multiple rules which specify the same output fuzzy set. The following sections address these issues by detailing the methods and mechanisms which allow fuzzy systems to reason and move from crisp inputs to crisp actions.

Combining Antecedents

The IF-side of a fuzzy rule is a conditional statement of one or more antecedents. Given a set of crisp inputs, the simple fuzzification procedure calculates the individual degrees of membership for each fuzzy proposition in the conditional statement. Fuzzification turns a crisp input into a fuzzy input, but does not tell us how to combine the individual degrees to form a degree of belief in the compound statement. The task of aggregating an overall truth value for the entire conditional statement (IF-side of the rule) from its individual degrees of membership is performed by the so-called Zadeh and compensatory operators.

Zadeh Operators

Adding an extra input variable, Humidity, to the rule set presented in the previous section gives:

Rule_1: IF Temperature IS Cool AND Humidity IS High THEN Fan_Speed IS Off,

Rule_2: IF Temperature IS Warm AND Humidity IS Low THEN Fan_Speed IS Low,

Rule_3 IF Temperature IS Hot AND Humidity IS Moderate THEN Fan_Speed IS High.

Fuzzy Logic for Real World Design

For reference, the fuzzy sets defining *Temperature*, *Humidity* and *Fan_Speed* are shown in Figure 3.6 (a), (b), and (c), respectively.

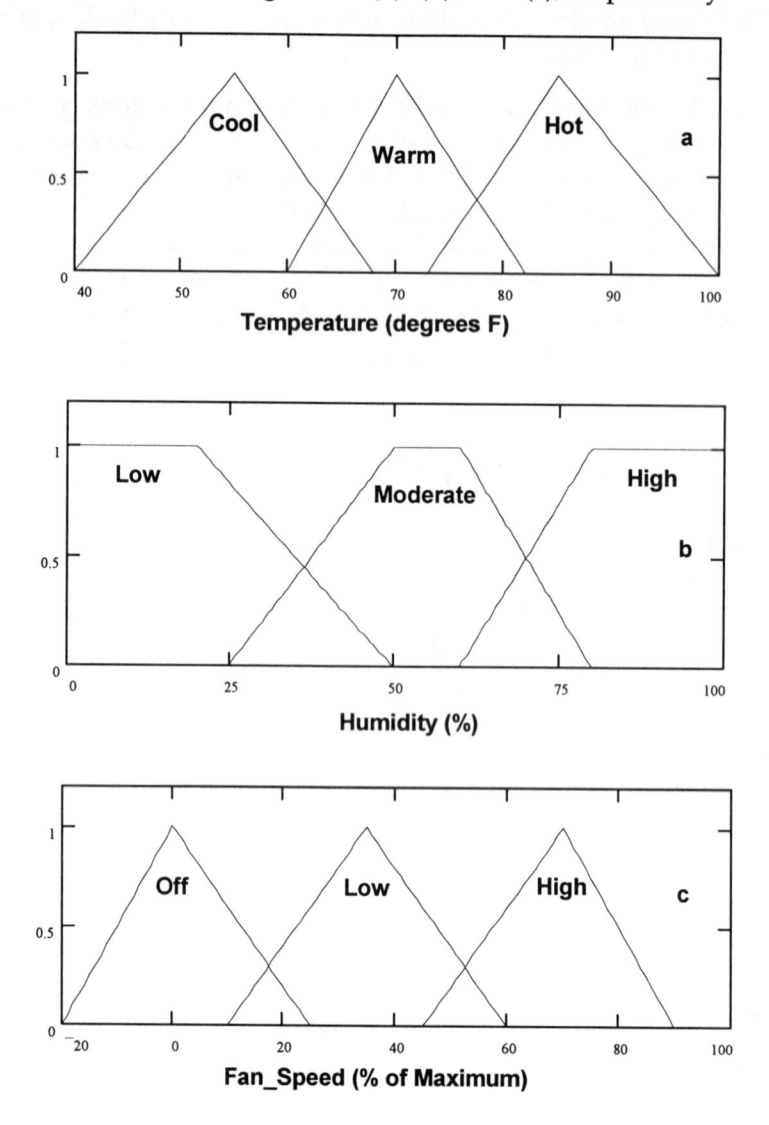

Figure 3.6: Membership functions for (a) Temperature, (b) Humidity, and (c) Fan_Speed.

Chapter 3: Fuzzy Rules and Operations

When the IF-side comprises more than one antecedent, the Zadeh or other compensatory operators are employed to determine the degree of truth of the compound conditional statement. These operators constitute a family of operators which act by treating each occurrence of the keyword *AND* as an instance of a mathematical operator. The operator takes the degrees of membership on both sides of the *AND* and combines them to yield a single truth value. The earliest description of such an operator, by Zadeh, is that of the simple intersection[46].

Mathematically, the *INTERSECTION OPERATOR* takes the degree of membership of two antecedents and returns the minimum of those two values. Mathematically, the action of this operator[47] is described by: $T(x,y) = \min[x,y]$, where $T(x,y)$ is interpreted as the degree of truth of the compound statement x AND y. In terms of the three rules defined above, the degree of truth of the IF-side of Rule_1 is: $D(Rule\_1) = \min[\mu_{Cool}(t), \mu_{High}(h)]$. Using the Zadeh intersection again for the remaining rules yields:

$$D(Rule\_2) = \min[\mu_{Warm}(t), \mu_{Low}(h)]$$
$$D(Rule\_3) = \min[\mu_{Hot}(t), \mu_{Moderate}(h)]$$

where t and h are expected to be crisp values of temperature and humidity, respectively. Table 3.1 summarizes the action of this operator for a temperature of 78 degrees F and humidity of 40%.

Rule	Temperature	Humidity	Result
Rule 1	$\mu_{Cool}(78) = 0.0$	$\mu_{High}(40) = 0.0$	$\min(0.0, 0.0) = 0$
Rule 2	$\mu_{Warm}(78) = .33$	$\mu_{Low}(40) = .33$	$\min(.33, .33) = .33$
Rule 3	$\mu_{Hot}(78) = .42$	$\mu_{Moderate}(40) = .60$	$\min(.42, .60) = .42$

Table 3.1 Evaluating IF-side of three rules for inputs 78 degrees and 40% humidity.

---

[46] Zadeh, "Fuzzy Sets," *Information and Control* 8, pp 338-353.
[47] The notation T(x,y) identifies the Zadeh intersection with the mathematical family of T-norms, which generalize the intersection function.

Any number of antecedents may be combined using the intersection operator by successive application of the operator.

Figure 3.7a shows degree of truth produced by this operator plotted against the degree of two separate antecedents. Notice that in using the intersection operator, ultimately, one antecedent determines the degree of truth of the entire IF-side of a rule. Because the $\min[x,y]$ function is used, the antecedent with the lowest degree of membership determines the truth of the compound conditional statement.

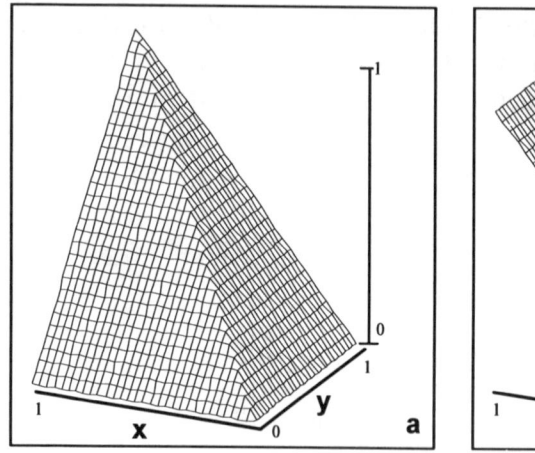

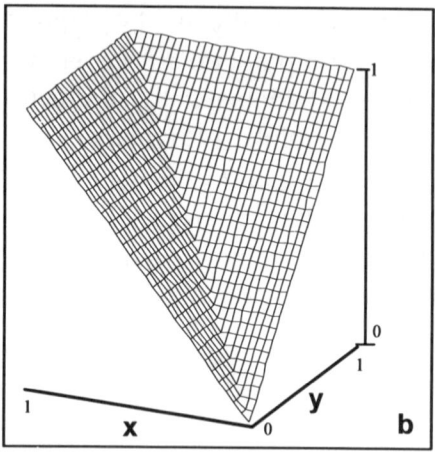

Figure 3.7: (a) Zadeh intersection x AND y, (b) Zadeh union x OR y.

Although the OR connective, which indicates a union operation, is not typically used in the evaluation of the compound degree of truth of a rule's IF-side, it provides a helpful comparison to the intersection operator. The Zadeh union[48] is defined as: $C(x,y) = \max[x,y]$, where $C(x,y)$ is interpreted as the degree of truth of the compound statement x OR y. The arguments x and y are expected to be crisp values of their corresponding variables. Any number of antecedents may be combined by successive application of the union operator. Figure 3.7b shows how

---

[48]The notation C(x,y) identifies the Zadeh union with the mathematical family of T co-norms, which generalize the union function.

the outcome of the union operation varies with the truth value of two antecedents.

Compensatory Operators

The MIN intersection operator is not the only way to interpret the *AND* connective found in fuzzy rules. As we pointed out in the preceding section, the intersection operator may be considered as the most severe interpretation of *AND*, because the truth of an IF-side is determined by only the single antecedent with the lowest degree of truth. Table 3.2 shows this effect clearly for the MIN intersection. This interpretation throws away all information contained in the other antecedent truth values.

x AND y	y=0.0	y=0.2	y=0.4	y=0.6	y=0.8	y=1.0
x=0.0	0.0	0.0	0.0	0.0	0.0	0.0
x=0.2	0.0	0.2	0.2	0.2	0.2	0.2
x=0.4	0.0	0.2	0.4	0.4	0.4	0.4
x=0.6	0.0	0.2	0.4	0.6	0.6	0.6
x=0.8	0.0	0.2	0.4	0.6	0.8	0.8
x=1.0	0.0	0.2	0.4	0.6	0.8	1.0

Table 3.2: MIN intersection is determined by the smallest truth value.

Similarly, the MAX union operator is not the only way to interpret the *OR* connective. The MAX union may be viewed as the most severe interpretation of the union operator because the truth of a compound *OR* statement is determined by only the single antecedent with the highest degree of truth. Just as in the case of the MIN intersection, the MAX union discards all information contained in the other antecedent truths values.

In response to the severity of the MIN and MAX operators, we have the COMPENSATORY OPERATORS. The COMPENSATORY OPERATORS are so called because their alternate interpretations of the *AND* and *OR* connectives compensate for the severity of the MIN intersection and MAX union operators. The dominant impact of the compensatory operators is the use of information contained in the truth values of more than just one of the antecedents.

Bounded Intersection

The BOUNDED INTERSECTION operator, sometimes abbreviated as BINTER, is defined as: $T(x,y) = BINTER(x,y) = \max[0, x+y-1]$. For an IF-side consisting of more than two antecedents, this equation is applied like a chain rule. The IF-side is evaluated by applying the operator to the first two antecedents, then applying the operator to that result and next antecedent. Thus,

$$BINTER(x,y,z) = BINTER(BINTER(x,y), z).$$

The final result is obtained when the last antecedent is used. Figure 3.8a graphically depicts the action of the bounded intersection operator. The figure shows that BINTER is insensitive to small truth values, but responsive to large truth values.

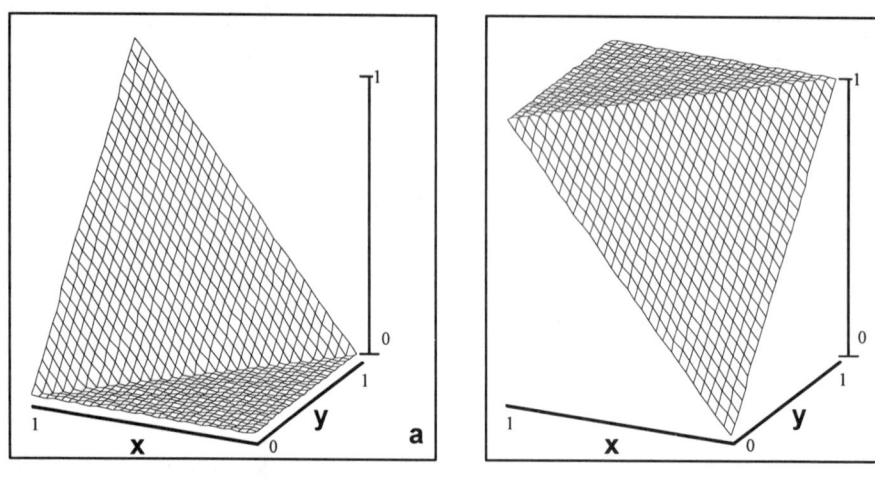

Figure 3.8: (a) BINTER intersection x AND y, (b) BUNION union x OR y.

Bounded Union

The *BOUNDED UNION* operator, sometimes abbreviated as *BUNION*, is the bounded form of the union operator. Union operators are typically associated with the *OR* connective, so it would be unusual to use a union operator in the evaluation of the IF-side of a rule. The mathematical form of this operator is:

$$C(x,y) = BUNION(x,y) = \min[1, x+y].$$

The action of this operator is shown in Figure 3.8b. The figure shows that *BUNION* is insensitive to high truth values, but sensitive to low truth values.

Product

As an alternative intersection operator, the *PRODUCT* operator simply multiplies all of the antecedent truth values of an IF-side together to arrive at a truth value for the compound statement. Figure 3.9a and Table 3.3 shows the product operator in two different ways.

x AND y	y=0.0	y=0.2	y=0.4	y=0.6	y=0.8	y=1.0
x=0.0	0.0	0.0	0.0	0.0	0.0	0.0
x=0.2	0.0	0.04	0.08	0.12	0.16	0.2
x=0.4	0.0	0.08	0.16	0.24	0.32	0.4
x=0.6	0.0	0.12	0.24	0.36	0.48	0.6
x=0.8	0.0	0.16	0.32	0.48	0.64	0.8
x=1.0	0.0	0.20	0.40	0.60	0.80	1.0

Table 3.3: Product intersection considers all antecedents.

Generalizing the action of this operator yields: $T(x,y) = [x \times y]$, where all of the arguments *x, y, etc.* are expected to be degrees of membership from their corresponding variables. The figure clearly displays the smooth nature of product intersection.

Fuzzy Logic for Real World Design

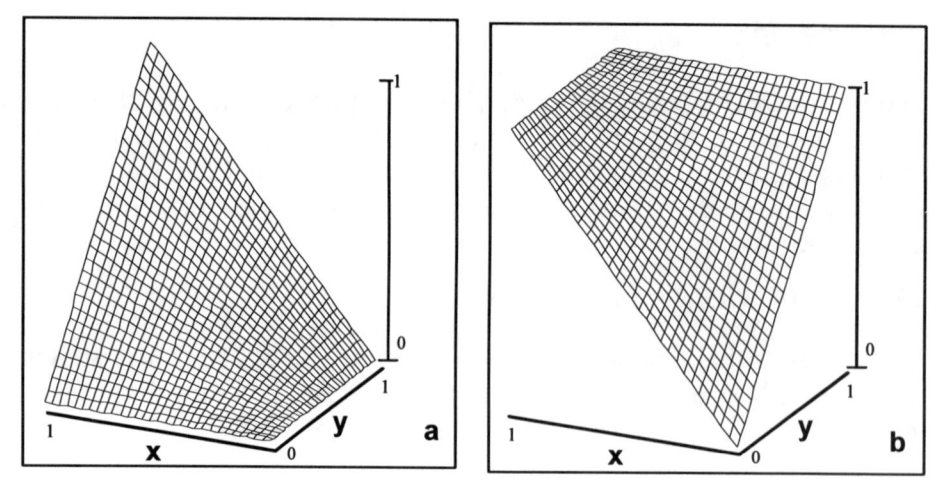

Figure 3.9: (a) PROD intersection x AND y, *(b) PSUM union* x OR y.

Probabilistic Sum

An alternative union operator, the *PROBABILISTIC SUM* is defined as: $C(x,y) = [x + y - x \times y]$. Like the bounded intersection operator, this equation is applied in chain rule fashion on an IF-side that consists of more than two antecedents. Figure 3.9b graphs the action of the probabilistic sum operator on two antecedents.

Although it's not immediately apparent, a little mathematical manipulation demonstrates that the probabilistic sum and product operators function as DeMorgan conjugates. A similar mathematical exercise will show that the binter and bunion operators also function as DeMorgan conjugates. Refer to Chapter 2 for details of DeMorgan's Theorem applied to fuzzy sets.

Yager Functions

All of the preceding compensatory operators have involved simple algebraic manipulations. The Yager class of compensatory operators specifies a parameterized family of *AND* and *OR* operators. The family of Yager intersection operators is given by the equation:

Chapter 3: Fuzzy Rules and Operations

$$T(x,y) = 1 - \min\left[1, [(1-x)^p + (1-y)^p]^{1/p}\right], p > 0,$$

where the parameter *p* controls the action of the operator. Figures 3.10a, b, c, and d show the behavior of this operator at the extremes of *p*, and two values in between.

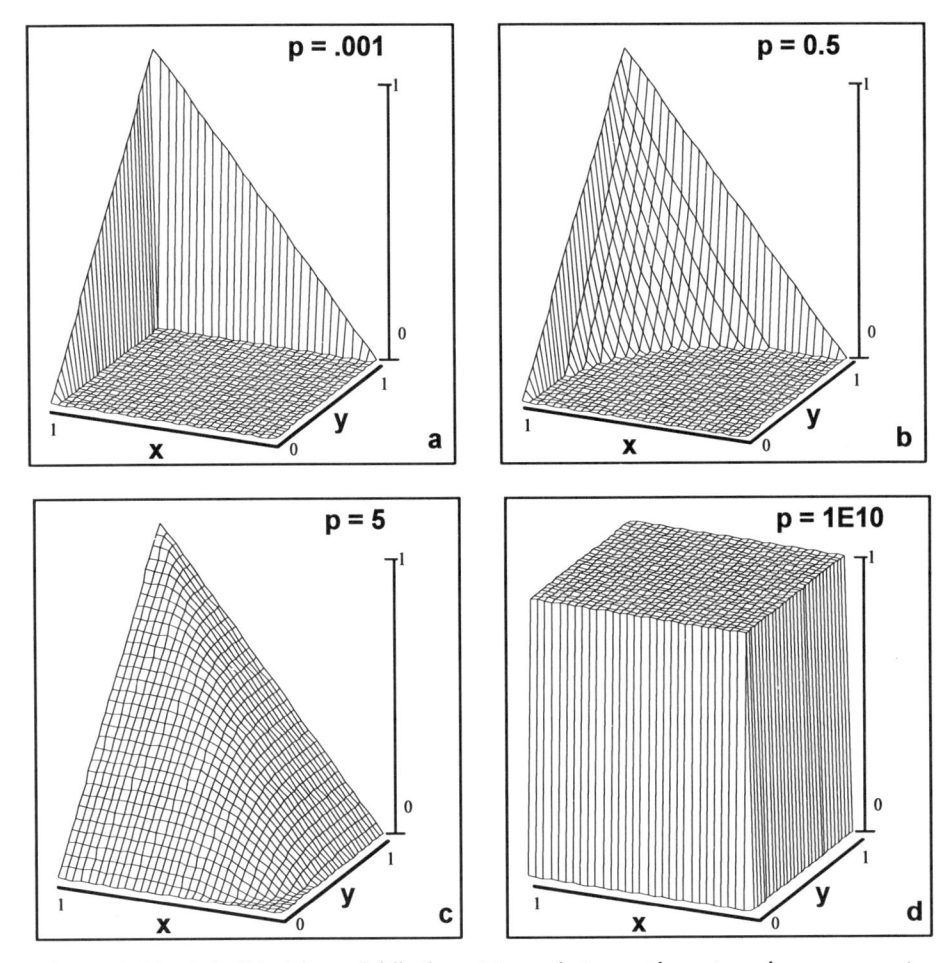

Figure 3.10: (a), (b), (c), and (d) show Yager intersection at various parameter values.

Each member of the Yager family of intersection operators has a DeMorgan conjugate union operator given by the equation:

$$C(x,y) = \min\left[1, [(x)^p + (y)^p]^{1/p}\right], p > 0.$$

Figures 3.11a, b, c and d show the behavior of this operator at the extremes of p, and two values in between.

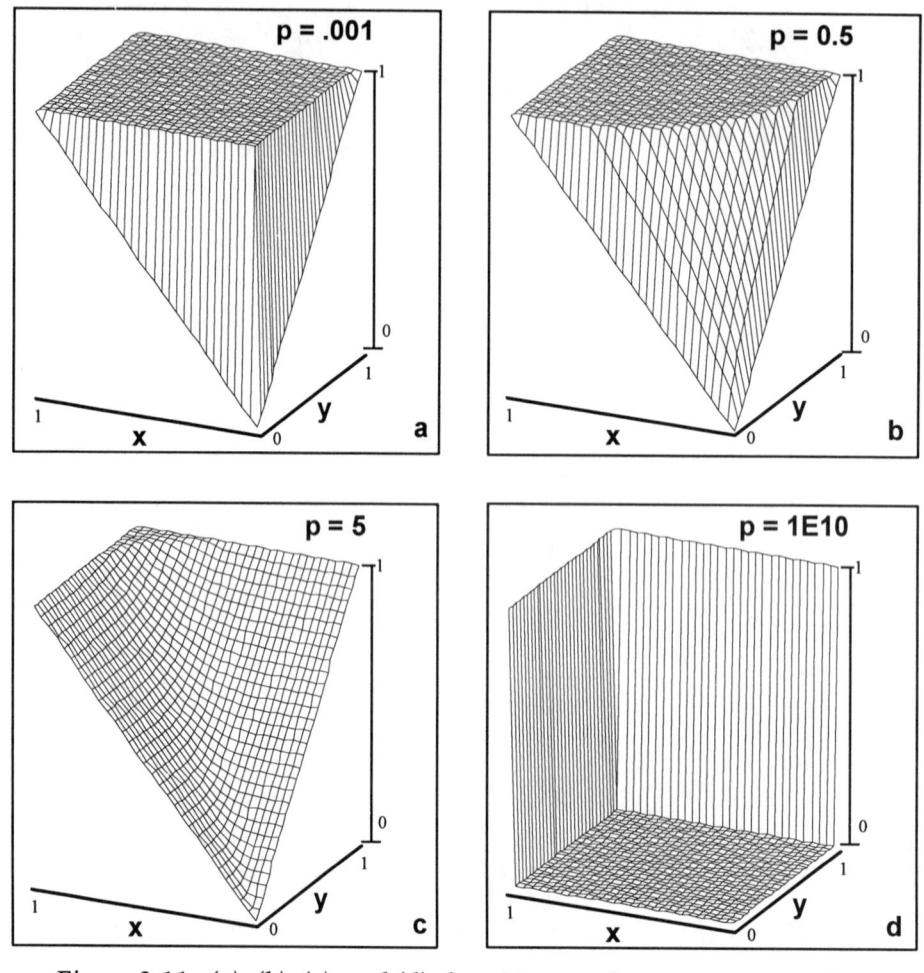

Figure 3.11: (a), (b), (c), and (d) show Yager union operator at various parameter values.

Although there exists an infinite variety of intersection and union operators, here we presented a representative sample that covers 99% of the real world applications of fuzzy logic. As with the definition of hedging operators, there is not one ultimate operator that proves all others inferior. The motivation in selecting operators should always be based in our common sense understanding and on the particulars of the problem we want to solve.

To this point, we have looked at each step in the process of evaluating the degree of truth of the conditional side of a fuzzy rule. That evaluation takes crisp inputs and produces a single degree of truth for the IF-side of a single rule. We now turn our attention to the THEN-side of a rule.

Inference

INFERENCE is the act of drawing a conclusion based on a premise. In the case of fuzzy logic rulebased systems, premises are spelled out as a combination of antecedents on the IF-side of rules. The consequents of a fuzzy rule can be interpreted as the conclusions drawn if the premises of a fuzzy rule are satisfied. Combining antecedents with any of the methods presented in the previous section will result in a single degree of truth for the entire IF-side of a rule. The fuzzy INFERENCE procedure specifies how the IF-side truth value applies to the consequents specified in a rule.

The strength of a conclusion should not be confused with the vbalue or action specified by the conclusion. The strength of the concusion simply indicates a degree of belief in the actual value or action specified in the rule consequent.

Characteristically, the degree to which the premise of a fuzzy rule is satisfied lies on the continuum of values [0,1]. What we expect from any inference procedure is that the strength of the conclusion should track with the strength of its associated premises. A premise truth of zero should lead to a conclusion with zero strength, and a premise truth of one should lead to a conclusion with maximal strength. In between the extremes of premise truth, the strength of a conclusion should increase as the strength of its premise increases. The inference methods presented below obey these simple guidelines.

Correlation Minimum

A rule consequent specifies the particular fuzzy set that results from satisfying the premise conditions of that rule's IF-side. The inference procedure adjusts the consequent fuzzy set based on the degree of truth of the IF-side of its associated rule. The CORRELATION MINIMUM INFERENCE method uses the truth value of the IF-side to truncate the consequent fuzzy sets. The truncation is performed at the level equal to the degree of truth of the premise.

As an example of this method, we'll consider the temperature control rules cited when we discussed intersection and union operators. Using crisp inputs 78 F temperature and 40% humidity, and the MIN intersection operator, the fuzzy regions produced by correlation minimum inference are shown in Figure 3.12.

Regardless of the initial shape of the consequent fuzzy set, the truncation step produces a flat topped fuzzy region. In practice, this truncation is viewed as a loss of information because the shape of the original fuzzy set is not preserved. As we'll see in the next section, the correlation product inferencing method overcomes this drawback.

Correlation Product

CORRELATION PRODUCT INFERENCE uses the truth value of the IF-side to scale the consequent fuzzy set. The consequent fuzzy set scaling works analogously to the NORM operator presented in Chapter 2. The consequent fuzzy set is multiplied by the degree of truth of the IF-side so that the highest membership grade of the resulting fuzzy set is less than or equal to the truth value of the IF-side.

Chapter 3: Fuzzy Rules and Operations

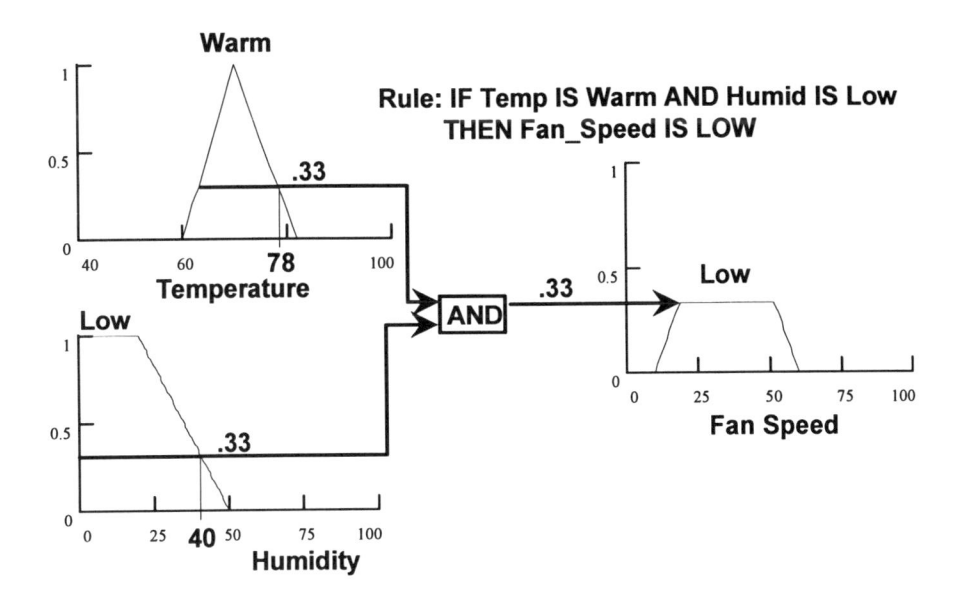

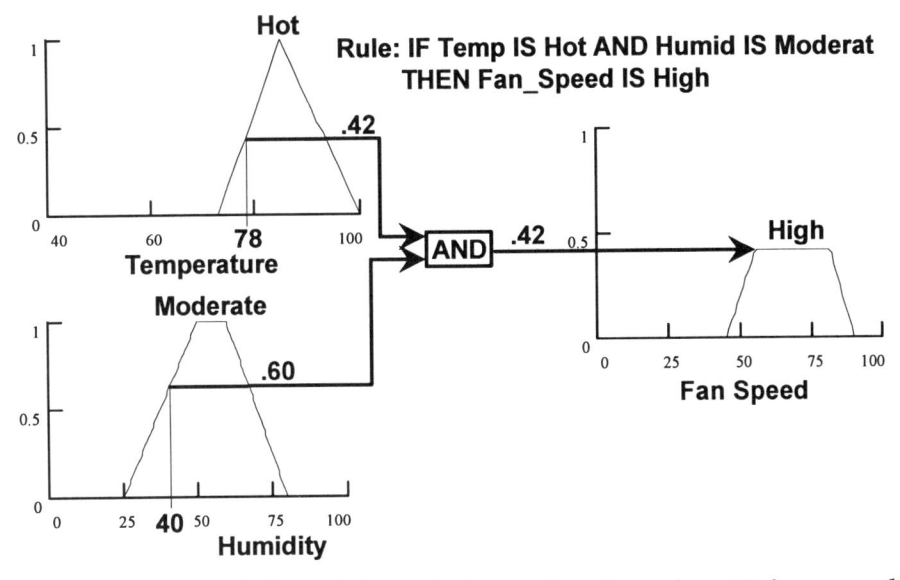

Figure 3.12: Correlation minimum inference for crisp values 78 degrees and 40% humidity, using MIN intersection operator.

Fuzzy Logic for Real World Design

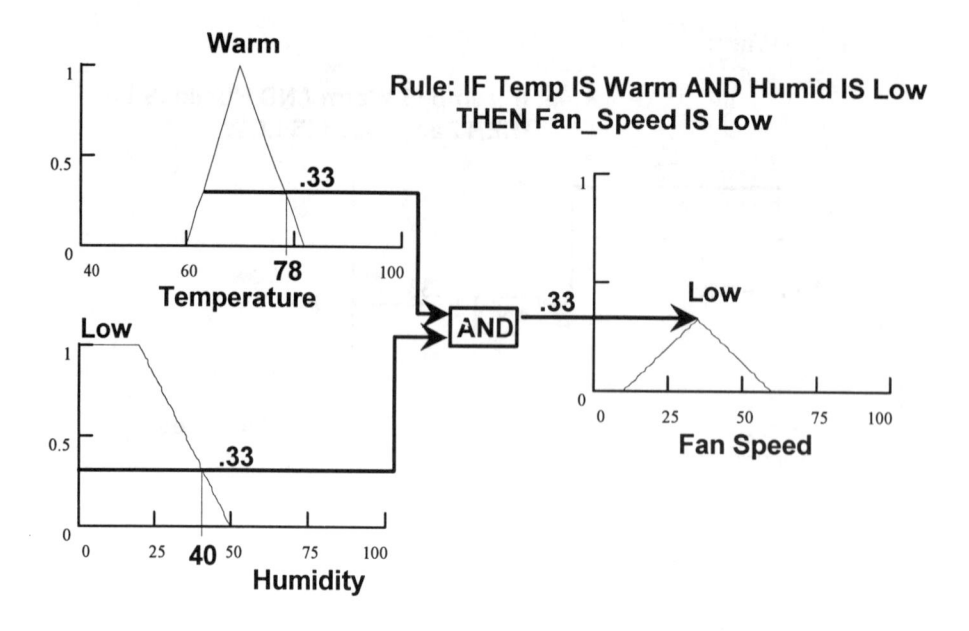

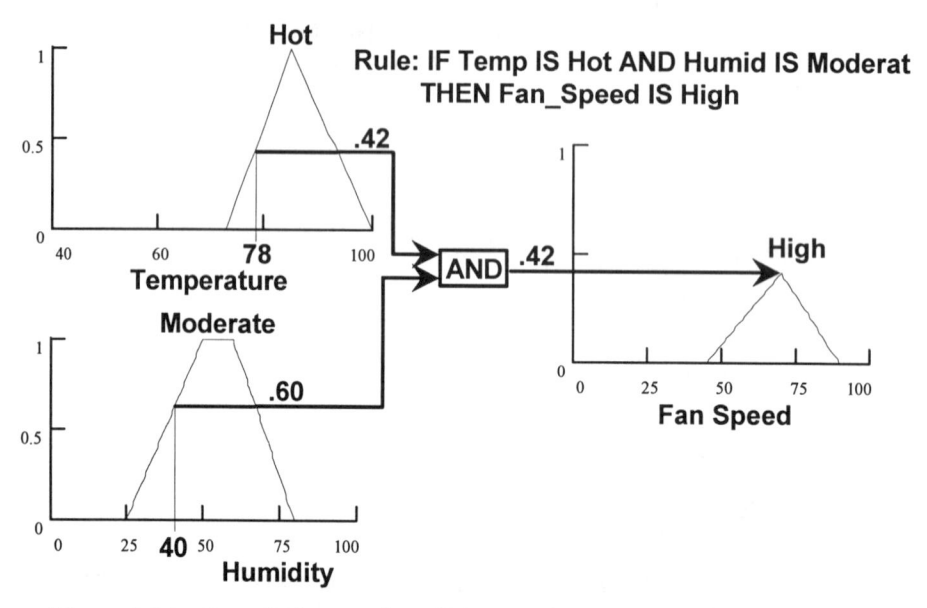

Figure 3.13: Correlation product inference for crisp values 78 degrees and 40% humidity, using MIN intersection operator.

Figure 3.13 shows how correlation product inference alters the consequent fuzzy sets involved in the temperature control example. In contrast to the correlation minimum inference technique, notice how correlation product inference preserves the shape of the consequent fuzzy set.

Min-Max

MIN-MAX INFERENCE takes the most shortcuts of any of the inference proceedures widely used. Similar to Correlation Minimum Inference, *MIN-MAX INFERENCE* uses the truth vaule of the IF-side of a given rule to truncate the consequent fuzzy sets on the THEN-side of the rule. *MIN* refers to the use of the MIN intersection operator, which we've already noted throws much of the antecedent information away.

As we will see in a following section, *MAX* refers to the way that the various consequent fuzzy sets produced by a rulebase are aggregated. For a given consequent membership function, only the strongest (*MAX*) rule consequent is used. For instance, if five different rules specify the same consequent, like *Fan_Speed IS Low*, only the rule that yields the highest IF-side truth value is used. In the same sense as the MIN intersection, the Max procedure throws away the information contained in the remaining four rule consequents.

Min-Max inference differs from the Correlation Minimum Inference in two important ways. First, correlation minimum inference does not restrict the AND operator to just the MIN intersection operator. Second, Correlation Mimimum Inference uses all consequents produced, not just the largest contributors.

Rulebase Evaluation

When considered together, the rules of a fuzzy model are called a *RULEBASE*. The rulebase, together with the membership functions completely encode the knowledge in a fuzzy system. We have already seen how crisp inputs percolate through the system to generate a consequent fuzzy set for each applicable rule. Two additional steps remain before we arrive at a crisp output. *AGGREGATION* sums the fuzzy sets produced by individual rules to determine an overall fuzzy region that indicates the total effect of all rules. The final step, *DEFUZZIFICATION*, computes a single crisp output from the aggregated fuzzy space.

Aggregation

For any given set of crisp inputs, several rules typically yield non-zero contributions to the output. *AGGREGATION* describes the process of taking the fuzzy sets produced by each individual rule and summing them into an overall fuzzy set. The set that is produced is interpreted as the total support provided to the output by the consequent fuzzy regions.

Additive Aggregation

The first, and most intuitive, approach to aggregating fuzzy consequents is a simple additive process. Figures 3.14 and 3.15 show the additive aggregation of our fan control rules for a temperature of 78 degrees and humidity of 40%. Although only two consequents were produced in our example, the aggregation process is the same regardless of the number of non-zero consequent fuzzy sets a system produces. As a result of aggregating many rules, this additive process may produce truth values greater than 1.

Chapter 3: Fuzzy Rules and Operations

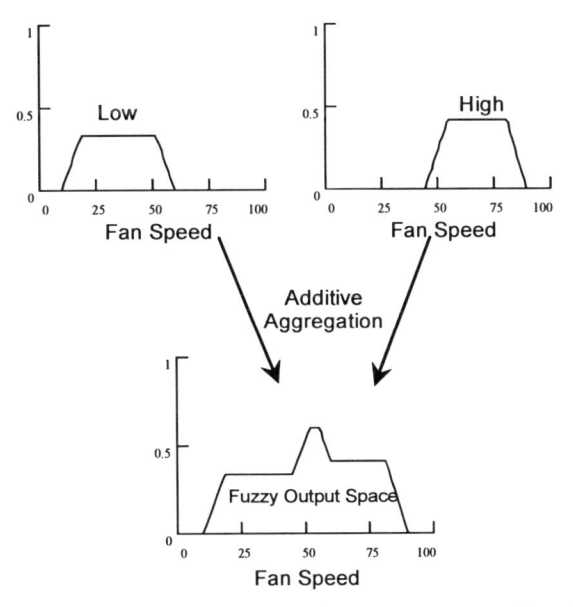

Figure 3.14: Additive aggregation of consequents produced by MIN intersection and correlation minimum inference.

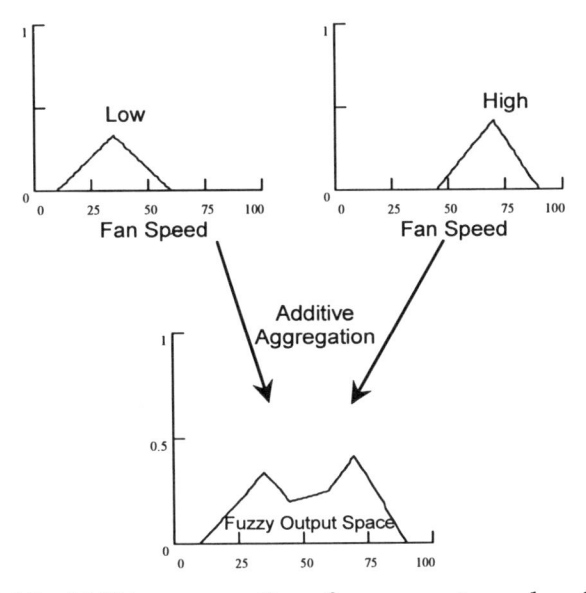

Figure 3.15: Additive aggregation of consequents produced by MIN intersection and correlation product inference.

95

Max Aggregation

The second, and more restrictive approach to aggregation is Max aggregation. We previewed Max aggregation while describing the workings of Min-Max Inference. Although many rules can simultaneously yield non-zero consequents, Max aggregation considers only the consequents with the highest truth values. Given the following rules:

Rule A: IF Temp IS Low AND Humidity IS Very High THEN Fan_Speed IS Moderate

   : Truth Value=0.30

Rule B: IF Temp IS Moderate AND Humidity IS High THEN Fan_Speed IS Moderate:

   Truth Value=0.60

Rule C: IF Temp IS Moderate AND Humidity IS Very High THEN Fan_Speed IS High

   Truth Value=0.20

Rule D: IF Temp IS Somewhat Low AND Humdity IS High THEN Fan_Speed IS High

   Truth Value=0.25

Max aggregation uses only rules *B* and *D*. While rules *A* (truth value 0.30) and *B* (truth value 0.60) both specify the consequent *Fan_Speed IS Moderate*, only the consequent having the highest truth value (rule *B*) is used. Similarly, the truth value of rule *C* (0.20) is lower than that of rule *D* (0.25), so rule *C*'s consequent is disregarded.

Summing the consequents produced by rules *B* and *D* produces the aggregated fuzzy output space. While disregarding all but the strongest contributions to the output certainly discards information, it speeds up the execution of the fuzzy model.

Defuzzification

DEFUZZIFICATION of the aggregated fuzzy output space yields a single crisp value. Defuzzification is the final payoff from the fuzzy system in the sense that in return for one set of crisp inputs, the whole fuzzy machinery produces a crisp output. In deriving a single value from the aggregated fuzzy space, the goal is to find the single crisp value that best represents the whole.

Chapter 3: Fuzzy Rules and Operations

One nearly unavoidable consequence of distilling the entire fuzzy output space down to a single crisp value is that information is lost. One crisp number cannot exactly represent the intricacy of the fuzzy output space. In practice, the lost information does not restrict the useful application of fuzzy logic[49]. Different defuzzification methods try to address the question of how to best represent the whole. We'll look at the most commonly used methods now.

Centroid

The most common defuzzification algorithm is the centroid, also called the center of mass or center of moments. As inputs vary, centroid defuzzification leads to smooth output transitions from one region to the next. Since smooth output transitions are desirable, especially in control system applications, and the calculation is straightforward, centroid defuzzification is applied widely.

The calculation takes its motivation from the physicists first moment of mass equation[50]. The fuzzy logic form of the algorithm calculates the weighted average of the fuzzy output space. Truth values supply the weighting. The equation takes the form: $Centroid = \dfrac{\int x\mu(x)dx}{\int \mu(x)}$, where the limits of integration are the range of the output variable x. In practice, we only need calculate a discrete sum of points. For a sum of points, the equation looks like: $Centroid = \dfrac{\sum_{i=1}^{N} x_i \mu(x_i)}{\sum_{i=1}^{N} \mu(x_i)}$, where N is the total number of points in the summation. Figure 3.16 shows the centroid of the fuzzy output produced by our fan control rules.

---

[49] This is especially true when fuzzy rulebased systems are used in feedback controllers.

[50] Robert Resnick, David Halliday, *Physics*, p. 165.

97

The centroid can also be viewed as the point where the area under the curve to the left exactly balances the area under the curve to the right.

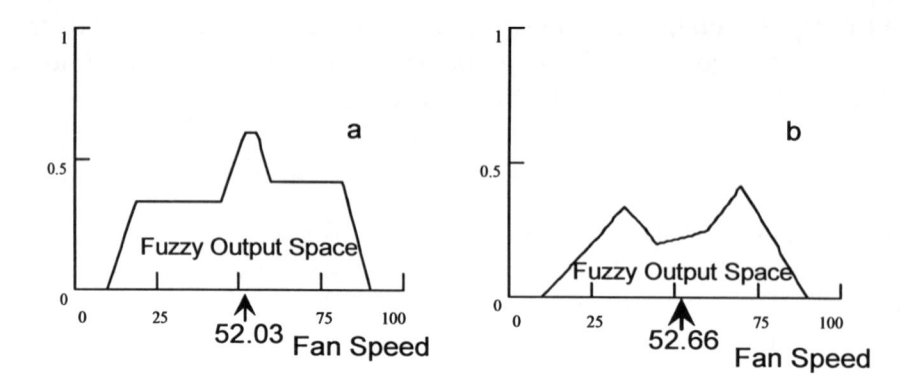

Figure 3.16: (a) Centroid of region in Figure 3.14, (b) Centroid of region in Figure 3.15.

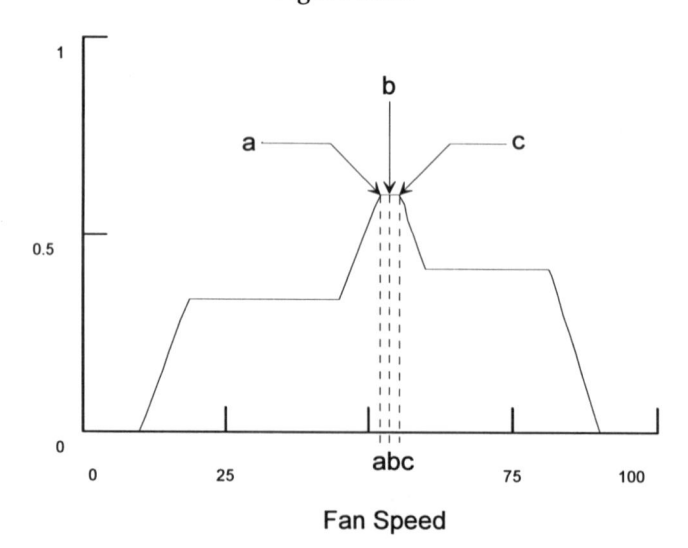

Figure 3.17: (a)Left-most Maximizer, (b)Composite Maximum, (c)Right-most Maximizer.

Chapter 3: Fuzzy Rules and Operations

Maximizers

The maximizer family of defuzzification methods includes *Composite Maximum, Leftmost Maximizer,* and *Rightmost Maximizer*. Each of these methods key off of the point or points with the greatest truth value in the fuzzy output space. The crisp value associated with that maximum truth value becomes the crisp output from the fuzzy system. For all versions of maximizer defuzzification, if one point in the fuzzy output space has a higher truth grade than any other point in the fuzzy output space, then defuzzification produces the crisp value associated with that point. Figure 3.17 shows all three of the maximizers applied to the fuzzy output space of Figure 3.14.

If, as in Figure 3.17, a maximum plateau exists in the fuzzy output space, the left-most point of the plateau defines the left-most maximizer defuzzification. Similarly, the right-most point of the plateau defines the right-most maximizer defuzzification. The composite maximum is just the average of the left-most and right-most maximizer functions. Maximizer defuzzification works well with correlation minimum inference because that inference method usually produces shifting plateaus in the fuzzy output space. Figure 3.18 shows the composite maximum applied to correlation minimum inference for two different sets of crisp input for our fan controller example.

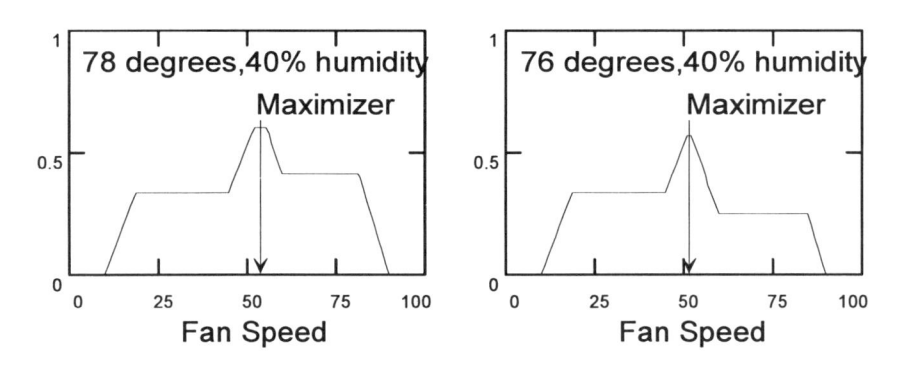

Figure 3.18: Maximizer vs. Centroid defuzzification for correlation minimum inference.

99

Fuzzy Logic for Real World Design

The strongest advantage of maximizer defuzzification is its speed of computation. However, the performance of maximizer defuzzification is sensitive to the inference method used. Specifically, maximizer defuzzification of a fuzzy output space produced by correlation product inference lacks the desirable characteristic of smooth output transitions. Correlation product inference usually produces a fuzzy output space devoid of truth plateaus. Because of this, and as illustrated by Figure 3.19, a small change in input can lead to abrupt changes in defuzzified output.

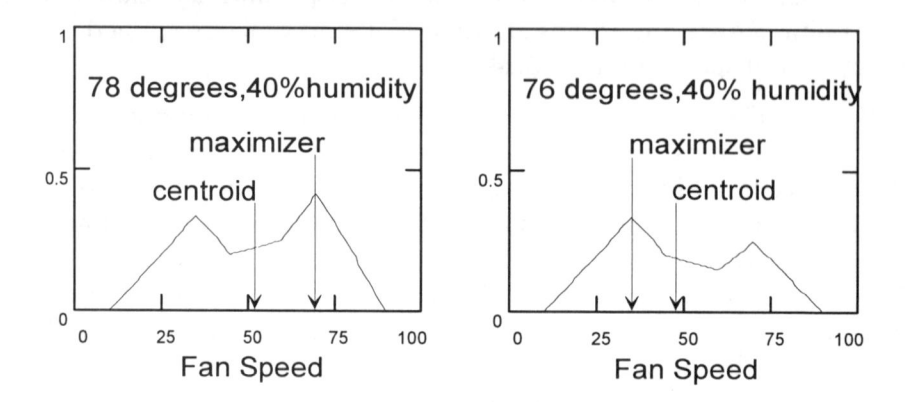

Figure 3.19: Maximizer vs. Centroid defuzzification for correlation product inference.

Singleton Defuzzification

Let us answer a few questions on singletons and their defuzzification at this point. A singleton, as was noted previously, is not a fuzzy set. So why is the singleton useful as an output fuzzy set? One reason is that under centroid defuzzification, singletons produce identical defuzzified outputs to symmetric fuzzy sets. This is easy to see since a symmetric fuzzy set (which looks like an isoceles triangle) has a centroid location that remains unchanged under both minimum correlation and product correlation inference. Figure 3.20 illustrates this connection.

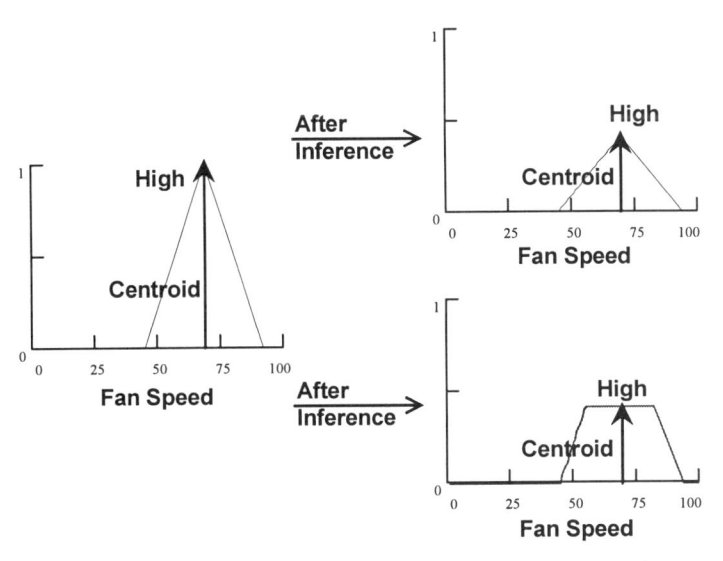

Figure 3.20: For an output variable, a singleton located at the the centroid of a symmetric membership function yields identical results under both correlation minimum and correlation product inference.

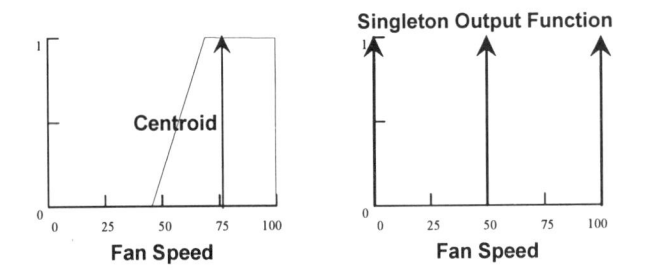

Figure 3.21: The centroid of a ordinary fuzzy set must be located inside its area, never at its edge.

Beneficially, using singletons instead of symmetric fuzzy sets reduces the computation needed to defuzzify into a crisp output. Individual centroids need not be computed for each inference loop, thus improving the speed of inference. Another reason to use singletons as output fuzzy sets stems from a limitation of the centroid calculation. When using fuzzy sets with finite areas, the crisp output of

Fuzzy Logic for Real World Design

the system can never reach the limits of its particular scope. For example, a fuzzy set that extends to lower limit of its scope, as depicted in Figure 3.21, has a centroid that will always be located inside the limits of its the variable's domain, regardless of the inference mechanism used.

The singleton may be located without the geometric constraint of the computed centroid.

A singleton may be located anywhere within the fuzzy variable's universe of discourse, including at the upper or lower bounds of that universe. This means that when defuzzifying singletons the full universe of discourse of the crisp output can be reached.

Assertions

An *ASSERTION* is a rule with no IF-side. Alternately, an assertion can be viewed as a rule whose IF-side is always true to maximal degree one. An assertion like, *Fan_Speed is Low*, acts to bias the output of a fuzzy system. With that assertion, the fuzzy set *Low* will always be part of the output space of the fuzzy system. Adding this assertion to the three rules in our temperature control example changes the output of our system as demonstrated by Figure 3.22.

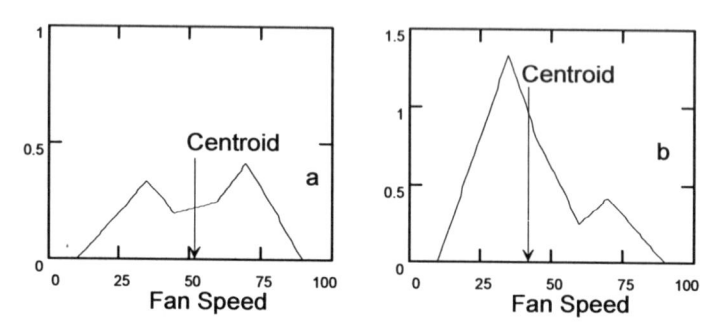

Figure 3.22: (a)Fuzzy output space for (78 degrees, 40%), (b)Effect of assertion Fan_Speed is Low *on fuzzy output space shown in (a).*

Alpha Cuts

In Chapter 2 we introduced the concept of an alpha-cut as it applied to a fuzzy set. In the context of a fuzzy set, the alpha cut was viewed as a thresholding operation. An *ALPHA CUT* applied to a rule performs a similar thresholding operation. If the truth value of the IF-side of a rule is below the alpha cut threshold, then the truth value of the rule is set to zero. Thus the alpha cut establishes the minimum truth value that a premise must obtain before that particular rule can contribute to the output. It is possible to apply a different alpha cut to each rule in a rulebase.

The best use for a rule alpha cut involves exception conditions and alarms. A rule meant to signal an alarm uses a high alpha cut threshold to prevent false triggering. Such an alarm might be triggered by this rule premise: *IF Boiler Pressure IS very High AND Temperature is High*. In response to the alarm, the fuzzy system fires rules to alert an operator, shut down the system, or bring the system back to a safe operating region.

Rule level alpha cuts find few other uses in practical systems. Without rule alpha cuts, properly designed fuzzy systems smoothly vary their output in response to changing system inputs. Rule alpha cuts degrade the ability of the fuzzy system to produce smooth output transitions. Alpha cuts cause unsmooth steps in the output response of the fuzzy system which tend to be detrimental to overall system performance.

Summary

This chapter presented fuzzy IF/THEN rules as the second of two key constructs which define fuzzy logic. The first key construct, the fuzzy set, was covered in the previous chapter and was the foundation upon which we built our understanding of the fuzzy inference process. We introduced the idea of a fuzzy proposition and saw that rules were just conditional combinations of fuzzy propositions. Fuzzy rules express a linkage between fuzzy regions of multiple variables. That linkage is, by its nature, approximate.

Fuzzy Logic for Real World Design

Whereas membership functions and rules can be viewed as the data structures of a fuzzy system, methods of manipulating those data are necessary to achieve a working system. To that end we examined, in detail, the mathematical operations involved in fuzzification, rule evaluation, and defuzzification. Table 3.4 summarizes the methods we explored.

INTERSECTION	INFERENCE	AGGREGATION	DEFUZZIFICATION
MIN BINTER Product Yager Functions	Correlation Minimum Correlation Product	MAX Additive	Left-most Maximizer Right-Most Maximizer Composite Maximum Weighted Average Centroid

Table 3.4 Summary of operators and methods covered in this chapter.

The next chapter explains how the elements of fuzzy logic may be be combined in the context of sovling real world problems.

Chapter 4:
Fuzzy Engineering

The term *FUZZY ENGINEERING* describes the application of fuzzy logic to the solution of real world problems. This chapter introduces a systematic approach to the fuzzy engineering process.

When to Use Fuzzy Logic

Fuzzy logic adds to and expands upon existing problem-solving techniques in powerful and revolutionary ways. In augmenting, rather than replacing traditional methods, fuzzy logic makes it possible to attack problems which were previously beyond our capabilities. While fuzzy logic is already responsible for an entire generation of successful, intelligent consumer products, it is destined to become just one more tool in an engineer's hands. As such, it becomes important to know what types of problems lend themselves naturally to a fuzzy approach.

Although the decision to use fuzzy logic techniques on a particular problem is itself fuzzy, Tables 4.1 and 4.2 attempt to give reliable guidelines. No doubt even these tables are incomplete, as newer and more diverse uses of fuzzy logic are announced every week.

Fuzzy Logic for Real World Design

FUZZY	NOT FUZZY
Mathematical model not feasible	Problem can't be represented numerically
Non-linear problem dynamics	Strictly Boolean situation
Approximate	Not measurable
Imprecise	Not quantifiable
Ambiguous	Unknown
Range of meanings	Exclusive meanings
Human expertise	Mathematical equations
Quick prototyping for approximate results	High precision requirements
Higher speed or smaller implementation size	Existing solution good enough
Too complex for Boolean logic-combinatorial explosion	Infrastructure impediments prevent the adoption of fuzzy methods

Table 4.1: Where is the fuzziness?

Classification	Fault identification
	Selection of one action or policy from many
Evaluation and Modeling	Risk and criticality assessment
	Cost estimation
	Asset valuation
	Credit scoring
	Forecasting
Control of Dynamic Systems	Replacement of human operator
	Non-linear control
	Pattern recognition
	Intelligent consumer appliances
Ranking of Alternatives	Investments
	Sales leads
	Sites
	Target markets

Table 4.2: Candidate applications of fuzzy engineering.

How to Use Fuzzy Logic

The previous chapters have laid out the meaning and math behind fuzzy rulebase modeling. We explored two key constructs, membership functions and IF/THEN rules, which together encapsulate the knowledge of the fuzzy model. We also saw how the three key operations of fuzzification, rule evaluation, and defuzzification permitted the fuzzy model to reason from crisp input to crisp output. This chapter builds upon that foundation by examining each step of the fuzzy engineering process.

Important aspects of the fuzzy design effort include: description of system operation, definition of input variables, output variables and membership functions, elicitation of control rules, simulation of fuzzy model and tuning of system parameters, translation to target system, and integration into user application. Each step of the process will be illustrated by working through a simple inverted pendulum control problem.

Describe the System

The first step is to describe how the system should work in plain language. Of course, this first step is common to all problem solving. The system description needs only to express the general workings of the system. The point is to capture the essence of the problem, identify inputs and outputs of the system, and understand how the inputs generally relate to the outputs. Figure 4.1 shows the high level perspective at this stage of the process.

Figure 4.1: Highest level block diagram.

Describing the desired system operation requires an understanding of what inputs to, and what outputs from the total system are necessary and possible. Describing the entire system will typically identify elements like input sensors, decision making elements, and output effectors. While fuzzy models significantly enhance the intelligence of machines, fuzzy models tend not to be stand-alone solutions. Fuzzy models comprise only part of the overall solution. Partitioning your system down into smaller scale pieces may produce functional elements like those depicted in Figure 4.2.

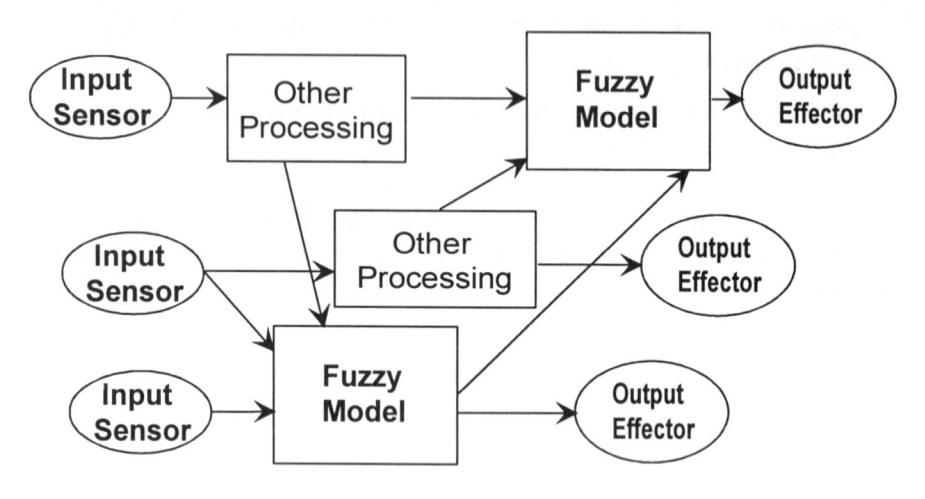

Figure 4.2: Fuzzy models as components of the total solution.

Once the desired system operation is outlined, the role of the fuzzy model will be clear. Identification of intelligent control or decision making elements in a system suggests the application of fuzzy modeling.

We will use the simplified inverted pendulum control problem throughout this chapter to illustrate each step of the process. Although the simplified inverted pendulum is, in a sense, a toy problem, the fuzzy engineering process follows the same steps as are necessary for more complex, real-world problems. Figure 4.3 illustrates the inverted pendulum apparatus. The objective, which requires active real-time control, is to balance the pendulum on its head.

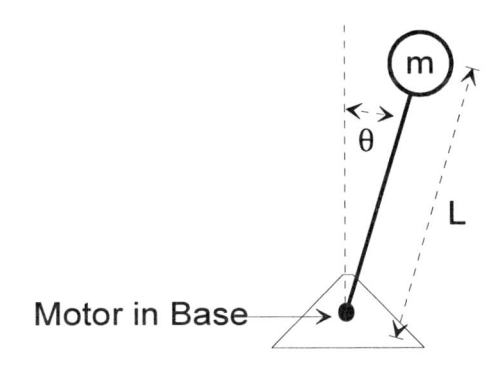

Figure 4.3: Inverted pendulum apparatus.

The description of the system depicted in Figure 4.3 follows. The pendulum consists of a rigid rod of length **L** attached to a bob of mass **m**. The pendulum is constrained to move only in the plane of angle **θ**. The base of the rod is connected to the shaft of a motor such that the angle of the rod with respect to a completely vertical position is described by angle **θ**. When powered, the motor applies a torque which can rotate the rod and mass combination to the left or the right. We want to develop a fuzzy model which will balance the pendulum in an upright position (where **θ=0**). In addition, the fuzzy controller will bring the pendulum into balance for any initial angle or speed.

Successfully controlling dynamical physical systems and processes generally rests upon the understanding and effective control of the degrees of freedom naturally embedded in the system of interest. Since the pendulum is constrained to move only in the plane of **θ**, the physical system of Figure 4.3 has only two degrees of freedom to consider, namely angle **θ**, and angular speed $\dot{\theta}$. These degrees of freedom dictate our fuzzy model inputs. The two inputs to the fuzzy model will be angle θ and angular speed $\dot{\theta}$[51].

---

[51] In practice, the ∫θdt term is quite powerful for the condition $\theta \cong \phi$ and $\dot{\theta} \cong \phi$, and is often the only way to achieve a comfortable gain margin. However, for instructive purposes we wish to keep this example less complex.

The output from the fuzzy model must relate to the output effector we have chosen, namely the motor. We want to control the voltage to the motor such that a positive voltage applies a clockwise torque (in the direction of positive θ), and a negative voltage applies a counter-clockwise torque (in the direction of negative θ). So the single output from the fuzzy model will be motor voltage v.

Specify Ranges of All Inputs and Outputs

Once the inputs to and the outputs from the fuzzy model have been identified, select the universe of discourse for each fuzzy variable. Recall that the universe of discourse for a fuzzy variable is defined as a range of crisp values. The universe of discourse should be wide enough to capture all of the meaningful range of that variable. The motivation in specifying a universe of discourse is practicality. The meaningful range of a given variable seldom spans $-\infty$ to $+\infty$.

The ranges of inputs and outputs are problem dependent. For instance, if we use an 8 bit analog-to-digital (A/D) converter to acquire the angle data in our inverted pendulum example, we have at most 2^8, or 256, distinct crisp values of θ. A 12 bit A/D converter would yield at most 2^{12}, or 4096 distinct crisp values. On the other hand, we may simulate the fuzzy model in software, using floating point variables which, in practice, yield a near continuum of crisp values.

It is usually possible to identify extreme values of a particular fuzzy variable, beyond which system operation is not altered. Those extreme values provide good limits for a fuzzy variable's universe of discourse. Simulating our example problem in a high level language on the computer allows us to use floating point variables, which eliminates, for now, the concerns of digitizing to 256, 4096, or more discrete values. Furthermore, the angle θ is limited by the pendulum bob hitting the table surface. We define the meaningful range of θ in our problem as -100 degrees to +100 degrees. The universe of discourse for the angular speed $\dot{\theta}$ is more a matter of judgment than an intrinsic limitation of the apparatus. We must decide on the meaningful range for $\dot{\theta}$ based on either experience or guesswork. The development and refinement of a fuzzy model is an iterative process, so specification of the universe of discourse does not need to be exactly optimal at this stage. Hence, we choose the range -50 degrees/second to +50 degrees/second for angular speed $\dot{\theta}$.

Specifying the universe of discourse for a fuzzy model output depends on how the output will be applied to the system. In a physical control system, the output effector device tends to dictate the meaningful range of its associated fuzzy output variable. For example, an output from a fuzzy model which is intended to control a chemical reactor may operate a valve. The meaningful range for that output variable depends on the mechanism of operating that valve. Since our example problem is a physical control system, the actuation of the output effector (the motor) dictates a universe of discourse from -5 volts to +5volts.

In an information system type application like inventory forecasting, the fuzzy output variable can be more naturally expressed. In such problems, the meaningful range of the fuzzy variable simply follows the semantics of the model. For instance, if the fuzzy output variable is *Number_of_Widgets* then its meaningful range spans the minimum to the maximum number of widgets in inventory.

Partition into Membership Functions

Traditional fuzzy model development methods rely on a human expert to develop membership functions from experience and introspection. When relying on human expertise, fuzzy rulebased modeling is intrinsically suited to the job. The ease of expressing human expertise with fuzzy membership functions and fuzzy IF/THEN rules, and the fuzzy nature of human expertise team together to make a superior modeling tool.

The ease of expressing expertise with fuzzy logic should not be confused with actually possessing expertise in a particular area. The expertise referred to here is expertise in a specific area of application, like process control or inventory forecasting. An important point to remember is that fuzzy logic in the hands of the application expert will produce better results, faster, than in the hands of a non-expert.

Partitioning each fuzzy variable into overlapping membership functions starts the infusion of real knowledge into the fuzzy model. Several decisions must be made at this stage regarding the number, shape, and placement of membership functions along a given universe of discourse.

Number of Membership Functions

When translating human expertise[52] into fuzzy sets, most fuzzy variables end up with between two and seven membership functions. While more than seven membership functions certainly can be used, especially in fuzzy function approximation, most applications use seven or less. Too few membership functions will result in inadequate and ineffective control. Too many membership functions, in addition to being cumbersome, add an uneeded computational burden.

Psychology provides a partial explanation for the number of membership functions typically used. In our language we seldom identify more classes than *Small, Medium,* and *Large*. If a fuzzy variable spans negative to positive values, then the natural representation might be *Negative Large, Negative Medium, Negative Small, Zero, Positive Small, Positive Medium,* and *Positive Large*. This scheme amounts to seven membership functions.

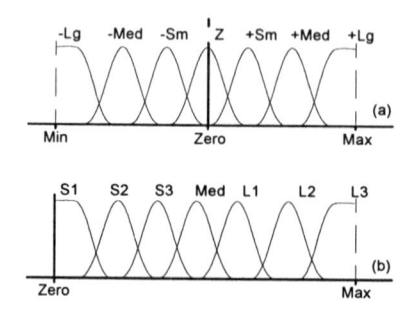

Figure 4.4: (a) Bipolar fuzzy variable, (b) Unipolar fuzzy variable.

---

[52]Current research continues to explore intelligent, automated methods of partitioning fuzzy variables into membership functions. Such methods rely on data from the system of interest, rather than human expertise, to generate membership functions. Promising methods include neural network techniques and genetic algorithms.

We can generalize all fuzzy variables to just two types: bipolar, and unipolar. Bipolar variables will range from negative values, through zero, to positive values. It is often useful to define a membership function that straddles zero, especially in process control applications. In process control applications, corrective action must be taken when the system deviates from a specified set-point. When the system already rests at the set-point, no action is needed. The membership function straddling zero allows the representation of "at the set-point." Equal numbers of membership functions on either side of zero allows a symmetrical representation of deviations from the set-point. Partitioning bipolar fuzzy variables into an odd number of membership functions allows the representation of the *zero* point as well as symmetry around *zero*.

Unipolar variables range from zero to either positive or negative values. A good example of a unipolar variable is *Humidity*, which we used in the previous chapter. *Humidity* spans the range [0,100]. Since no provision for straddling zero is needed for a unipolar variable, either even or odd numbers of membership functions work well.

The inverted pendulum uses three bipolar fuzzy variables. Angle θ spans the interval [-100,+100], angular speed, $\dot{\theta}$, spans the range [-50,+50], and the output fuzzy variable voltage, v, spans the range [-5,+5]. We will use five membership functions for each fuzzy variable.

Shape of Membership Functions

Membership functions approximate fuzzy classes and concepts. Since fuzzy classes and concepts defy exact description, any membership curves we draw must be regarded as approximations. Far from being a drawback, however, the approximate nature of membership function shapes results in a broader set of choices for those shapes. Fuzzy systems are tolerant of the variety of approximations to the underlying fuzzy concepts that we make. This tolerance frees the fuzzy model designer to select membership function shapes based not only on the underlying fuzzy classes but also based on implementation concerns like memory constraints, execution speed, etc.

In embedded control applications, where a fuzzy model executes on an 8 bit or 16 bit microcontroller, triangular and trapezoidal membership function representations predominate. Frequently in such applications output fuzzy sets are singletons. The proliferation of embedded fuzzy applications indicates that even the limited triangular and singleton membership function shapes allow adequate modeling of fuzzy classes and concepts.

Fuzzy models aimed at information systems application, like an intelligent sales advisor, seldom require the instantaneous response necessary in embedded control. Such fuzzy models are freer to exploit the entire range of shapes including S-, Z-, and Pi-curves, Gaussian shapes, and multipoint, piecewise linear shapes.

The inverted pendulum example uses triangular shaped membership functions.

Placement and Overlap of Membership Functions

Placement and overlap receive most of the effort and attention devoted to defining the membership functions of a fuzzy model. Also, when refining the performance of a fuzzy model, most of the effort is devoted to tuning the placement and overlap of those membership functions. While it is true that the best performance will not be obtained from a fuzzy model without tuning, it is also true that following a few simple guidelines can yield significant initial results. Here are a few guidelines.

Placement

The placement, and more specifically the density, of membership functions across a universe of discourse strongly influences the degree of control achievable in each region. The general rule to follow in placement is that more membership functions should be used in the areas where finer control is needed. Fewer membership functions should be used in regions of the universe of discourse where a small change in input does not require a large change in output.

Figure 4.5 shows how different partitioning can be used to enhance control in certain regions. As the density of membership functions

Chapter 4: Fuzzy Engineering

varies across a universe of discourse, the density of control rules derived from those membership functions also varies. In Figure 4.5a the universe of discourse is divided equally among seven membership functions. The scope of each membership function is denoted by label R1, R2, etc. The arrows indicate the region over which each membership function is non-zero. The uniform partitioning resulting from R1, R2, etc., causes control rules using those membership functions to be equally distributed across the universe of discourse. Figure 4.5b, while using the same number of membership functions, shows an unequally divided universe of discourse. The high density region surrounding R3 allows more rules to effect the model output in that region. The low density regions R6 and R7 are affected by fewer rules.

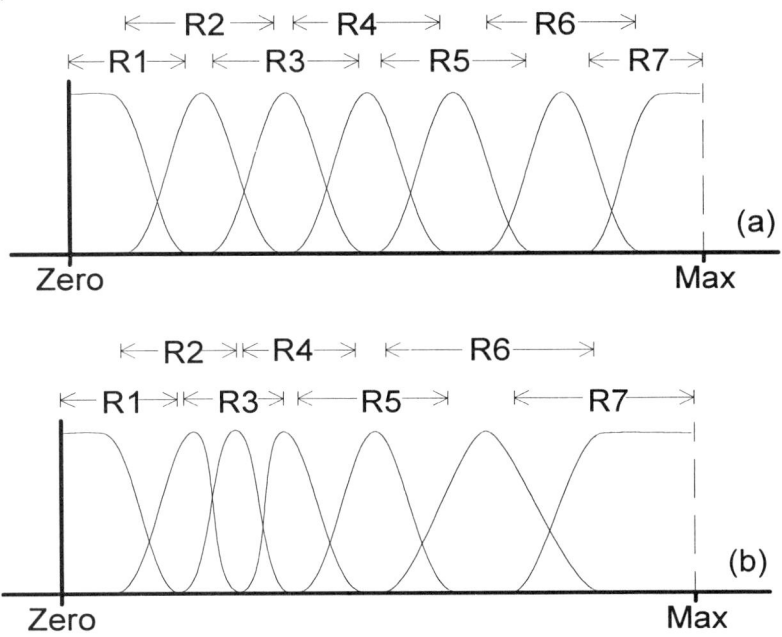

Figure 4.5: (a) Equal partitioning, (b) Unequal partitioning.

In the inverted pendulum example, the regions of greatest control should be the regions centered around $\theta = 0$, and $\dot{\theta} = 0$. Why? The goal is to balance the pendulum with $\theta = 0$ and $\dot{\theta} = 0$. Clustering membership functions around $\theta = 0$ and $\dot{\theta} = 0$ allows more control rules to affect those local regions. A higher density of membership functions in a region leads to a higher density of control rules in that region. In dense regions, small changes in input can cause large changes in the fuzzy model output. Such non-linear response is an inherent benefit of fuzzy modeling.

Overlap

Overlap refers to the degree to which adjacent membership functions share the same scope. Membership functions with no overlap reduce to a Boolean system. Excessive overlap can cause instability due to a large and changing number of control rules which contribute to the output. Luckily, the right amount of overlap is easily found between these two extremes.

The following guidelines address the vast majority of applications.

1. *Every point in the universe of discourse should be a member of at least one fuzzy set.*

2. *Most points in the universe of discourse should be a member of two fuzzy sets.*

3. *Only one fuzzy set should have the maximum degree of truth (1.0) at any point in the universe of discourse.*

4. *Where membership functions overlap, the sum of membership degrees for all fuzzy sets should be between .5 and 1.0.*

5. *In special cases, small regions of the universe of discourse may be members of at most three fuzzy sets.*

Use of an objective metric to assess membership function overlap can assist in the design of a fuzzy model. The *OVERLAP RATIO*[53] is one such metric. Figure 4.6 illustrates the two quantities which make up the metric. As shown in that figure, the Overlap Ratio is computed by dividing the scope of two adjacent membership functions into the amount of the overlap.

---

[53]Proposed by Steve Marsh et. al. in Motorola Fuzzy Logic Education Kit.

Chapter 4: Fuzzy Engineering

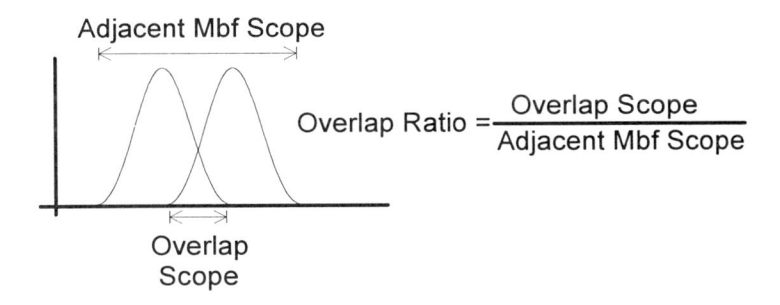

Figure 4.6: How to calculate Overlap Ratio.

Most fuzzy models use Overlap Ratios between 0.15 and 0.6. An Overlap Ratio of 0.33 corresponds to Figure 4.7a. An Overlap Ratio of 0.15 corresponds to Figure 4.7b. Low Overlap Ratios characterize a fuzzy model having strong Boolean qualities. In general, higher Overlap Ratios allow a fuzzy model to tolerate noisier or more ambiguous inputs.

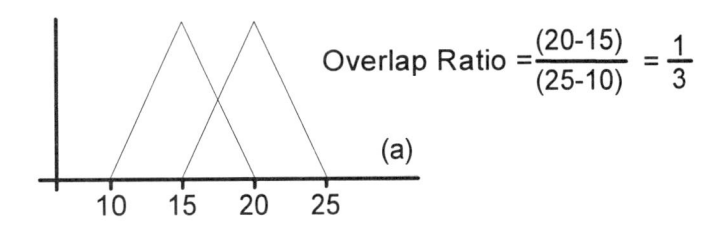

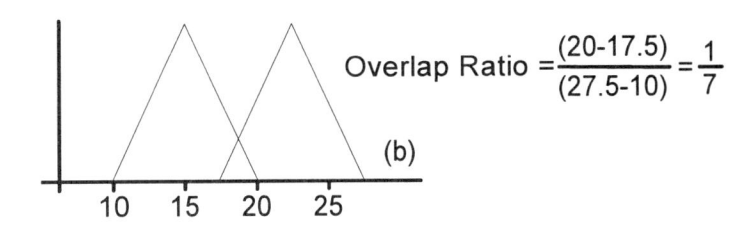

Figure 4.7: (a) Overlap Ratio of 1/3, (b) Overlap Ratio of 1/7.

117

Symmetry

There are two symmetry aspects which impact membership function partitioning. One aspect of symmetry involves all of the fuzzy sets of a particular variable. The other aspect of symmetry applies only to a single membership function. Understanding both aspects will aid in the construction of a good fuzzy model.

The first symmetry guideline applies to bipolar fuzzy variables. Often, the best approach to partitioning a bipolar variable is to make the right and left halves of the universe of discourse mirror images. Mirror image partitioning implies that the system being modeled or controlled behaves symmetrically with respect to positive and negative values of that fuzzy variable. This is the dominant approach in process control environments, where the ideas of set-point and distance from set-point drive the model. If the underlying model behaves asymmetrically about the set-point, fuzzy set partitioning should reflect that. See Figure 4.4a for an example of mirror image partitioning.

Symmetry of an individual membership function refers to the shape of that membership function with respect to its point of maximal truth. Although most of the membership functions graphed to this point have been symmetric, asymmetric functions are frequently used in applications. An asymmetric membership function has sides with different slopes. The steeper sloped side changes in degree more than the shallow sloped side for the same change in crisp value of the fuzzy variable. The the fuzzy sets labelled *NSml* and *PSml* in Figure 4.9a demonstrate this asymmetry.

Using all of the guidelines presented thus far, the membership function partitioning for the variables of the example problem is shown in Figures 4.9a, b, and c.

Chapter 4: Fuzzy Engineering

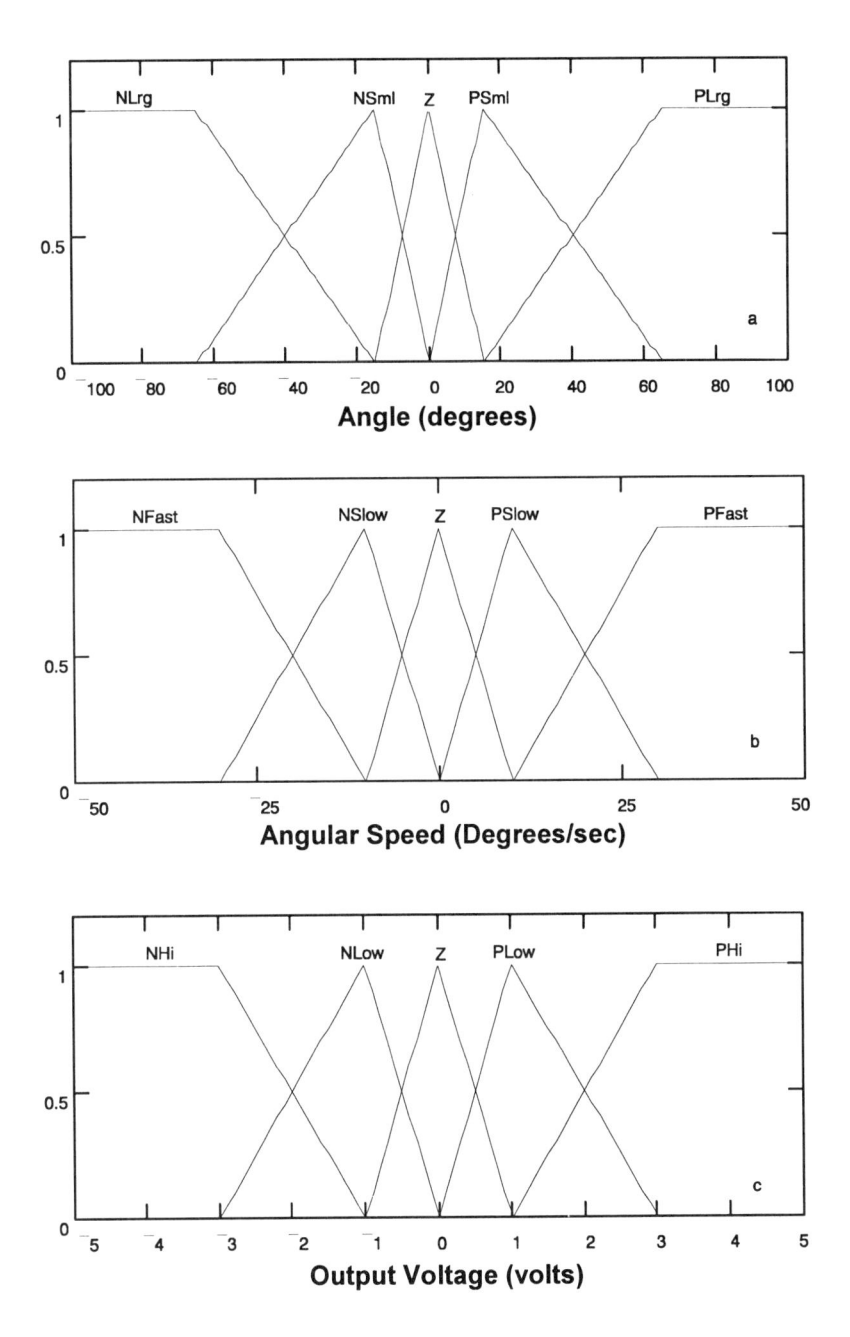

Figure 4.9(a), (b), (c): Membership functions for inverted pendulum control.

119

Construct IF/THEN Rules

While membership function creation begins the infusion of knowledge into the fuzzy model, rule writing conveys the bulk knowledge. Rules capture the relationships between input and output fuzzy variables in the form of IF/THEN statements. IF/THEN rules easily model complex system behavior and non-linear relationships. Selection of intersection operator, inference, aggregation, and defuzzification methods complete the infusion of knowledge. Ahead, we will look at each of these steps.

Rule Writing 1

At the same time that fuzzy rules model complex relationships, their natural language syntax makes them easy to read, write, and understand. The traditional fuzzy model development process relies on a human expert to write rules describing system operation.[54] When process and system knowledge resides with an expert, rule writing involves just a few steps.

An important question to ask is: how many rules are needed? No single answer to that question could satisfy the diverse requirements of fuzzy modeling. However, some general guidelines are helpful. The first guideline is to avoid writing conflicting rules. Two or more rules are said to conflict if they have the same IF-sides but different THEN-sides. Assuming all possible combinations of input fuzzy sets are used to write a complete set of non-conflicting rules, the equation $Max\_Rules = N_1 * N_2 * ... * N_i$ calculates the maximum number of rules available. The subscript *i* indicates the number of input variables, N_1 is the number of membership functions in the first input variable, N_2 is the number of membership functions in the second input variable, etc. The inverted pendulum uses two input variables, each of which is divided into five membership functions. Using the previous equation shows that a maximum of 25 non-conflicting rules are possible.

---

[54] In the same vein as the prior footnote regarding automatic generation of membership functions, much promising work continues in the area of automatic generation of rules. Promising methods include neural network techniques, genetic algorithms, and the Wang-Mendel algorithm.

Another useful guideline is that at least one rule must have a non-zero degree for any point in the multi-dimensional space specified by the universes of discourse of the input variables in the fuzzy model. The danger in not using enough rules to cover the universes of discourse of the fuzzy model inputs is that the fuzzy model output will be undefined for certain combinations of crisp inputs. Put another way, an insufficient number or poor distribution of rules leaves holes in the fuzzy model.

First, write the obvious rules. Obvious rules exist in almost any fuzzy model. A rule might be considered obvious if an expert is not needed to identify it. Often, the easiest rules to write govern the behavior of the system at the extremes of its variables. In the inverted pendulum example we can identify several obvious rules. Consider these rules which act when the inverted pendulum system is far from the ideal (the ideal being angle = 0, and angular speed = 0).

Rule 01: IF Angle IS Neg_Lrg AND Speed IS Neg_Fast THEN Voltage IS Pos_Hi

Rule 02: IF Angle IS Pos_Lrg AND Speed IS Pos_Fast THEN Voltage IS Neg_Hi

These two rules express the knowledge that if the angle is large (either positive or negative) and the pendulum is moving quickly away from the ideal system state, then the appropriate action is to apply a large torque in the opposite direction. A positive voltage applies a torque to the pendulum in the direction of positive θ and a negative voltage applies a torque in the direction of $-\theta$.

The next two rules could also be considered obvious by extending rules 01 and 02:

Rule 03: IF Angle IS Neg_Sml AND Speed IS Neg_Fast THEN Voltage IS Pos_Hi

Rule 04: IF Angle IS Pos_Sml AND Speed IS Pos_Fast THEN Voltage IS Neg_Hi

Rules 03 and 04 say that if angle is incorrect to a small extent, and the pendulum is moving fast in the wrong direction, then a large torque must be applied in the opposite direction to restore the pendulum to equilibrium. As demonstrated by the first four rules, rules involving bipolar fuzzy variables often appear in symmetric pairs, equal but opposite in effect. If an underlying system behaves symmetrically around a set-point, then symmetric fuzzy rules naturally express that feature of the system.

Another obvious rule governs system operation at the set-point:

Rule 05: IF Angle IS Zero AND Speed IS Zero THEN Voltage IS Zero

This rule says that if the pendulum is at the ideal state, then no action is needed.

Next, write the less obvious rules. This is the step where experience and intuition are used to shape the finer areas of the fuzzy model. An expert's knowledge of the subtleties in a system are freely expressed through the grammar of IF/THEN rules. At this stage in the example problem we can write:

Rule 06: IF Angle IS Neg_Sml AND Speed IS Pos_Slow THEN Voltage IS Zero

Rule 07: IF Angle IS Pos_Sml AND Speed IS Neg_Slow THEN Voltage IS Zero

Rules 06 and 07 express the knowledge that if the pendulum is close to the ideal angle, and it is moving in the direction of the ideal state, then do nothing. The next four rules express similar subtleties of the pendulum control.

Rule 08: IF Angle IS Zero AND Speed IS Pos_Slow THEN Voltage IS Neg_Low

Rule 09: IF Angle IS Zero AND Speed IS Neg_Slow THEN Voltage IS Pos_Low

Rule 10: IF Angle IS Zero and Speed IS Pos_Fast THEN Voltage IS Neg_Low

Rule 11: IF Angle IS Zero and Speed IS Neg_Fast THEN Voltage IS Pos_Low

A convenient device for visualizing an entire rulebase is shown in Table 4.3. The table shows the rule matrix corresponding to a two input, one output system. The two inputs define the rows and columns of the matrix while the entries in the matrix define the output for that specific combination of inputs. In the fuzzy logic literature this rule matrix is also known as a *FUZZY ASSOCIATIVE MEMORY* (or *FAM*). The 11 rules defined to this point partially fill the rule matrix of Table 4.3.

Chapter 4: Fuzzy Engineering

Speed\Angle	Neg_Lrg	Neg_Sml	Zero	Pos_Sml	Pos_Lrg
Pos_Fast			Neg_Low		Neg_Hi
Pos_Slow		Zero	Neg_Low		Neg_Hi
Zero			Zero		
Neg_Slow	Pos_Hi		Pos_Low	Zero	
Neg_Fast	Pos_Hi		Pos_Low		

Table 4.3: Rulebase matrix relating two inputs and one output.

Do the rules summarized in Table 4.3 sufficiently cover the possibilities? In order to answer that question positively, the fuzzy model must not have any holes in it. Figure 4.10 shows how the existing rules cover the two dimensional input space defined by the membership functions for Angle and Angular Speed. The labels **0, 1, 2,** or **3** indicate the number of overlapping rules that apply to each region.

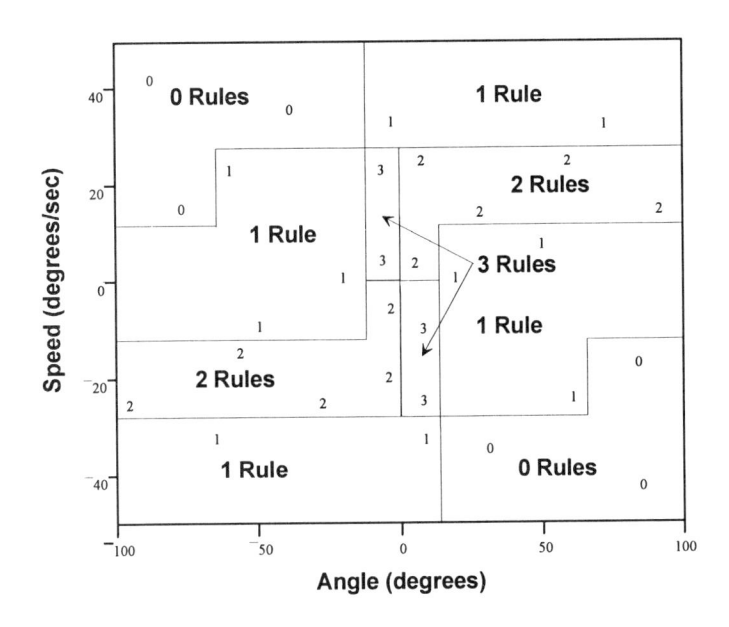

Figure 4.10: Rule coverage map for the 11 rules defined in Table 4.3.

123

The coverage map of Figure 4.10 shows several regions that need more rule coverage. The regions labeled **0 Rules** are regions of the fuzzy model that have been left undefined by the eleven rules written thus far. In addition, the regions labeled **1 Rule** need more coverage to enhance the smoothness of the fuzzy model output.

Speed\Angle	Neg_Lrg	Neg_Sml	Zero	Pos_Sml	Pos_Lrg
Pos_Fast	Pos_Low	Zero	Neg_Low	Neg_Hi	Neg_Hi
Pos_Slow	Pos_Hi	Zero	Neg_Low	Neg_Hi	Neg_Hi
Zero	Pos_Hi	Pos_Low	Zero	Neg_Low	Neg_Hi
Neg_Slow	Pos_Hi	Pos_Hi	Pos_Low	Zero	Neg_Hi
Neg_Fast	Pos_Hi	Pos_Hi	Pos_Low	Zero	Neg_Low

Table 4.4: Completed rule matrix for the pendulum model defines 25 unique rules.

By further application of our expertise we write rules to complete the rulebase. A complete rulebase is shown in Table 4.4. Before we began writing rules, we noted that the physical pendulum should behave symmetrically around the ideal state (angle = 0, angular speed = 0). The rules now reflect that expectation. To see how that symmetric behavior is encoded in the rulebase, notice how pairs of equal but opposite inputs specify an equal but opposite output. The following two rules demonstrate this property.

Rule_12: IF Angle IS Neg_Lrg AND Speed IS Pos_Fast THEN Voltage IS Pos_Low

Rule_13: IF Angle IS Pos_Lrg AND Speed IS Neg_Fast THEN Voltage IS Neg_Low

As Figure 4.11 illustrates, the additional rules in Table 4.4 eliminated the holes left by the eleven rules of Table 4.3. Also, the vicinity of the ideal system state (angle = 0, angular speed = 0) exhibits the highest density of overlapping rules, thus enhancing control in that region.

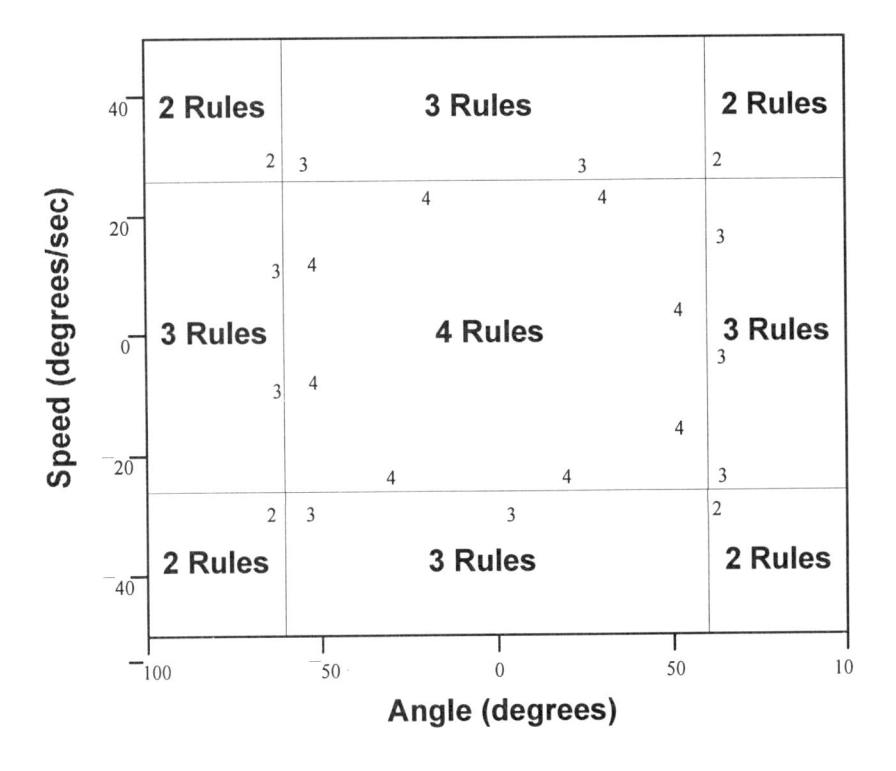

Figure 4.11: Rule coverage map for the 25 rules defined in Table 4.4.

The 25 rules of Table 4.4 embody the extent of our expertise in pendulum control. However, the performance of the fuzzy model as it balances the pendulum will determine the ultimate quality of those rules.

Rule Writing 2

While for small systems, constructing and using a complete FAM like the one in Table 4.4 works well, large systems require another approach. The number of unique rules possible in a large system grows quickly as the number of input variables and input membership functions grows. For instance, while the pendulum example (two inputs with five membership functions each) yielded 25 non-conflicting rules, a slightly larger system (four inputs with five membership functions each) makes possible 625 non-conflicting rules. A larger system still, (five inputs with seven membership functions each) allows up to 16,807 non-conflicting rules. It is unlikely that a human expert would write more than 200 rules, much lesss 16,807.[55]

Luckily for the designer of a complex fuzzy model, large systems seldom require anywhere near the maximal number of rules. The key to writing rules for a large system is writing the rules which govern the most important aspects of the system. Then write the less obvious, but intuitively correct rules. And finally, write rules which cover all other possibilities. This last step is easier to do than it sounds. For instance, consider two rules from the credit scoring task,

IF Debt_Ratio IS Border_Line AND Current_Income IS Border_Line THEN Risk IS Moderate

IF Debt_Ratio IS Very Bad THEN Risk IS High

The first rule expresses the finer grained knowledge necessary to treat a borderline credit applicant. The second rule flatly states that a bad debt ratio is sufficient grounds to deny credit. *Current_Income* is unimportant if *Debt_Ratio* is *Very Bad*. The second rule is the equivalent of writing many rules which include both inputs, *Debt_Ratio*, and *Current_Income*. In general, more opportunities for this type of short cut will be found as a fuzzy model grows in complexity.

Selection of Operators and Other Methods

In addition to writing rules, the designer must make several choices at this stage of fuzzy model development which impact the evaluation and interpretation of rules. Those choices include the functional form of the AND operator, inference method, aggregation method, and defuzzification strategy. A number of factors bear upon these choices including, but not limited to, implementation constraints, execution

---

[55] One well known fuzzy logic application involves a large and complex information system. Yamaichi Securities devised a fuzzy based expert system for trading stocks trading. Ten industry experts generated approximately 600 rules related to stocks, companies, and banks. Notably, the system lost nothing in the American mini-crash of 1987.

speed requirements, and underlying model semantics. A summary of choices is provided in Table 4.5.

INTERSECTION	INFERENCE	AGGREGATION	DEFUZZIFICATION
MIN BINTER PRODUCT Yager Functions	Correlation Minimum Correlation Product	Max Additive	Left-most Maximizer Right-most Maximizer Composite Maximum Weighted Average Centroid
Other	Other	Other	Other

Table 4.5: Choices for the fuzzy modeller.

The choices explicitly listed in Table 4.5 are detailed in prior chapters. The table entry 'Other' is intended to remind the reader that, although the lists cover the vast majority of all fuzzy systems in use today, alternative functional forms exist, and others can be invented, as needed, to support unusual or special modeling requirements.

Of course, not every choice listed in Table 4.5 is appropriate for every application. Table 4.6 ranks the various choices as to their computational efficiency and memory requirements.

While memory and speed trade-offs help define the right combination of operators and methods for a given application, one other concern supersedes them in importance. This prime concern is: how well does our fuzzy model handle the fuzziness in the problem we are trying to solve?

In order to be effective, a fuzzy model must be capable of handling the underlying model semantics. In other words, the fuzzy model must be able to adequately represent and reason with the fuzziness inherent in the underlying system. In general, the greater the fuzziness in the underlying system, the more powerful must be our operators and methods to achieve adequate fuzzy model performance. Looking at two different cases will help illustrate the varied implications of fuzziness in a model.

INTERSECTION	COMPUTATIONAL SPEED	MEMORY USE
MIN	High	Low
BINTER	Medium	Low
PRODUCT	Medium	Low
Yager Functions	Low	Medium
INFERENCE		
Correlation Minimum	High	Low
Correlation Product	Medium	Medium
AGGREGATION		
MAX	High	Low
Additive	Low	High
DEFUZZIFICATION		
Left-most Maximizer	High	Low
Right-most Maximizer	High	Low
Composite Maximum	Medium	Medium
Weighted Average	Medium	Medium
Centroid	Low	High

Table 4.6: Tradeoffs in computational speed and memory usage.

The first case considers the example problem for this chapter, balancing the inverted pendulum. What are the sources of fuzziness in the underlying system? The equations of motion, i.e. the physics which determine the behavior of the pendulum, are well known and can be written with near exactitude. The equations of motion specify the behavior of the pendulum both in the qualitative and quantitative senses. In addition, we expect such things as friction and aerodynamic drag to insignificantly affect the pendulum's motion. Thus, little fuzziness exists in our understanding of the behavior of the pendulum.

In attempting to control the motion of the pendulum, the fuzzy model must encode the equations of motion in the approximate reasoning framework of fuzzy logic. Unless an unlimited number of membership functions is used, such an encoding of the equations of motion demands a degree of approximation. However, good approximations are possible even with a small number (3 to 7) of membership functions. Adjusting the fuzzy model to accurately, though not exactly, represent the equations of motion is possible precisely because there exists an exact mathematical model of the pendulum's behavior.

Fuzziness also enters the inverted pendulum system at the point where the input data angle and angular speed are read from the input sensors. This type of fuzziness is called measurement uncertainty. Poor sensors cause higher uncertainty while better sensors reduce uncertainty.

For the second case, consider the credit scoring problem. In commercial banking, the credit scoring process rates prospective borrowers on their ability to make payments and their future potential to default on a loan. The majority of credit applications are scored by a human expert, who uses experience (the rulebase), judgment (the operators and methods), and line items from the credit application (the inputs) to determine credit risk (the output).

Credit scoring is far from an exact science. Not even the credit scoring expert can write down exact equations which govern their assessment and decision making processes. The credit scoring process is intrinsically imprecise. So, regardless of the precision of input data like income, debt to equity ratio, accounts receivable, etc., the inherent fuzziness of the decision model limits our ability to make precise statements about credit worthiness. Yet, fuzzy logic does allow us to make meaningful statements about credit worthiness.

This first case is an example of the kind of system of which we possess knowledge of a more precise nature. In contrast, the second case is an example of the kind of system where exact knowledge and mathematical equations are non-existent or meaningless. Fuzzy models addressing systems similar to the first case tolerate operators and methods which discard information, like the MIN intersection operator or MAX aggregation, better than fuzzy models which address systems similar to the second case. The crux of this issue is information preservation and usage. A system which is inherently fuzzier requires more robust operators and methods to preserve and reason with the available information. Information preservation must guide the selection of operators and methods used in the fuzzy model. Table 4.7

rates the various operators and methods with regard to information preservation.

INTERSECTION	INFORMATION PRESERVATION
MIN	Low
BINTER	Low
PRODUCT	High
Yager Functions	Low to High
INFERENCE	
Correlation Minimum	Low
Correlation Product	Medium
AGGREGATION	
MAX	Low
Additive	High
DEFUZZIFICATION	
Left-most Maximizer	Low
Right-most Maximizer	Low
Composite Maximum	Low
Weighted Average	Medium
Centroid	High

Table 4.7: Implications of various operators and methods on information usage.

Simulate Fuzzy Model and Tune System

Although the performance of the initial fuzzy model is usually surprisingly good, adjustment of rules and membership function numbers and shapes to achieve desired system performance will be necessary. The heuristics used so far in the development of both membership functions and rules must be regarded as the first loop in fuzzy model construction. An iterative process of adjusting membership functions and rules after observing system behavior helps drive the fuzzy model toward optimal performance.

In practice, membership functions receive more attention in the tuning phase than the rules. It is easier to express accurate rules than to optimally partition each membership function on the first pass of a design. Bad rules make a greater impact on model output, and are easier to identify, than a membership function which is shifted slightly from its ideal location.

Simulation

The most convenient approach to tuning fuzzy models involves computer simulation. The heart of a computer simulation is a model which mimics the behavior of the real system in response to changing outputs of the fuzzy model. In order to use simulation as a tuning aide we must be able to construct a computer model of the rest of the system. For the inverted pendulum example, and many other control problems, physics gives us the equations of motion which form the basis of a computer simulation. Importantly, computer simulation provides a low cost route to design iteration, still an unavoidable feature of fuzzy model development.

In the inverted pendulum example, two torques affect the acceleration of the pendulum mass. The torque supplied by the motor, T_M, is produced by the fuzzy model to counteract the torque produced by gravity. In terms of the variables mass, length, and angle described in Figure 4.3, the equations of motions for the pendulum are:

$$\alpha = \frac{T_M}{l^2 M} + \frac{g \sin \theta}{l}$$

$$\omega = \omega_0 + \alpha t$$

$$\theta = \theta_0 + \omega_0 t + \frac{\alpha t^2}{2}$$

where the angular acceleration is denoted $\alpha \equiv \ddot{\theta}$, and the angular speed is denoted $\omega \equiv \dot{\theta}$. Once the initial conditions θ_0 and ω_0 are specified, the three equations of motion determine the evolution of the pendulum. In order to simulate the system completely, we must know how the motor behaves in response to changing fuzzy model outputs. We assume a permanent magnet DC motor with characteristic torque given by the expression: $T_M = K_M v$, where the constant K_M depends upon the motor, and the voltage v is the output of the fuzzy model.

To write a more faithful simulation, we must consider the effects of the *CONTROL LOOP* timing. What is control loop timing? Imagine driving a car around a curve in the road. By adjusting the steering wheel a few times each second, the typical driver easily negotiates the changes in a curving roadway. What would happen if steering adjustments could only be made once per second? Would the car stay on the road if steering adjustments came at two, four, or eight second intervals? The time between adjustments is described as the control loop timing. For now we only need to acknowledge that if control loop timing is too long, fuzzy model outputs change too slowly to effectively control a system.

Control loop timing contributes significantly to the stability of any sampled data control system. Many of the standard measures of stability were derived specifically for linear systems. Since fuzzy logic models exhibit non-linearities, alternate stability measures must be used. Chapter 6 explores the non-linear stability concerns faced in fuzzy control.

We simulate the effect of finite control loop timing for the pendulum example in the following way. First we select a minimum level of granularity of Δt for the simulation. Choosing a $\Delta t = 1ms$ means that the equations of motion will be iterated once for each millisecond. If we had chosen a minimum granularity of 5ms, the equations of motion would be iterated once for every 5ms interval. We then define the control loop timing as an integer number of Δt intervals.

The fuzzy model outputs are updated at the end of each control loop interval. The model outputs are then held constant until the end of the next control loop interval. Using a finite Δt, requires a slight reformulation of the previously listed equations of motion:

$$\alpha = \frac{T_M}{l^2 M} + \frac{g \sin \theta}{l}$$

$$\omega_{i+1} = \omega_i + \alpha(\Delta t)$$

$$\theta_{i+1} = \theta_i + \omega_i(\Delta t) + \frac{\alpha(\Delta t)^2}{2}$$

where the previous angle and speed are given by θ_i, ω_i, and the current angle and speed are given by $\theta_{i+1}, \omega_{i+1}$. The simulation updates θ_i, ω_i

every Δt, but the fuzzy model, and hence T_M, updates only once per control loop.

Now that the mechanics of the simulation have been established, we can concentrate again on the inverted pendulum example. Using the previously defined membership functions and rules, Figure 4.12 shows the time response of the fuzzy model to a perturbation of the pendulum from the set point. Notice that the model does indeed drive the pendulum back to the set point. Of course, this only marks the starting point of our analysis. A complete analysis must consider the performance of the fuzzy model over the entire universe of discourse of each of the input fuzzy variables.

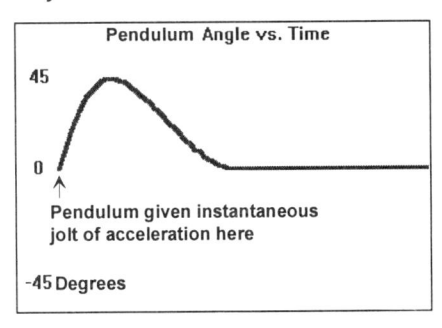

Figure 4.12: Response of the initial fuzzy controller to bring the pendulum back to equilibrium.

One powerful way to understand the behavior of a fuzzy model invokes the human ability to process volumes of visual information. The rules of Table 4.4 and membership functions of Figure 4.9 define the action applied to bring the pendulum back to the set point from any arbitrary state. Figure 4.13 visualizes the output action of the fuzzy model in response to all allowable inputs. This type of graph is often referred to as a control surface, however, as a visualization technique it easily applies to a vast spectrum of problems.

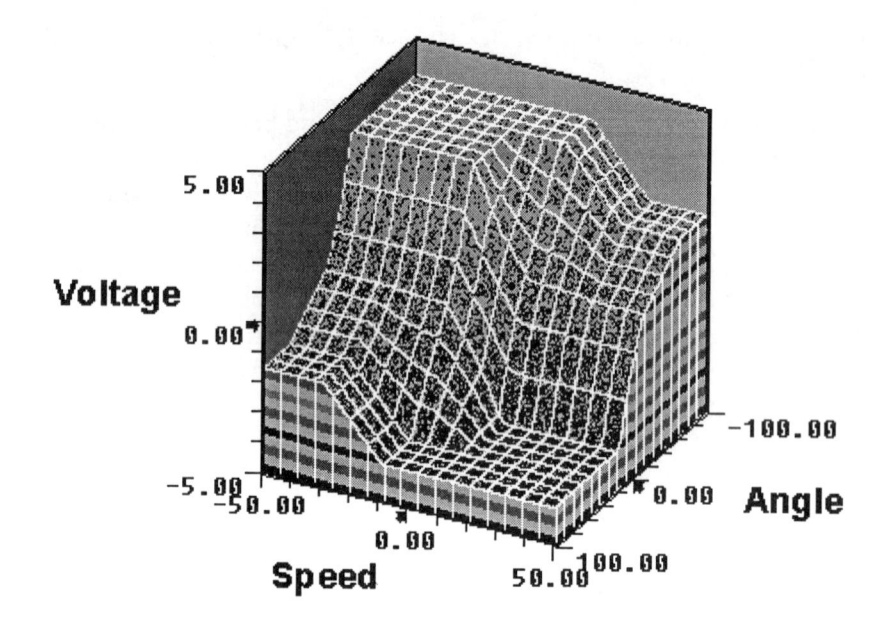

Figure 4.13: Fuzzy model control surface relating angle, speed and output voltage.

Translate Fuzzy Model to Target System

The central question to answer in this section is: how do we implement the fuzzy model in the final application? Implementation concerns include hardware vs. software, speed vs. cost, information system vs. embedded control, etc. On the hardware side, concerns include quantization levels of inputs and outputs, microcontroller selection, possible need for high speed, dedicated fuzzy chips, and availability of development tools. On the software side, concerns include maintainability, reusability, portability, size, and speed. In practice, hardware and software issues intermix, so the designer must satisfy the various concerns concurrently.

Ideally, to avoid unforeseen limitations caused by implementation constraints, performance criteria for the final system should be gauged

Chapter 4: Fuzzy Engineering

before simulating and tuning the fuzzy model. Also, be aware that the variety of specialized fuzzy hardware and software development environments do not support a uniform set of membership functions shapes, intersection operators, aggregation methods, or defuzzification algorithms.

Figures 4.14 and 4.15 put performance needs and capabilities in perspective.

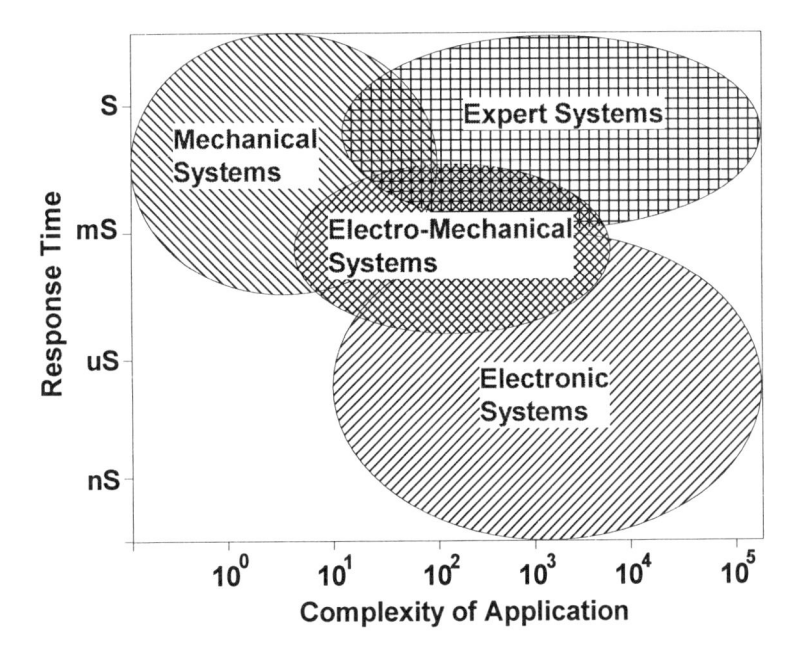

Figure 4.14: Complexity vs. response time for various problem domains.[56]

---

[56] Rinaldo Poluzzi, "Hardware Structure and Design Methodology for a Dedicated VLSI Fuzzy Controller (WARP)", Proceedings of Fuzzy Logic '93.

135

Fuzzy Logic for Real World Design

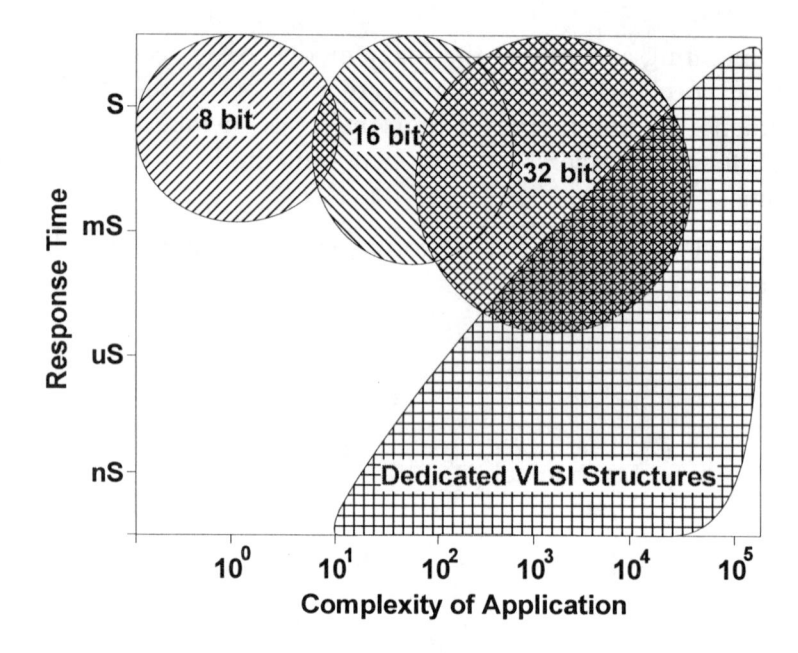

Figure 4.15: Fuzzy computational capabilities of various technologies.[57]

Integrate Fuzzy Model into User Application

By now the fuzzy model has been tweaked and tuned. However it is still not a stand-alone solution. Strictly speaking, a fuzzy model only contains the knowledge and decision making elements of the total solution. Integrating the model into a larger application completes the fuzzy engineering process. The typical application will need some or all of the following elements: interface to databases, interface to input sensors, interface to output actuators, exception handling, error flags, operator alerting, user interface, collection of performance data, and other pre- and post-processing hardware and software.

---

[57]Ibid.

Chapter 4: Fuzzy Engineering

Interfaces

Interfaces are the most crucial feature dealt with in integrating fuzzy models into larger applications. Interfaces carry crisp values into and out of the fuzzy model. The inputs to the model may come from any number of sources including: database, electronic spreadsheet, digital signal processing hardware, analog-to-digital converter, user supplied keyboard entry, etc. Paying careful attention to the format of the crisp data, and the speed with which such data can be acquired and used will contribute to a successful fuzzy model implementation.

Information Systems

Information system fuzzy models like medical diagnosis, failure analysis, and inventory forecasting link to databases for inputs and may also directly solicit input from users. In general, information systems deal with the higher levels of complexity and many inputs. Although tremendous inferencing speed may not be required, greater flexibility in membership function shapes, intersection operators, aggregation, and defuzzification methods must be supported.

Figure 4.16 illustrates another type of complexity encountered in information system models. A modeling problem often can be broken down into several smaller problems, which individually require a fuzzy model. In a hierarchical system, the output of one fuzzy model may be used in a downstream fuzzy model. In the case of Figure 4.16, the interface must direct the output from fuzzy model A to the input of fuzzy model B.

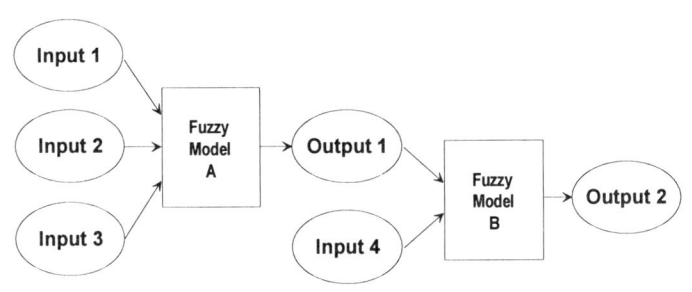

Figure 4.16: Hierarchical fuzzy model.

Control Systems

The critical nature of the interface between a control system and its fuzzy model derives from the control system's sensitivity to control loop timing and sensor resolution. Integration of a fuzzy model into a control system application follows guidelines similar to those used in the integration of digital signal processing hardware into an application. Concerns over control loop timing revolve around the total time needed to process inputs, execute the model, and effect outputs. Concerns over resolution revolve around having enough bits to generate as fine a control surface as needed. We will look more closely at resolution issues and control loop issues in Chapter 6.

Error and Exception Handling

As part of a larger system, the fuzzy model can incorporate exception handling and error flags to notify the user application of unusual or dangerous events. Rule level alpha-cuts, described in Chapter 3, generate alarms like the following:

Alpha-cut level of 0.90

IF Reactor_Temp IS Too_High AND Coolant_Flow IS Low THEN Melt_Down_Avoidance

When the IF-side of this rule produces a truth value greater than the alpha-cut level, the *Melt_Down_Avoidance* action takes immediate effect. In this case, *Melt_Down_Avoidance* acts like a Boolean output variable to initiate corrective action. Since this may be the only rule which affects the *Melt_Down_Avoidance* variable, fuzzy model operation is unaffected under safe operating conditions.

The same approach can generate indicator and error flags to the user application for problems less severe than nuclear meltdown. The following two rules handle problems of varying severity:

Alpha-cut level of 0.70

IF Boiler_Pressure IS Too_High AND Pressure IS Increasing THEN Open_Safety_Valve

Alpha-cut level of 0.90
IF Safety_Valve IS Open AND Pressure IS Increasing THEN Shut_Down_Boiler

Collection of Performance Data

When a fuzzy model executes on a platform with computing power to spare, one of the most useful tasks that can be performed with the excess computing capability is the collection of performance data. Performance data shows how well the fuzzy system accomplishes its goal. Performance data can then be used off-line as a diagnostic and tuning tool, or with the appropriate algorithms, on-line use can lead to self-tuning of the fuzzy model. Research into architectures and methods of implementing adaptive fuzzy systems promises mastery over an ever increasing spectrum of problems. As such, a complete treatment of this fertile topic is beyond the scope of this volume and must be left to other authors and future books.

Summary

In this chapter we presented a structured approach to fuzzy rulebased engineering. Determining the need for and potential usefulness of a fuzzy model in an application is the first step of the process. Both likely and unlikely candidates for fuzzy logic appear in Tables 4.1 and 4.2. While these tables should in no way be construed as complete and final, they serve as useful guides. After all, it is equally wasteful to bother with an unsuitable approach to a problem as it is to ignore a promising new approach.

Next, we covered in some detail how to develop a fuzzy model from concept to implementation. We broke fuzzy model development into seven manageable steps and illustrated the entire process utilizing a practical example.

1. *System description.* Define how the fuzzy model fits into the total system solution. Describe, in words, how the system should work. Capture the essence of the problem by identifying inputs, outputs, and their relationship to each other.

2. *Specification of input and output variable ranges.* Identify extreme ranges of all inputs and outputs. Define the universes of discourse for all fuzzy variables. Thought must be given to input resolution, sensors, and output format.

3. *Membership function partitioning.* Partition each fuzzy variable into overlapping membership functions. Decide on number, shape, location, symmetry, and overlap.

4. *Rule writing.* Construct IF/THEN rules. Write the obvious rules first. Then write the less obvious but intuitively correct rules next. Finally write the unobvious rules through introspection and experience. Select intersection, inference, aggregation, and defuzzification methods based on the constraints of computational speed, memory use, and information preservation.

5. *Simulation and tuning.* Adjust membership functions and rules to achieve desired model performance. Computer simulation and/or testing on the target system may be necessary. Visualization techniques and commercially available tools aide this area of fuzzy engineering significantly.

6. *Translation of model to target system.* This step deals with how the simulated and tuned fuzzy model is implemented in the user application. Design trade-offs consider speed, memory, and complexity. We also looked at variety of hardware and software implementation issues.

7. *Integration of fuzzy model into user application.* This final step pointed out that a fully developed fuzzy model is still not a stand-alone solution. Additional elements may be needed to complete the user application including: interfaces to input sensors and output effectors, user interfaces, exception handling, operator alerting, and collection of performance data.

Many of the nuances encountered while applying these seven steps will be explored in subsequent chapters.

Chapter 5:
Fuzzy Engineering II

This chapter expands on *Fuzzy Engineering I* by taking a closer look at the membership function partitioning, rule writing, simulation, and tuning needed to finalize fuzzy controllers and models. We illustrate different aspects of the fuzzy engineering process using several simple control and information systems applications.

The first example we will look at is a continuation of the simple pendulum example used in *Fuzzy Engineering I*. We will introduce graphical techniques useful for rule and fuzzy set tuning in the fuzzy controller. Although the application is a very simple one, the graphical analysis methods we develop will carry over into the analysis and tuning of real world fuzzy controllers.

The second example presents the fuzzy equivalent to proportional-integral-derivative (PID) control. PID control is a well established, even dominant, control algorithm for process and set-point type control systems. Because simulation dramatically enhances the test-and-tune phase of fuzzy engineering, this example also details the construction of a simulation model for the apparatus under control.

The third example presents an information system. This example leads us through a decision support system that helps you decide which US. city is the best place for you and your family to live.

Fuzzy Logic for Real World Design

Pendulum Revisited

It has been said of the fuzzy engineering process that while obtaining the first 80% of the performance objectives is fairly easy, obtaining the final 20% presents a real challenge. In this section we will illustrate several tools and techniques that help you obtain that final 20%. The simplicity of the inverted pendulum problem allows us to concentrate on the tools and techniques rather than the problem itself.

State Space Evolution

We start by recalling the pendulum equations of motion from the previous chapter: In terms of the variables mass, length, and angle described in Figure 4.3, the equations of motion for the pendulum are:

$$\alpha = \frac{T_M}{l^2 M} + \frac{g \sin \theta}{l}$$

$$\omega = \omega_0 + \alpha t$$

$$\theta = \theta_0 + \omega_0 t + \frac{\alpha t^2}{2}$$

The only influence of the fuzzy controller is to vary the torque, T_M applied to the pendulum. The pendulum responds to the combined effects of gravity and the fuzzy controller in the way specified by the equations of motion. The equations of motion for the pendulum will hereafter be referred to as the pendulum simulation. Once the initial conditions θ_0 and ω_0 are specified, the three equations of motion and the output of the fuzzy controller completely determine the evolution of the pendulum. At any time the variables θ and ω completely specify the state of the pendulum.[58] For this reason the variables angle and angular speed are called state variables of the pendulum system. The explicit purpose of the fuzzy controller is to govern the evolution of the system's state variables.

---

[58] In general, two state variables are needed for each degree of freedom in a system to completely specify its state. The pendulum of this example has only one degree of freedom and hence requires only two state variables.

Chapter 5: Fuzzy Engineering II

Because it uses the state variables as inputs, the fuzzy controller is directly tied to the state of the system. In order to understand the effects of the controller on the system, we must study the state space evolution of the combined pendulum-fuzzy controller system. The most intuitive way to study the state space evolution is graphically. Figure 5.1 shows one possible trajectory through the state space.

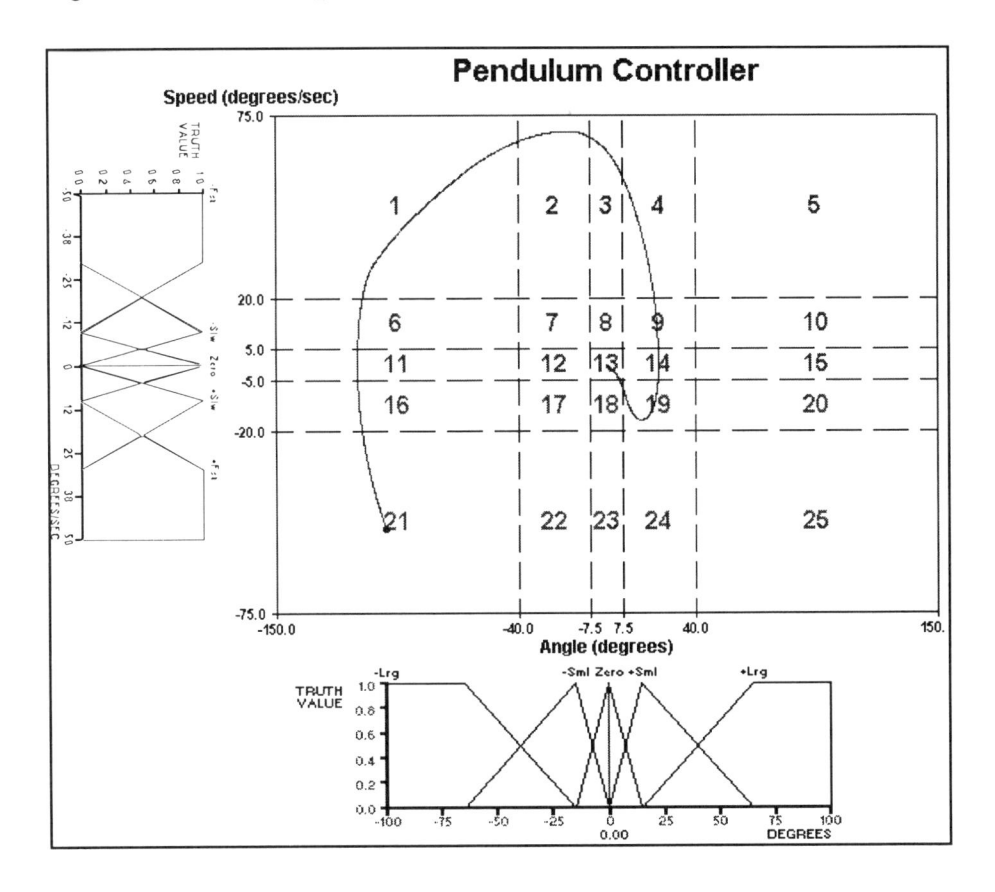

Figure 5.1: Representative state space trajectory for the pendulum controller. The number inside each dashed box indicates the strongest rule in that region. Rules are numbered and listed in Table 5.1.

The graph showing the pendulum state space evolution is generated by picking a starting point in the state space and then iterating through

143

the fuzzy controller and the pendulum simulation. In Figure 5.1, the starting point is indicated by a small round dot in the lower left corner of the graph. Obtaining successive points in the state space trajectory amounts only to executing the controller/simulator loop many times and plotting the changes in the pendulum's state variables. Each loop starts with the new state variables generated by the previous simulation loop.

The state space graph of Figure 5.1 presents plenty of useful information. The single curved line that starts with a small round dot (i.e. $\theta = -100, \omega = -50$) and ends at the center of the graph (i.e. $\theta = 0, \omega = 0$) traces the state of the pendulum as the controller drives to the set-point. The membership function definitions below the X-axis correspond to the *Angle* input while the membership functions for *Speed* are shown to the left of the graph's Y-axis. The labels printed just above each membership function help to identify the various regions of the state space graph. The dashed lines on the graph correspond to the points at which two adjacent membership functions intersect. The dashed lines divide the graph into numbered regions of operation. Identifying each region with one or more rules that operate in that region will aid in tuning the controller's performance.

For reference, Figure 5.2 shows the initial fuzzy set definitions for the pendulum controller. Table 5.1 lists the initial rulebase for the controller.

In this example, the fuzzy controller is tasked with maintaining the pendulum in an upright position described as the state $\theta = 0, \omega = 0$. We know that this state is an unstable equilibrium by looking at the open loop response of the pendulum simulation shown in Figure 5.3.

Chapter 5: Fuzzy Engineering II

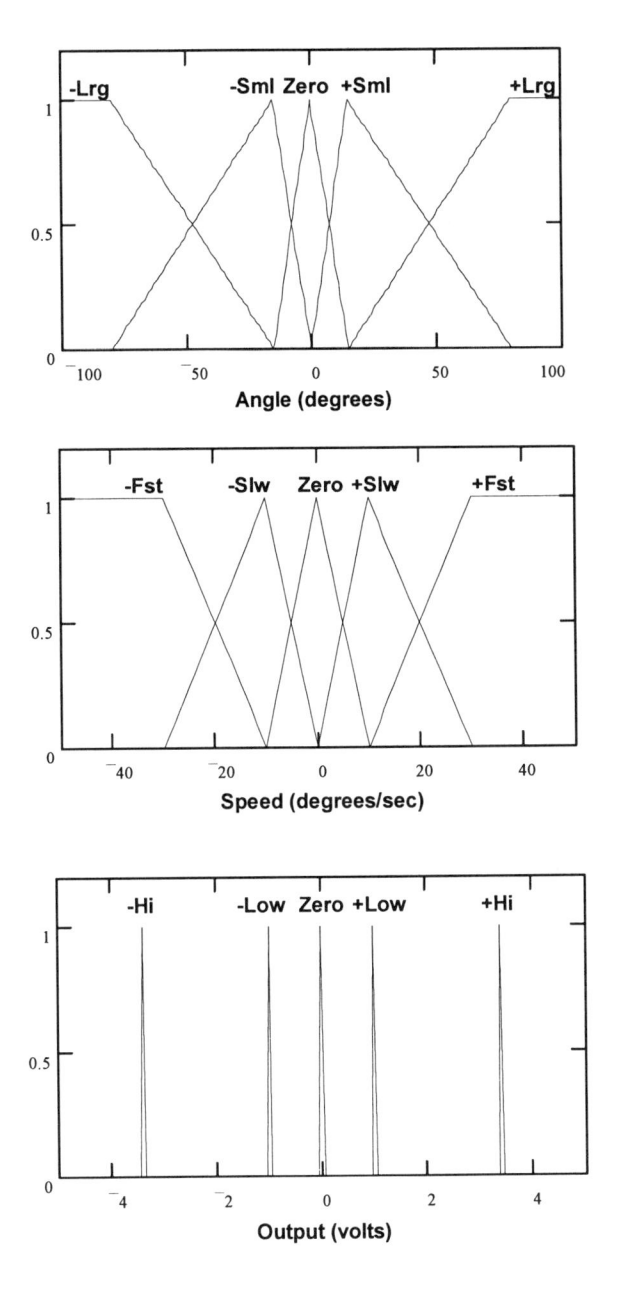

Figure 5.2: Initial function partitioning for inputs angle and speed, and output singletons for voltage.

145

```
01: IF Angle is -Lrg AND Speed is +Fst THEN Voltage is +Low
02: IF Angle is -Sml AND Speed is +Fst THEN Voltage is Zero
03: IF Angle is Zero AND Speed is +Fst THEN Voltage is -Low
04: IF Angle is +Sml AND Speed is +Fst THEN Voltage is -Hi
05: IF Angle is +Lrg AND Speed is +Fst THEN Voltage is -Hi
06: IF Angle is -Lrg AND Speed is +Slw THEN Voltage is +Hi
07: IF Angle is -Sml AND Speed is +Slw THEN Voltage is Zero
08: IF Angle is Zero AND Speed is +Slw THEN Voltage is -Low
09: IF Angle is +Sml AND Speed is +Slw THEN Voltage is -Hi
10: IF Angle is +Lrg AND Speed is +Slw THEN Voltage is -Hi
11: IF Angle is -Lrg AND Speed is Zero THEN Voltage is +Hi
12: IF Angle is -Sml AND Speed is Zero THEN Voltage is +Low
13: IF Angle is Zero AND Speed is Zero THEN Voltage is Zero
14: IF Angle is +Sml AND Speed is Zero THEN Voltage is -Low
15: IF Angle is +Lrg AND Speed is Zero THEN Voltage is -Hi
16: IF Angle is -Lrg AND Speed is -Slw THEN Voltage is +Hi
17: IF Angle is -Sml AND Speed is -Slw THEN Voltage is +Hi
18: IF Angle is Zero AND Speed is -Slw THEN Voltage is +Low
19: IF Angle is +Sml AND Speed is -Slw THEN Voltage is Zero
20: IF Angle is +Lrg AND Speed is -Slw THEN Voltage is -Hi
21: IF Angle is -Lrg AND Speed is -Fst THEN Voltage is +Hi
22: IF Angle is -Sml AND Speed is -Fst THEN Voltage is +Hi
23: IF Angle is Zero AND Speed is -Fst THEN Voltage is +Low
24: IF Angle is +Sml AND Speed is -Fst THEN Voltage is Zero
25: IF Angle is +Lrg AND Speed is -Fst THEN Voltage is -Low
```

Table 5.1: *Pendulum control rules used to generate Figures 5.1 and 5.3.*

In the figure, each small dot on the graph represents a new starting point for the simulation. A point is then plotted for each step of the simulation. Each simulation loop accounts for 0.012 seconds. After 100 iterations of the simulation are plotted, the next starting point is chosen. Figure 5.3[59] shows a grid of these starting points. The set-point for control is found in the box labeled 13, where $\dot{\theta} = 0, \theta = 0$. The open loop response of the pendulum is obtained by removing the controller from the loop. We generated the state space evolution of the open-loop system by iterating the pendulum simulation with the controller input $T_M = 0$. The instructiveness of studying the open loop response of any system that we wish to control cannot be overemphasized.

---

[59]The software shown here, **PhaseView** is a fuzzy logic development tool produced by Synerdaptix Inc., the same company that produced **FuzzyLab**, the software accompanying the book.

Chapter 5: Fuzzy Engineering II

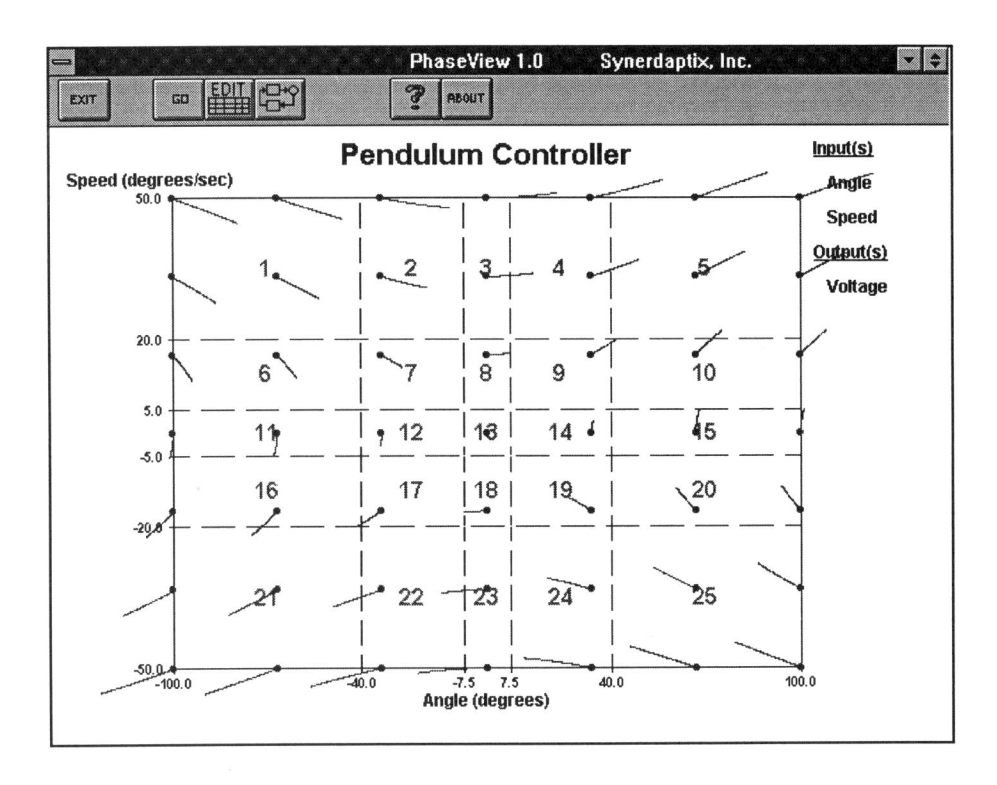

Figure 5.3: Pendulum open loop response. Each dot starts 100 simulation loops.

The open loop response of any system reveals information that is key to the controller design. For example, the open loop response of the pendulum state demonstrates that the set-point at $\theta = 0$ is an unstable equilibrium point. This is confirmed by the fact that all of the state space trajectories of Figure 5.3 diverge from the set-point. In addition, the symmetry of the state space trajectories of Figure 5.3 indicates that the pendulum responds symmetrically with respect to angular deviations from the set-point. Symmetry in the open loop response means that the controller can be designed to provide symmetrical control based on the input *Angle*. Simply stated, an angular input of +5 degrees will require an equal but opposite control action to the angular input of -5 degrees.

Although the time response of the system is already embedded in the state space graph, it may not be readily apparent if you are unfamiliar with this method of looking at the data. Oscillation, overshoot, convergence, and other properties commonly characterized in control systems are all recognizable in the state space graph. To illustrate this point, Figures 5.4 (a) and 5.5 (a) show the state space graphs that correspond to the time responses shown in Figures 5.4 (b) and 5.5 (b), respectively. Note that Figures 5.4-5.5 are general illustrations that have no direction correlation to our pendulum control system.

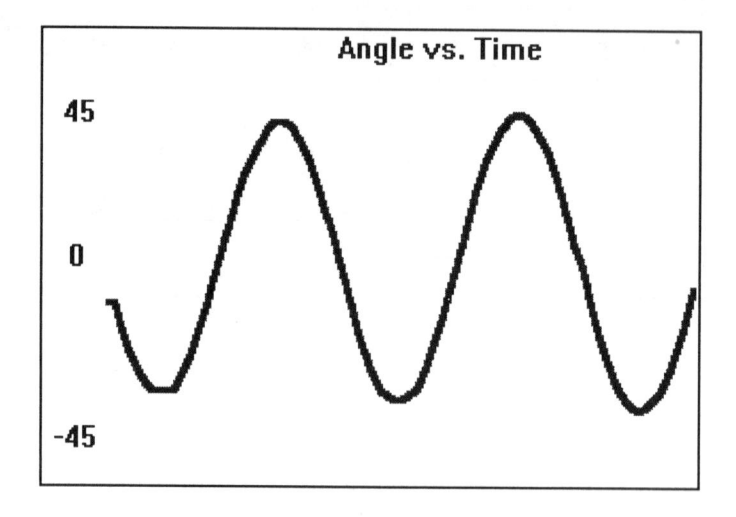

Figure 5.4 (a): Oscillatory time response.

Chapter 5: Fuzzy Engineering II

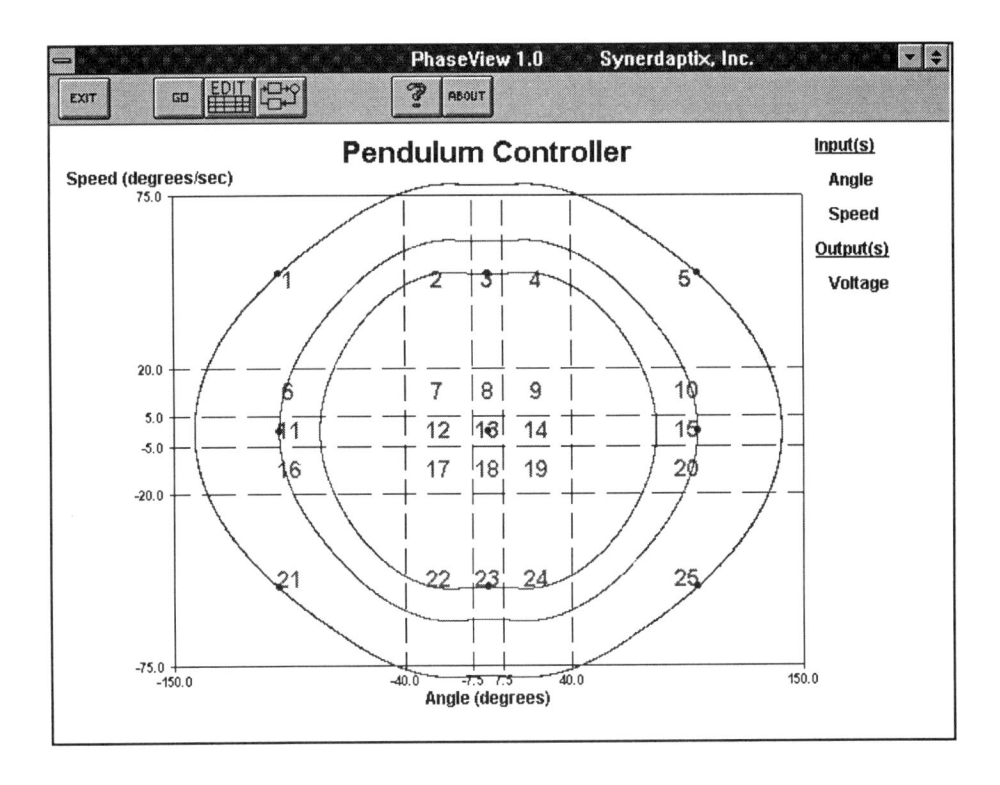

Figure 5.4 (b): State space representation of three oscillators with differing amplitudes.

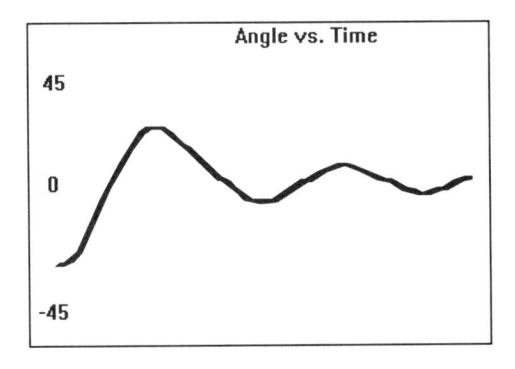

Figure 5.5 (a): Time response displaying overshoot and convergence.

Fuzzy Logic for Real World Design

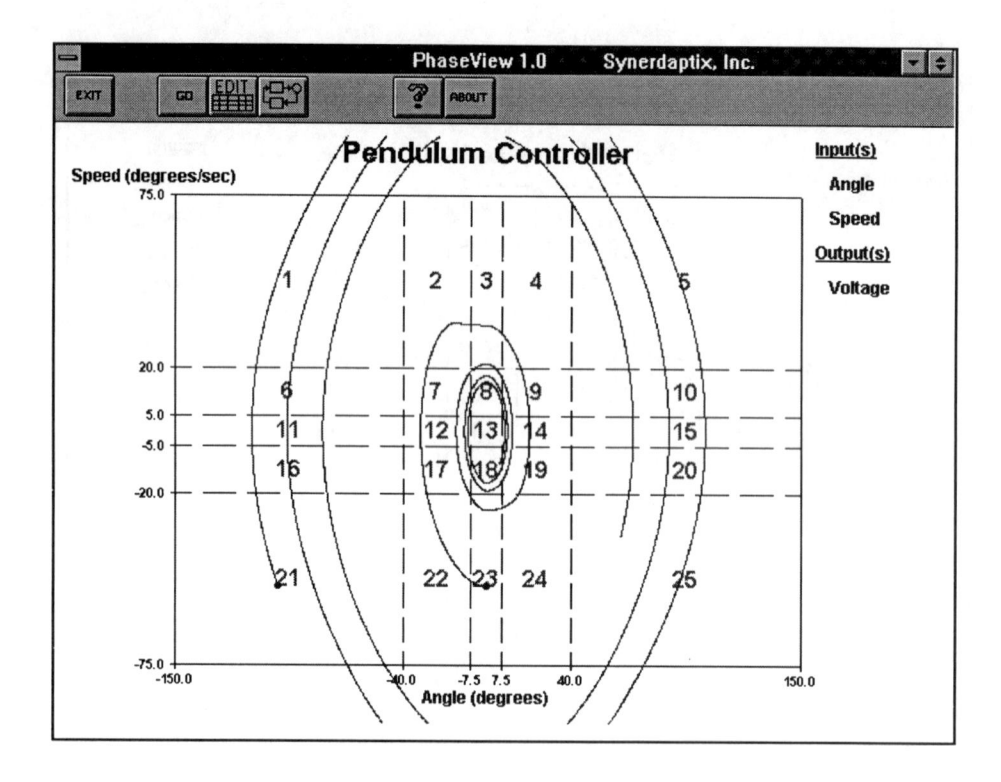

Figure 5.5 (b): Two state space traces showing overshoot and convergent oscillatory behavior.

Using the state space graph, we can investigate the behavior of the closed loop feedback system over the entire operating range of the fuzzy controller. A useful way to view the state space graph is illustrated in Figure 5.6. The figure shows the state space evolution by dividing the universes of discourse of the fuzzy inputs into a grid. Starting with each grid point, the controller and simulation were iterated 800 times and plotted. Thus the graph shows the way that the state of the pendulum evolves, over time, under the influence of the fuzzy controller. Notice that all paths shown in Figure 5.6 converge to the set-point located at $\dot{\theta} = 0, \theta = 0$. This is precisely the goal of the control system.

Chapter 5: Fuzzy Engineering II

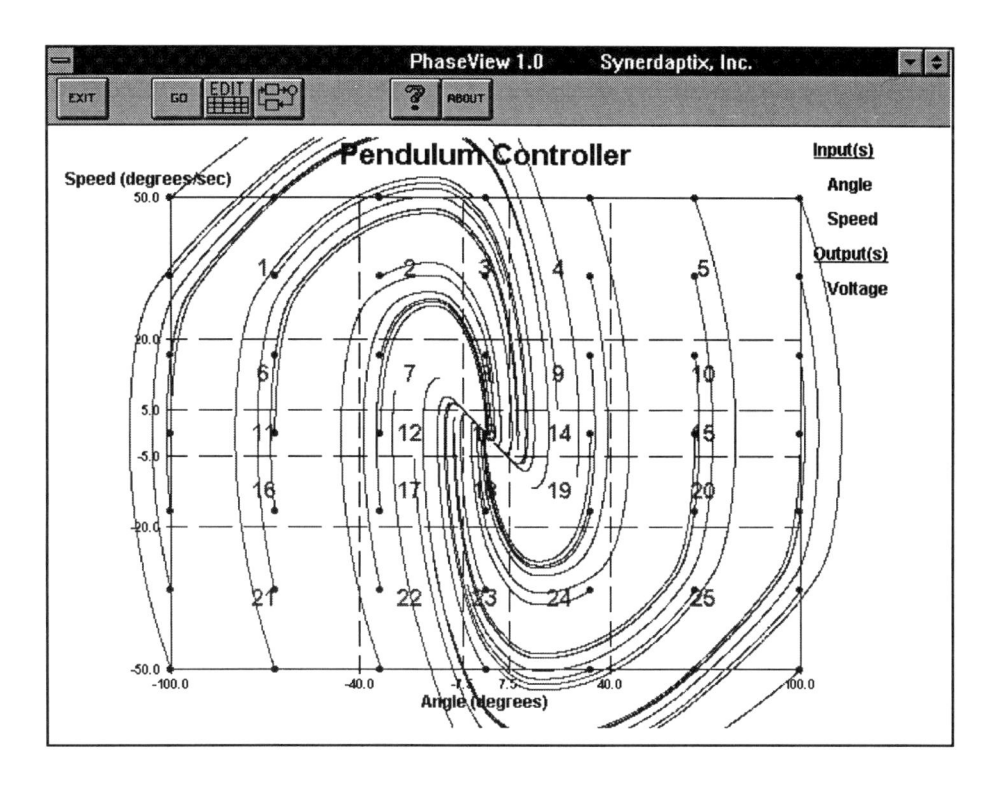

Figure 5.6: Pendulum-fuzzy controller closed loop response. All paths through state space converge to the set-point.

The numbered areas of the graph in Figure 5.6 correspond to the rules listed in Table 5.1. The intent of the numbered areas in Figure 5.6 is to indicate the rule that has the most effect in that particular segment of the state space. Due to the overlap of the membership functions for both inputs, up to four different rules apply to a non-zero degree at any point in the state space. To see that this is so, recall the membership function partitions shown in Figure 5.2. At most points in the domain of each input variable, two membership functions simultaneously have a non-zero degree of membership. Two membership functions for each input yields four possible combination of input fuzzy sets for most points in the domain. If the controller behaves unacceptably in a particular area, the corresponding rules can be easily identified. As an

151

Fuzzy Logic for Real World Design

example we will change rules 7 and 19, which affect control near the set-point. Change rules 7 and 19 to:

```
07: IF Angle is -Sml AND Speed is +Slw THEN Voltage is +Low
19: IF Angle is +Sml AND Speed is -Slw THEN Voltage is -Low
```

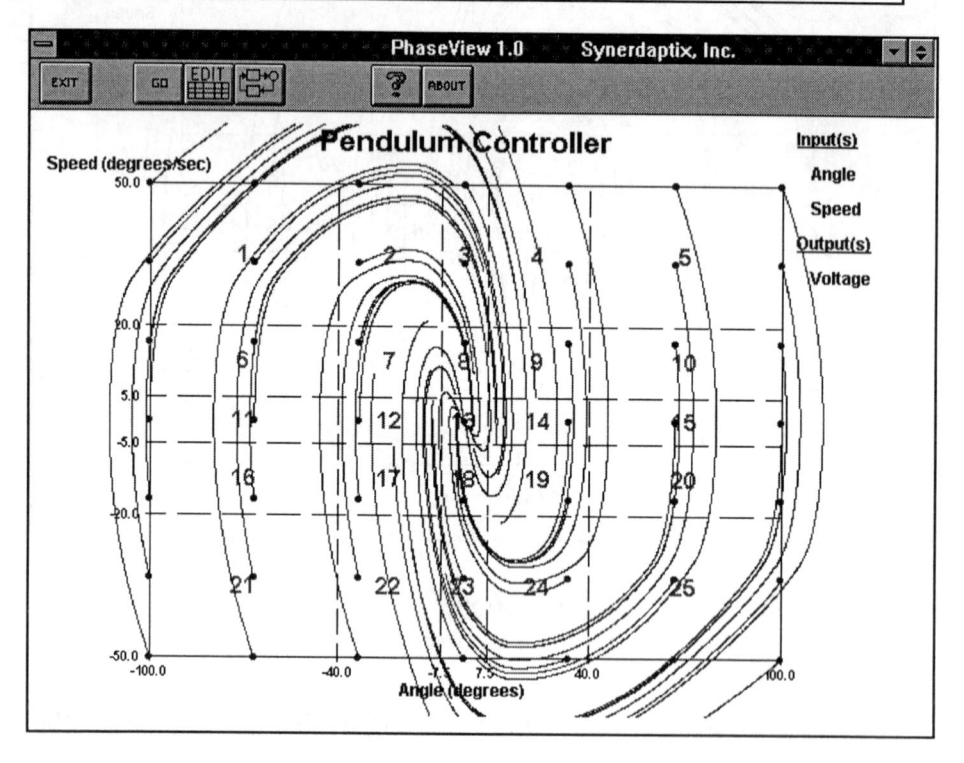

Figure 5.7: Pendulum-fuzzy controller closed loop response with alteration to rules 7 and 19.

In comparing Figures 5.6 and 5.7 notice that the changes to rules 7 and 19 most strongly affected the control regions labeled 7 and 19. The regions adjacent to 7 and 19 were weakly affected by the rule changes, while non-adjacent regions of Figure 5.6 were left totally unaffected. This example highlights a property of fuzzy controllers that eases the tuning effort. Fuzzy controllers with overlap in the range 0 to 50% display a strong degree of locality in their behavior. This allows local adjustments to the controller's behavior without affecting the entirety of the control space. This property can be further exploited in the tuning of membership functions.

Tuning the Fuzzy Controller

The qualitative performance objectives for this type of control system can be expressed as: 1) guiding the system to the set-point from any point in the state space, and 2) maintaining the system at the set-point once it has been reached. In addition to the qualitative criteria, a host of quantitative performance criteria may be established including: instantaneous error e, absolute error $|e|$, error squared e^2, average error over time $\bar{e}$, average absolute error $|\bar{e}|$, etc. An important point to remember is that optimizing the fuzzy controller for one of these performance measures will not necessarily optimize the controller's performance against the other measures. This is true not only of fuzzy-based controllers, but of all controllers regardless of their underlying technologies. Our goal in this section is to provide insight into the general problem of tuning a fuzzy controller that can then be applied to any specific problem.

Viewing the state space evolution of the controller/simulation makes clear how the system converges to the set-point. The state space graph also reveals information useful in further tuning the controller. To simplify our examination of controller tuning, consider the single trajectory of Figure 5.1. The trajectory shows overshoot into the right half-plane of the state space graph that we wish to reduce. By inspecting rules 02 and 03 we can gain some insight into possible causes of the overshoot:

```
02: IF Angle is -Sml AND Speed is +Fst THEN Voltage is Zero
03: IF Angle is Zero AND Speed is +Fst THEN Voltage is -Low
```

These rules are responsible for damping out the high speed of the pendulum around $\theta = 0$. The presence of overshoot indicates that the speed has not been damped sufficiently. To dampen the speed more aggressively, the motor must apply more torque in opposition to the pendulum's motion. To correct this we change rules 02 and 03 to:

Fuzzy Logic for Real World Design

```
02: IF Angle is -Sml AND Speed is +Fst THEN Voltage is -Low
03: IF Angle is Zero AND Speed is +Fst THEN Voltage is -Hi
```

And, to maintain symmetrical controller response, we change rules 23 and 24 to:

```
23: IF Angle is Zero AND Speed is -Fst THEN Voltage is +Hi
24: IF Angle is +Sml AND Speed is -Fst THEN Voltage is +Low
```

Figure 5.8 demonstrates the effect of these rule changes on the state space evolution of the pendulum. Notice that the overshoot has been reduced significantly.

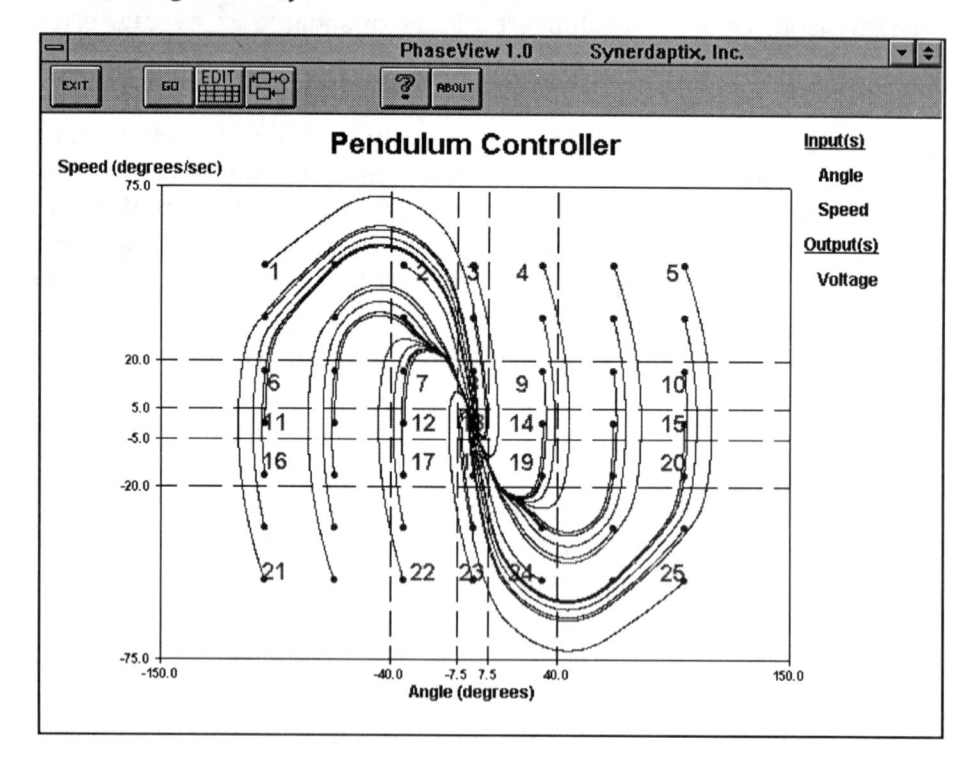

Figure 5.8: Pendulum response with rule corrections to reduce overshoot.

154

Chapter 5: Fuzzy Engineering II

In continuing our tuning efforts, we shift our focus to adjustment of the membership functions. Let us assume that the gross adjustments via rule changes have been exhausted. Fine tuning is possible by adjustments to the peaks and widths of the membership functions both on inputs Angle and Speed and on the output Voltage. In terms of the state space graph, adjustments to input membership functions shift the regions over which their corresponding rules exert influence.

Before we attempt to fine tune we must clearly understand the desired results of the tuning process. With regard to a generic set-point controller, the pendulum is a representative system. The first goal of a set-point controller is to restore the system to the set-point as quickly and accurately as possible. For the pendulum, this goal translates to driving the pendulum angle to zero as quickly as possible while minimizing overshoot. The second goal of the generic controller is to maintain the system at the set-point with no oscillation. For the pendulum, this goal amounts to staying at $\theta = 0, \omega = 0$ once it is reached.

As an example of finely tuning the controller via membership functions, consider the region where *Angle is Zero* (i.e. the fuzzy region around $\theta = 0$). That region is responsible for not only maintaining the system at the set-point once it has been achieved, but also for damping out excessive pendulum speed near the set point. Altering the widths and peaks of the *Angle* membership functions alters the way that the controller applies its various rules to damp out the pendulum's speed. Increasing the scope of *Zero* expands the region of damping, while decreasing its scope has the opposite effect. The trick is to adjust *Zero* to provide the fastest response possible while minimizing overshoot of the set-point.

Next, consider the regions covered by the *Angle* fuzzy sets *-Lrg* and *+Lrg*. In these regions the controller drives the pendulum toward the set-point. To obtain the fastest response, the controller drives the pendulum as hard as possible. Expanding the scopes of *-Lrg* and *+Lrg* toward *Zero* would the expand the regions over which the pendulum is driven hard, resulting in higher speeds during a return to the set-point. However, the adjustments to *-Lrg* and *+Lrg* must be limited to producing speeds that can be dampened out by the time *Angle* reaches zero.

155

Fuzzy Logic for Real World Design

Finally, consider the regions covered by the *Angle* fuzzy sets *-Sml* and *+Sml*. The controller blends both driving and damping behaviors in these regions. The driving and damping nature of the controller depends more on the pendulum speed here than in other regions. Tuning the overall system response via adjustments to both *-Sml* and *+Sml* must blend changes to damping and driving behavior appropriately.

Figure 5.9 is a synthesis of the tuning principles outlined above. The figure was generated with the same rulebase as the previous figure. Membership functions of both inputs *Angle* and *Speed* were altered to improved the system response.

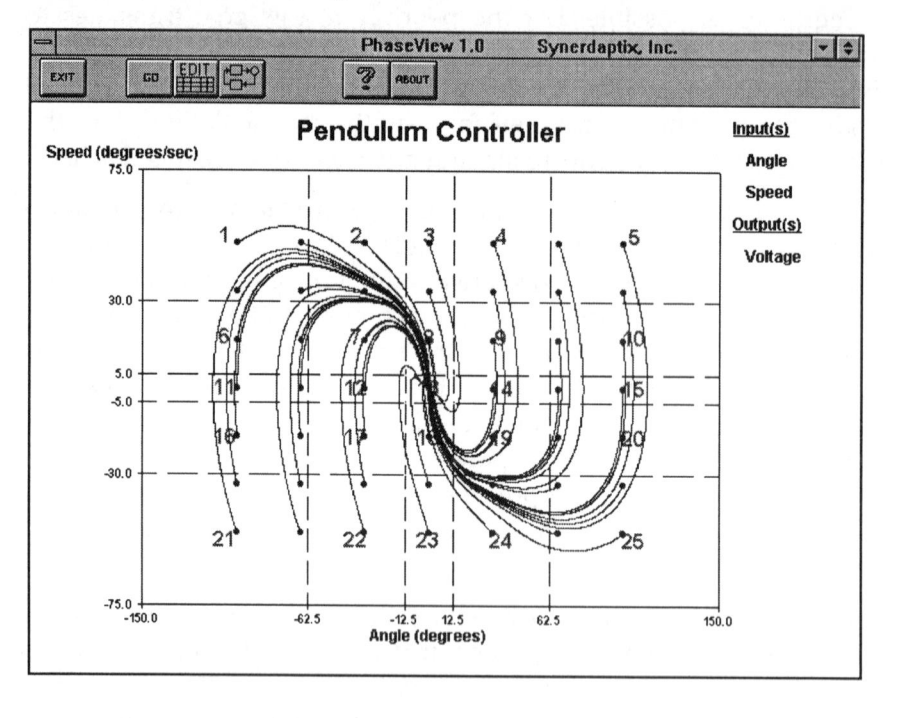

Figure 5.9: Improved response achieved by membership function tuning on both Angle *and* Speed

Adjustments to the output *Voltage* membership functions are also possible. Since the output functions are singletons, the single adjustable

Chapter 5: Fuzzy Engineering II

parameter for each function is its center location. Because an output membership function may appear in the consequent of several different rules, greater care must be exercised in adjusting output sets. Figure 5.10 illustrates the inter-relatedness of rules and consequent fuzzy sets by identifying the consequent *Voltage* fuzzy set associated with each state space region. Note that here again, the label that appears in each region is not the only rule consequent that applies to that region; rather it is just the strongest consequent in that region.

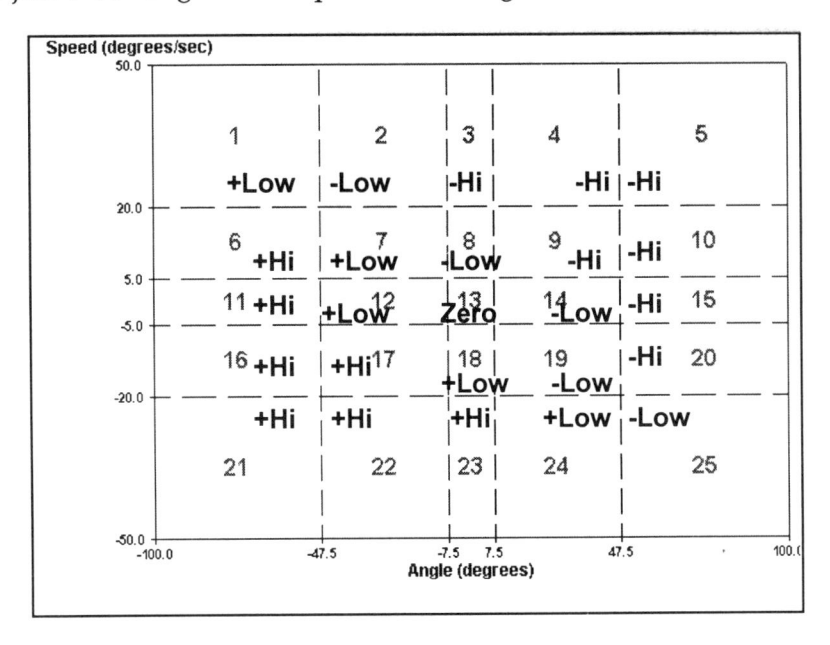

Figure 5.10: Output Voltage *fuzzy set associated with each state space region. Labels refer only to the strongest rule consequent in each region.*

As an example of this sensitivity, consider that a change to the *Voltage* membership function +*Hi* strongly affects the state space regions numbered 6, 11, 16, 17, 21, 22, and 23. The adjacent state space regions, numbered 1, 2, 7, 12, 13, 18, 19, and 24 are also affected, but to a lesser degree. Nonetheless, performance improvements can be obtained by a careful examination of the potential effects on the state space evolution. The final system response, depicted in Figure 5.12, was

157

Fuzzy Logic for Real World Design

obtained by increasing the magnitude of the output singletons +*Hi* and -*Hi*, and a small supplemental modification to the fuzzy sets covering the *Angle* universe of discourse. The final membership function partitionings are shown in Figure 5.11.

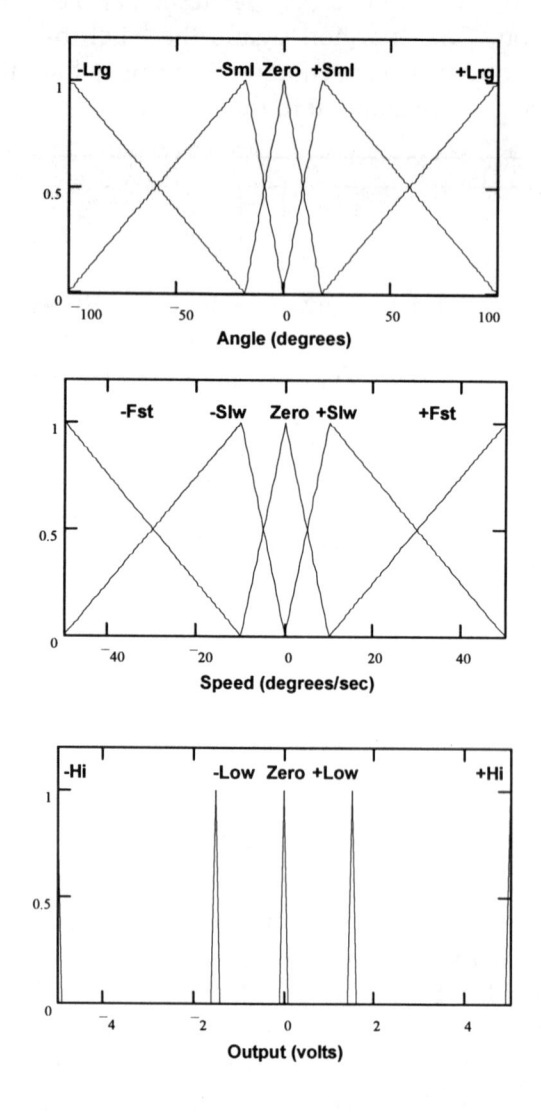

Figure 5.11: Final fuzzy set partitions for pendulum controller inputs angle and speed, and output voltage.

Chapter 5: Fuzzy Engineering II

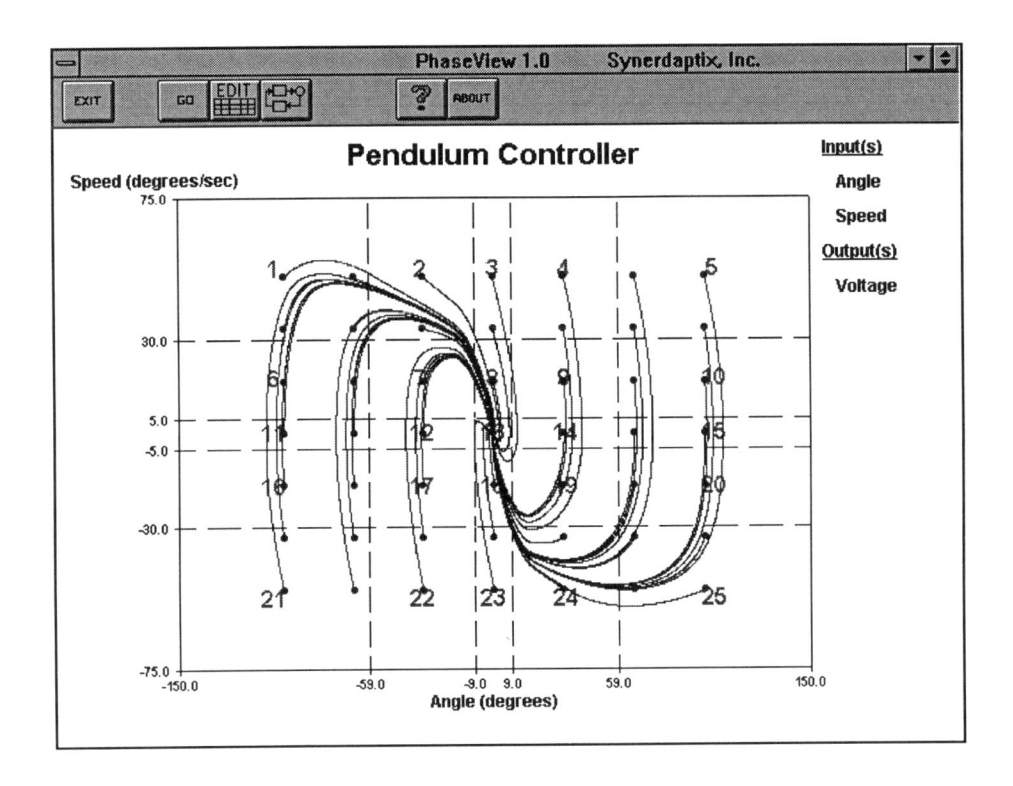

Figure 5.12: Increasing the magnitudes of output singletons +Hi and -Hi produces further improvement to the state space evolution.

Finally, Figures 5.13 (a) and (b) display the time response of pendulum angle and motor voltage for the completed controller.

Fuzzy-PID Control

The use of proportional-integral-derivative (PID) control and its relatives proportional-derivative (PD), and proportional-integral (PI) control is wide-spread in the realm of set-point control. The following equation captures the basic form of the PID control law:

$$u = K_p e + K_d \dot{e} + K_i \int e$$

Fuzzy Logic for Real World Design

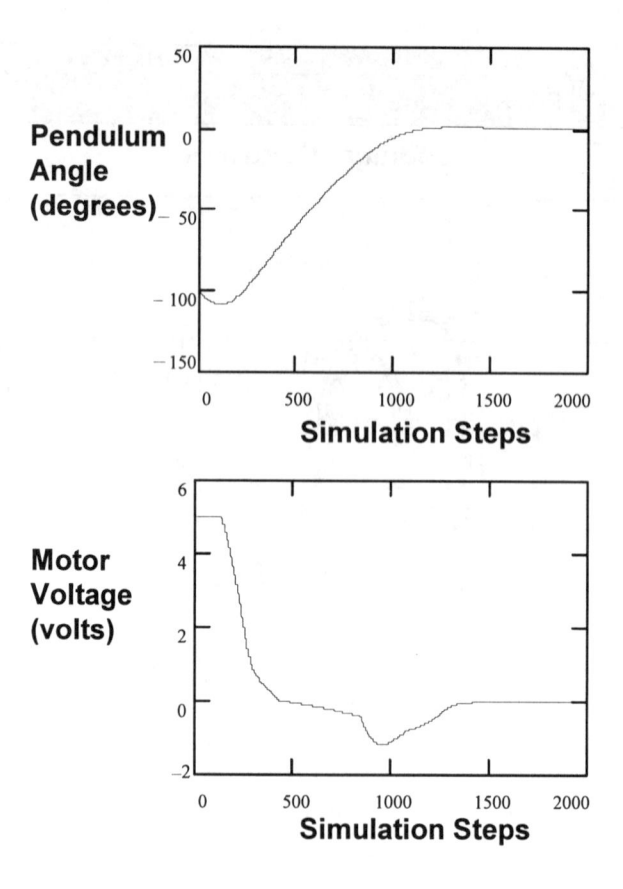

Figure 5.13 (a): Pendulum angle and (b) motor voltage, plotted over 2000 simulation steps. One simulation step = one control loop = 12 milliseconds.

The output of the controller, u, is a function of the present error, e, the rate of change in the error, $\dot{e}$, and the integral of the error over time, $\int e$. In a PI controller the K_d term is zero. In a PD controller the K_i term is zero. The designer's skill and art in selecting the constants associated with each of the PID scaling factors, K_p, K_d, and K_i, largely determines the ultimate performance of the final controller. As long as the parameters K_p, K_d, and K_i remain constant over the operation of the controller, the PID controller is capable of only a linear response.

160

A PID-like fuzzy controller can be designed that uses e, $\dot{e}$, and $\int e$ as fuzzy inputs. As it happens, the PID-like fuzzy controller is a frequently used topology. We now recognize that the pendulum controller discussed at the beginning of this chapter is in fact a fuzzy PD-like controller. The fuzzy-based controllers offer more flexibility by means of rules and membership functions, and also allow the use of non-linear control strategies. No real theoretical difficulties exist that would prevent the conversion of an existing PID controller to a fuzzy-PID controller. This is so because it has been shown[60, 61] that any continuous linear or non-linear control function can be approximated by a fuzzy system to any degree of accuracy required. We now look at a practical example of a fuzzy PI-like controller.

Automobile Cruise Control

A cruise control module on an automobile maintains the vehicle at the speed set by the driver until either the automobile's brakes are activated or until the driver deactivates the controller. The block diagram in Figure 5.14 represents a simplified speed control system. The simple safety interlocks on the brake pedal and the on/off switch do not impact the controller design. We will focus first on developing a representative simulation model for the vehicle and then on the fuzzy controller design.

---

[60]Kosko, B., *Neural Networks and Fuzzy Systems*, Englewood Cliffs, NJ, Prentice-Hall, 1992.
[61]Wang, L., *Adaptive Fuzzy Systems and Control*, Englewood Cliffs, NJ, Prentice-Hall, 1994.

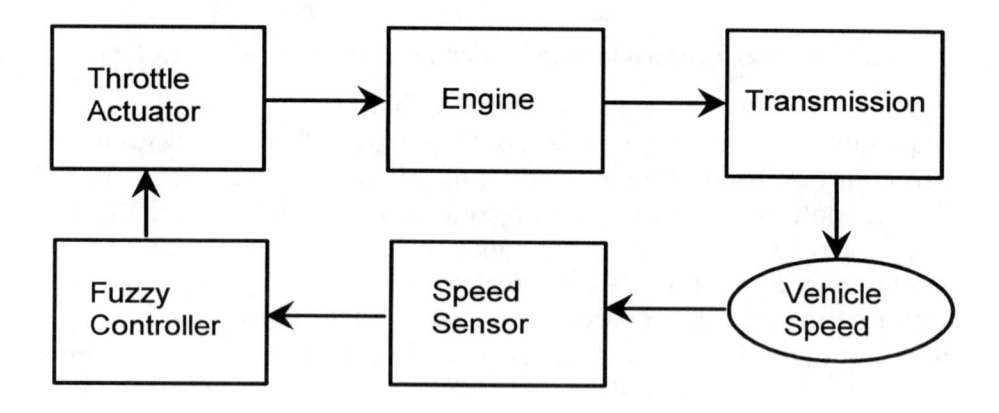

Figure 5.14: *Block diagram of the automobile cruise control. We will focus our attention on the control algorithm and on simulating the throttle/engine/transmission combination.*

Vehicle Simulation

The purpose of any simulation model used in controller design is to gain more insight into the dynamics of a particular system. A simulation model may be a tidy set of equations, or may itself be a fuzzy model representing the expert knowledge of a human operator. A simulation model is not absolutely necessary. However, having one available during controller design can speed up the tuning process and improve the results thereof. Without a simulation model, controller tuning must be performed in-situ on the target system.

Some features of the simulation are more important than others. To insure the basic usefulness of the simulation it should capture the system behavior both in the most likely region of operation and at the extremes of operation. Also, the simulation should capture the asymmetrical aspects of the system response. That is, we need to know if the system responds equally to perturbations in different directions. We now develop the automobile simulation as an illustration of these principles.

The throttle actuator and linkage, vehicle inertia, engine torque curve, transmission and gear-train, air resistance, and road grade all

Chapter 5: Fuzzy Engineering II

affect the response of the car to changes in the throttle setting. We will attempt to model the most important aspects of each of these factors.

The throttle actuator and linkage are modeled by a second order system given by the equation:

$$Throttle(a,b,t) = a + (b-a)\left(1 - \cos(c_1 t)e^{-t/c_2}\right)$$

t = time since last throttle change (seconds)
a = previous throttle position (% of maximum)
b = current throttle setting (% of maximum)
c1 = time constant 1 (0.3)
c2 = time constant 2 (8.0)

The actual throttle position lags behind the requested throttle position and is subject to moderate overshoot and oscillatory behavior. Figure 5.15 graphs the actual throttle setting in response to a commanded change from 0% to 100%.

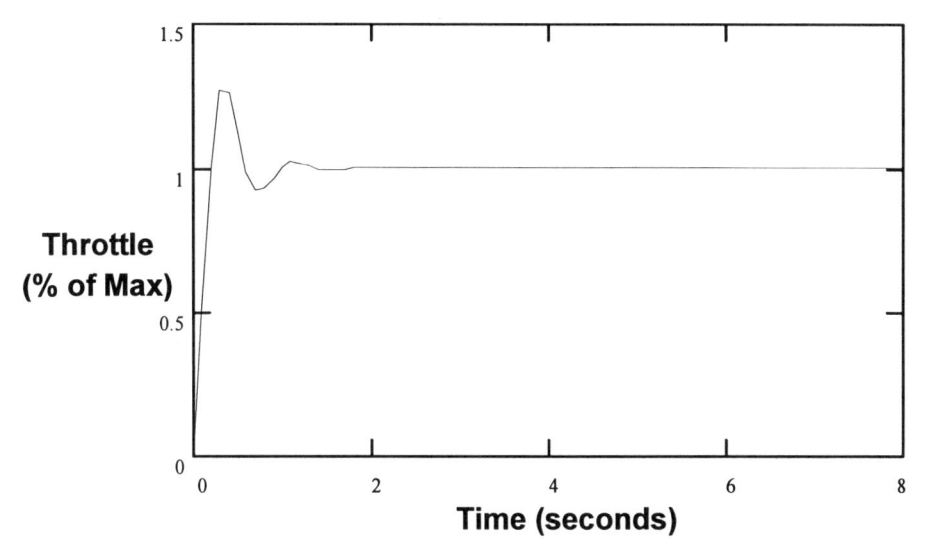

Figure 5.15: Second order throttle actuator and linkage response to full throttle(100%) command.

The engine power required to overcome gravity is related to the road grade, the vehicle's mass, and the current vehicle speed. That relationship is:

163

$$Gravity\_Power(m, v, \alpha) = mgv\sin(\alpha)$$
m = vehicle mass (100 slugs)
g = earth's gravitational acceleration (32.2 ft/s2)
v = vehicle road speed (ft/s)
α = road grade (radians) (positive is up-hill)

Traveling up-hill requires more power, while traveling down-hill releases gravitational potential energy which accelerates the vehicle. The previous equation captures all of the important effects of a changing road grade.

On a level roadway, air friction supplies the majority of resistance to increases in vehicle speed. The frictional force on the vehicle due to air resistance can be reasonably approximated as an exponential relationship. It is doubtful that an actual closed form solution exists that models the complex aerodynamic interaction between the vehicle body and the atmosphere, so an approximation must suffice. The power required to overcome air resistance takes the form of the following equation:

$$Air\_Power(v) = c_3 v \exp(c_4 v)$$
v = vehicle speed (ft/s)
c3 = constant (81.4)
c4 = constant (0.01)

The constants in this equation were derived from two points:
$$air\_power(100 Mph) = 100 Hp$$
$$air\_power(160 Mph) = 400 Hp$$
1 Mph = 1.467 ft/s
1 Hp = 550 ft-lb/s

Figure 5.16 graphs the power required to overcome air friction versus vehicle speed according to our approximation.

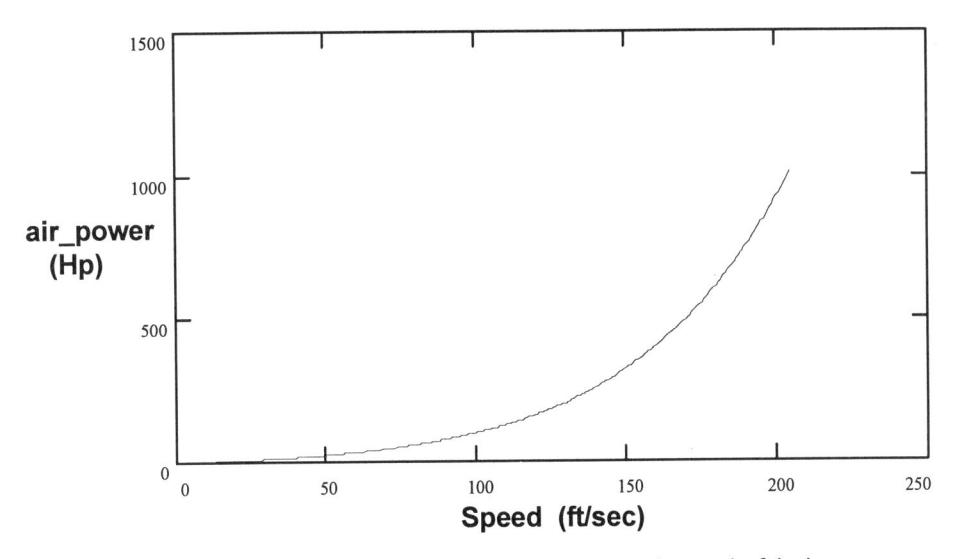

Figure 5.16: Power required to overcome aerodynamic friction.

The power output of an internal combustion engine is dependent on so many variables that an accurate analytical equation is impractical. Instead, we use empirical data for the engine's power output. The torque capability of the engine at maximum throttle applies most directly to our simulation. The raw data of torque versus revolutions-per-minute (RPM) for the engine is shown in Table 5.2, and a smoothed plot of the interpolated torque data is shown in Figure 5.17.

RPM	50	175	200	225	250	275	300	325	350
TQ	50	343	345	354	349	365	359	360	375
RPM	375	400	425	450	475	500	525	550	700
TQ	370	361	342	324	308	294	273	248	90

Table 5.2: Torque versus RPM at full throttle used in automobile engine simulation. Torque is in foot-lbs and RPM is in 10's.

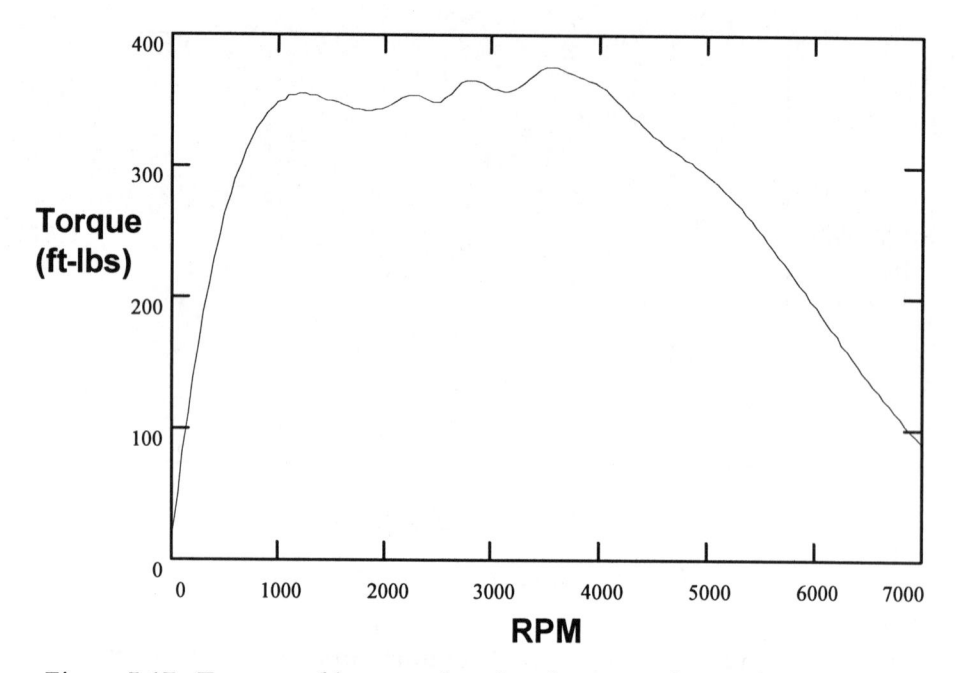

Figure 5.17: Torque making capacity of engine at maximum throttle plotted versus engine RPM.

The transmission and differential convey the engine torque to the tires and hence to the pavement. The transmission and differential end up multiplying the torque that reaches the ground by a gear ratio particular to that specific drive train. The power that reaches the pavement, as graphed in Figure 5.18, is given by the following equation:

$$Engine\_Power(throttle, v) = \frac{2 * v * gear\_ratio}{tire\_diam} * engine\_torque(v) * throttle$$

gear_ratio = final drive ratio for the combined transmission and differential (3.08)
tire_diam = diameter of the drive tires (2.25ft)

Chapter 5: Fuzzy Engineering II

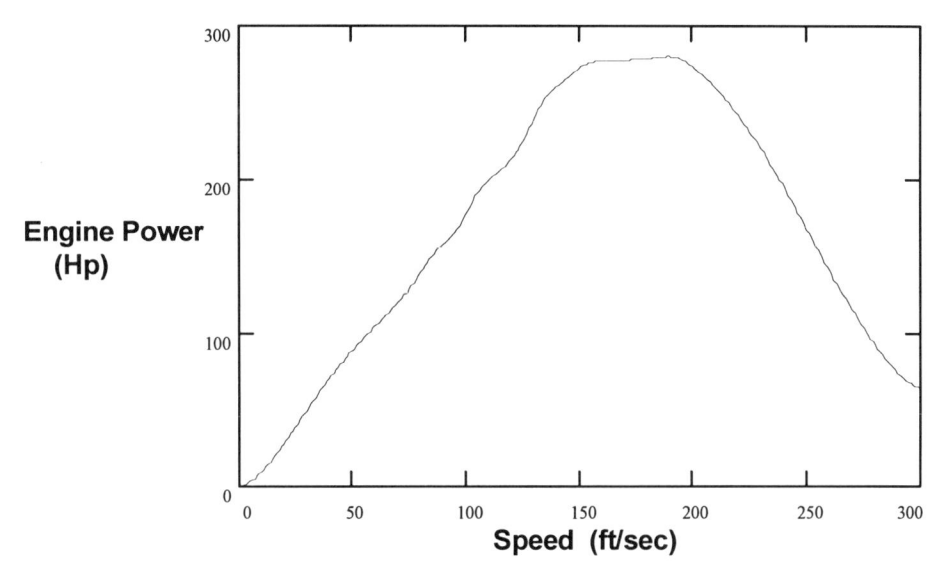

Figure 5.18: Engine power delivered to the road at maximum throttle plotted versus vehicle speed.

Finally, acceleration and deceleration of the vehicle results from the combined effects of the engine, air friction, and road grade, as shown by the following equation:

$$Acceleration(throttle, v, m, \alpha) = \frac{Engine\_Power(throttle, v) - Air\_Power(v) - Gravity\_Power(v, m, \alpha)}{mv}$$

To keep track of the vehicle's speed while the simulation proceeds we must use a numerical integration technique. The method used in the pendulum example is called Euler's method. Euler's method takes a simple differential equation, converts the infinitesimal differentials to finite intervals, and finally converts the original equation to a difference equation. Using Euler's method to track the vehicle's speed requires the following steps:

$$\frac{dv}{dt} = Acceleration(throttle, v, m, \alpha)$$

$$\Delta v = Acceleration(throttle, v, m, \alpha) * \Delta t$$

$$v_{i+1} = v_i + \Delta v$$

v_0 = initial speed
Δt is 'small'

167

Fuzzy Logic for Real World Design

While Euler's method handles simple situations like the pendulum and automobile cruise control well enough, it will not be as good for other, more complex simulations. The primary source of error in the Euler technique is that the change in the variable on the left side of the equation depends only on a calculation of the derivative (on the right side of the equation) made at the beginning of each time interval. More sophisticated numerical methods, like Runge-Kutta[62], improve on Euler's method by estimating more accurately the derivatives on the right side of the equation several times during one time interval.

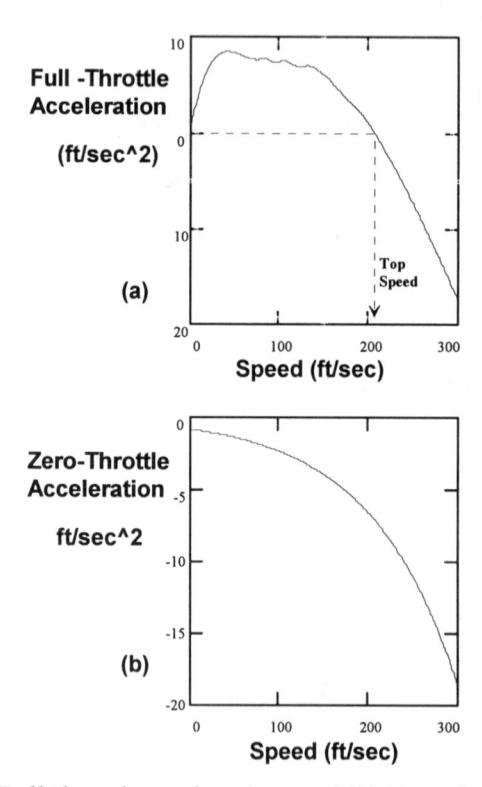

Figure 5.19: (a) Full throttle acceleration, and (b) Zero-throttle deceleration.

---

[62] An instructive practitioners guide to the mathematics of simulation can be found in W.H. Press et.al, *Numerical Recipes in C: The Art of Scientific Computing* Cambridge Univ. Press, Cambridge.

Having completed the simulation mathematics, we can examine the behavior of the vehicle under the range of conditions that the cruise controller is expected to handle. For this purpose, Figure 5.19 (a) shows full-throttle acceleration versus speed, and Figure 5.19 (b) shows zero-throttle acceleration versus speed.

The simulation reveals several interesting features. For example, at full-throttle on level ground, Figure 5.19 (a) clearly identifies the theoretical top speed of the vehicle as the point at which full-throttle acceleration crosses zero into negative territory. Furthermore, the vehicle responds asymmetrically to throttle increases versus throttle decreases. When increasing the vehicle speed, the engine supplies the power necessary to overcome the vehicle inertia and aerodynamic friction. However, when decreasing the vehicle's speed, the dominant factors are air friction and gear train losses. This asymmetric response will be used in designing and tuning the fuzzy controller.

Controller Design and Tuning

A PI-like fuzzy controller for automobile cruise control takes the following functional form:

$$u = \Phi(e, \int e)$$

u = output of the controller
e = speed deviation from the set-point
$\int e$ = integrated error from the set-point

The PI-control law can be recast into the form:

$$\frac{du}{dt} = F(\frac{de}{dt}, e)$$
$$\Delta u_i \equiv M(\Delta e_i, e_i)$$
$$u_i = u_{i-1} + \Delta u_i$$

In this final form, the output of the fuzzy system will be added to the existing throttle setting to determine the new throttle setting. The total throttle setting will be subject to an upper bound of 100% and a lower bound of 0%. For simplicity, the upper and lower bounds of the throttle setting will be imposed outside of the fuzzy system.

Fuzzy Logic for Real World Design

To start, each input will use five input membership functions, as depicted in Figures 5.20 (a) and (b), and the output will the five output singletons shown in Figure 5.20 (c).

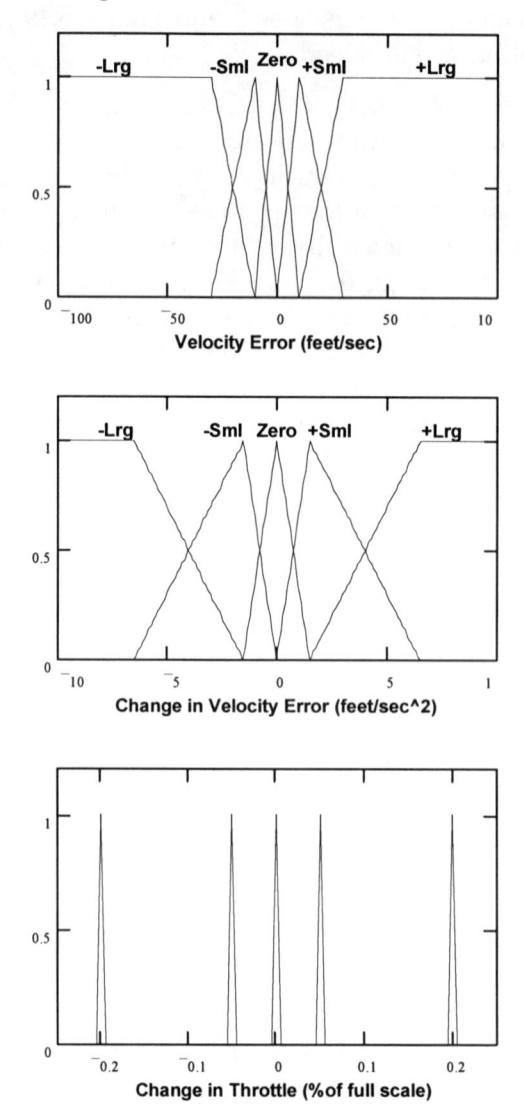

Figure 5.20: Initial membership functions for (a) input e, (b) input $\Delta e/\Delta t$, and (c) output Δu.

170

Chapter 5: Fuzzy Engineering II

Looking at the open loop response of Figure 5.21 shows again the zero-throttle response of the car. The vehicle deceleration is due primarily to air resistance.

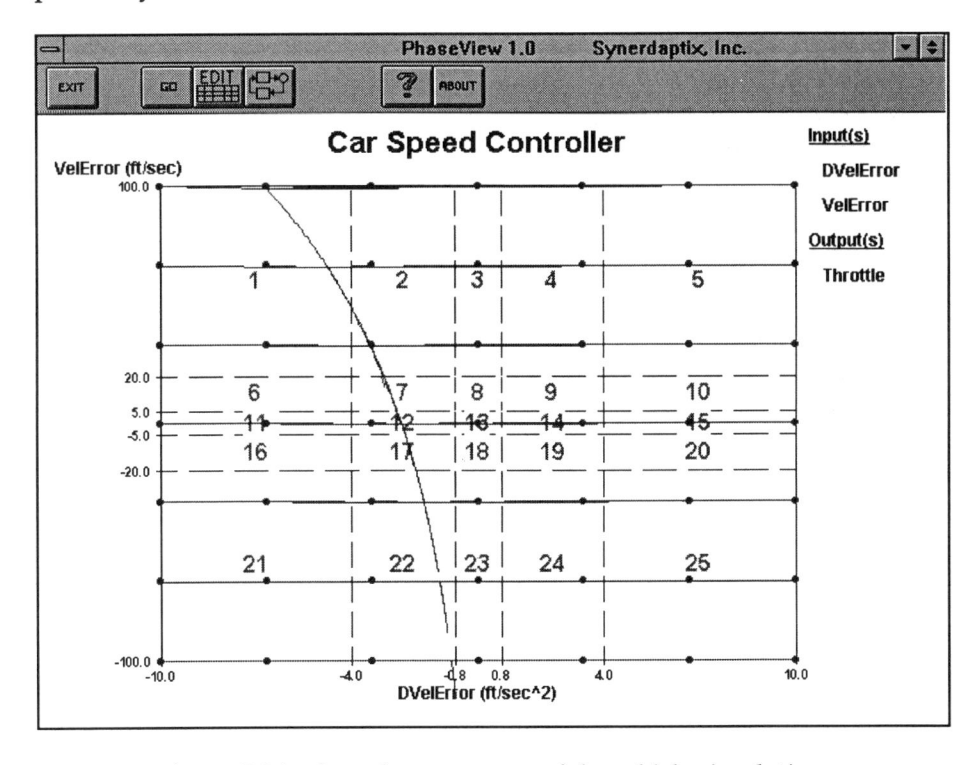

Figure 5.21: Open loop response of the vehicle simulation.

Using the open loop response as a guide, we can write a simple set of rules to start the design process. Table 5.3 lists the initial rule set. Figure 5.22 shows the state space performance of the speed controller corresponding to the initial rule set. Notice that only two rules are used to cover state space regions 1-5 and 21-25.

```
01-05: IF VelError is -Lrg THEN Throttle is +Lrg
06: IF VelError is -Sml AND DVelError is -Lrg THEN Throttle is +Lrg
07: IF VelError is -Sml AND DVelError is -Sml THEN Throttle is +Lrg
08: IF VelError is -Sml AND DVelError is Zero THEN Throttle is +Sml
09: IF VelError is -Sml AND DVelError is +Sml THEN Throttle is +Sml
10: IF VelError is -Sml AND DVelError is +Lrg THEN Throttle is Zero
11: IF VelError is Zero AND DVelError is -Lrg THEN Throttle is +Sml
12: IF VelError is Zero AND DVelError is -Sml THEN Throttle is Zero
13: IF VelError is Zero AND DVelError is Zero THEN Throttle is Zero
14: IF VelError is Zero AND DVelError is +Sml THEN Throttle is Zero
15: IF VelError is Zero AND DVelError is +Lrg THEN Throttle is -Sml
16: IF VelError is +Sml AND DVelError is -Lrg THEN Throttle is Zero
17: IF VelError is +Sml AND DVelError is -Sml THEN Throttle is -Sml
18: IF VelError is +Sml AND DVelError is Zero THEN Throttle is -Sml
19: IF VelError is +Sml AND DVelError is +Sml THEN Throttle is -Lrg
20: IF VelError is +Sml AND DVelError is +Lrg THEN Throttle is -Lrg
21-25: IF VelError is +Lrg THEN Throttle is -Lrg
```

Table 5.3 Initial rules for vehicle speed control.

Chapter 5: Fuzzy Engineering II

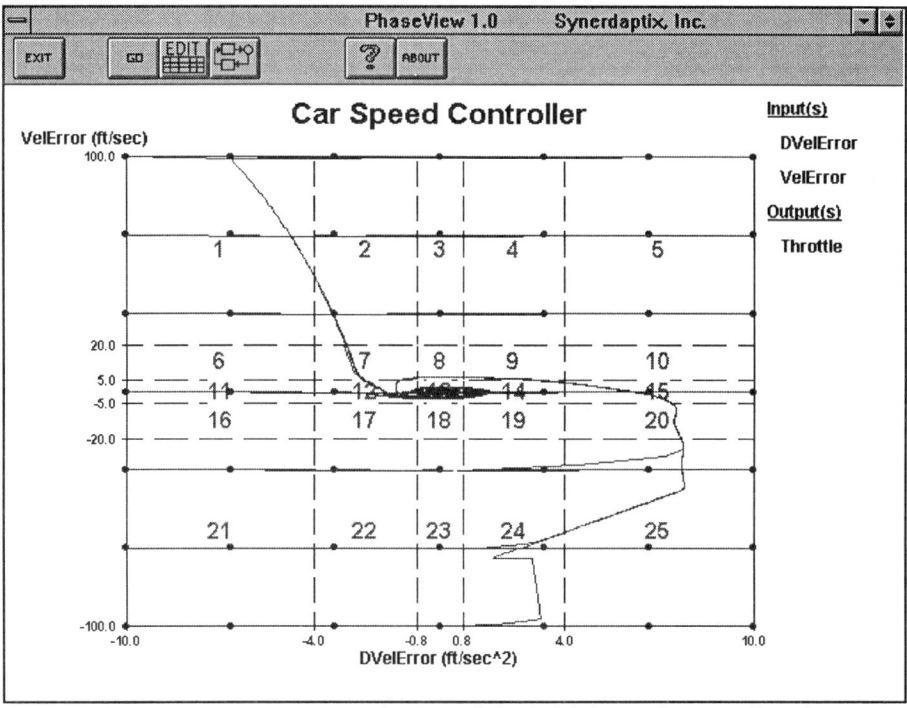

Figure 5.22: State space evolution of the initial speed controller shows oscillation and overshoot. Speed set-point is 110 ft/sec.

Recognizing that the state space evolution stays in the middle two-thirds of the input variable *DVelError*, that variable's universe of discourse can be reduced. In addition, the usable universe of discourse for *DVelError* is not quite symmetric. Mapping the membership functions to the useful range of a variable maximizes the precision of the fuzzy system for that variable.

The next step in the tuning effort is to make the necessary rulebase adjustments. The most obvious troubled region in the state space graph of Figure 5.22 appears at its center. The oscillations in the center regions of the graph are related to the rules for that region, numbered 12, 13, and 14. Larger output magnitude in both regions 12 and 14 will reduce the oscillation. Changing rules 12 and 14 from:

173

Fuzzy Logic for Real World Design

```
12: IF VelError is Zero AND DVelError is -Sml THEN Throttle is Zero
14: IF VelError is Zero AND DVelError is +Sml THEN Throttle is Zero
```
to:
```
12: IF VelError is Zero AND DVelError is -Sml THEN Throttle is +Sml
14: IF VelError is Zero AND DVelError is +Sml THEN Throttle is -Sml
```

dampens the oscillation as illustrated by Figure 5.23.

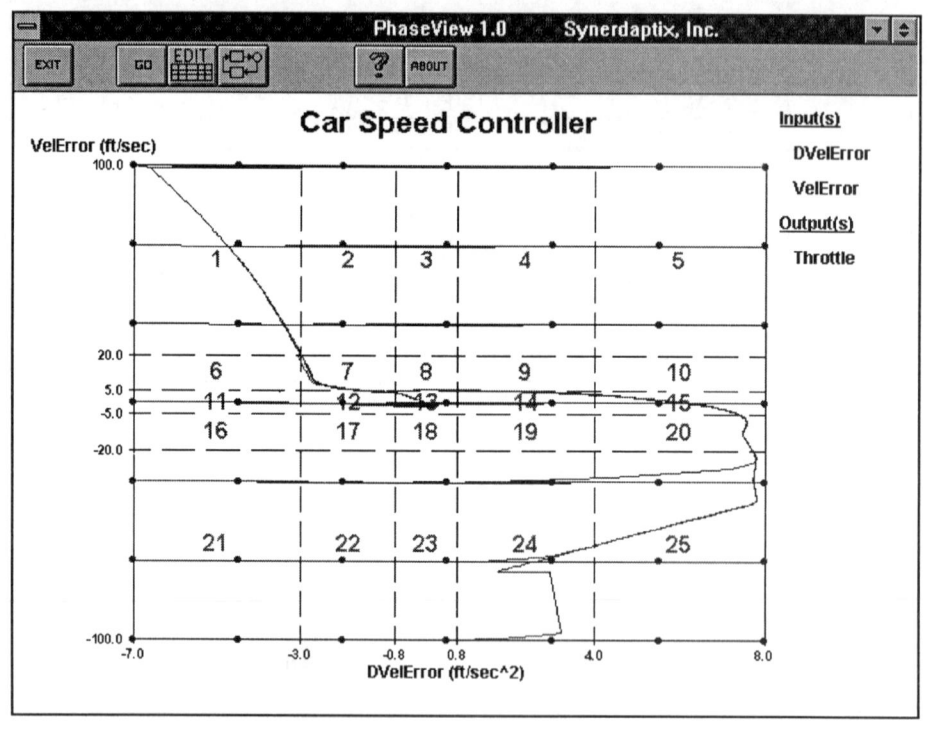

Figure 5.23: Reduced oscillation through refinement of rules 12 and 14.

The overshoot remaining during a transition from a slower to a faster speed can be reduced by more aggressively reducing the throttle when approaching the set-point. Rule 15 most impacts this situation, so its consequent may be changed to *-Lrg* to compensate.

Chapter 5: Fuzzy Engineering II

```
15: IF VelError is Zero AND DVelError is +Lrg THEN Throttle is -Lrg
```

An additional adjustment to the fuzzy model will compensate for the asymmetric response of the vehicle that we pointed out earlier. The only force operating to reduce a high speed is the force due to air friction. The effect of air friction is much weaker than the power supplied by the engine, so it is more important to avoid overshooting the desired speed. Increasing the magnitude of the *-Lrg* fuzzy set of the output *Throttle* will bias the output to more effectively avoid speed overshoot. The simulated state space evolution of the final fuzzy system is shown next in Figure 5.24.

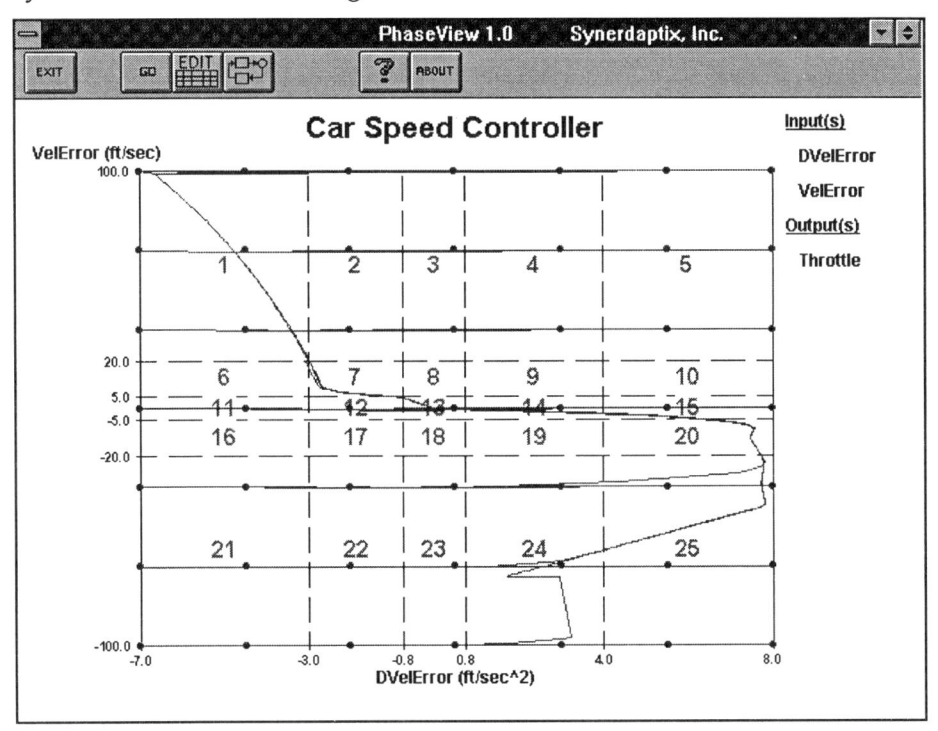

Figure 5.24: Reduced overshoot via asymmetry in output membership functions and changes to rules 11 and 15.

The final membership function partitions in Figure 5.25 show the asymetric adjustments made during the tuning process.

175

Fuzzy Logic for Real World Design

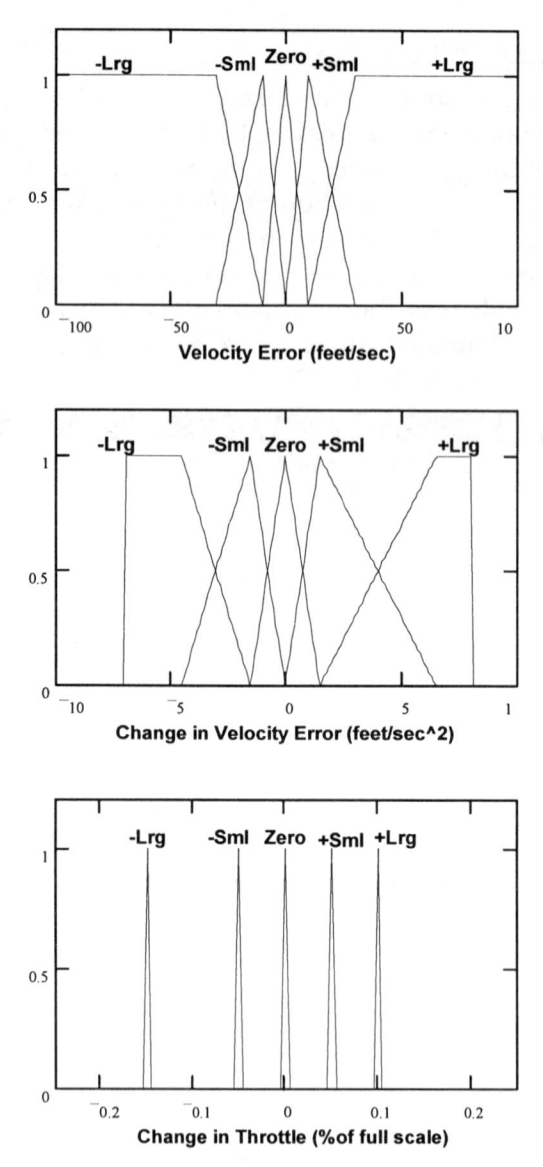

Figure 5.25: Final membership functions for (a) input e, (b) input $\Delta e/\Delta t$, and (c) output Δu.

To explore the effects of road grade on the tuned system, the simulation was expanded to include a fluctuating road grade. The time

Chapter 5: Fuzzy Engineering II

response in Figure 5.26 shows the behaviour of the simulated system and controller with a fluctuating road grade. At time=0 in the figure speed=10 ft/sec and the system has just been commanded to cruise at 110 ft/sec.

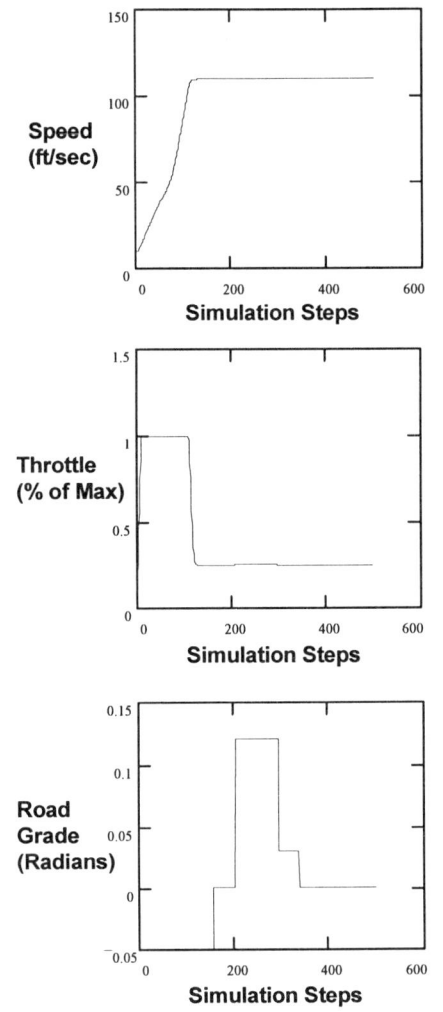

Figure 5.26: (a) Car speed, (b) throttle setting, and (c) road grade plotted over 400 simulation steps. One simulation step = 0.1 seconds.

177

From the speed response of the car shown in Figure 5.26 it appears not only that the changing road grade makes almost no impact on the controller's response, but also that the throttle setting commanded by the controller changes in a smooth and simple way. By looking at a smaller portion of the simulation we can see that the controller's behavior is more complex than it would first appear. Figure 5.27 shows a smaller portion of the simulation, at higher magnification. The figure reveals the finer structure of the controller's behavior as the speed and road grade vary.

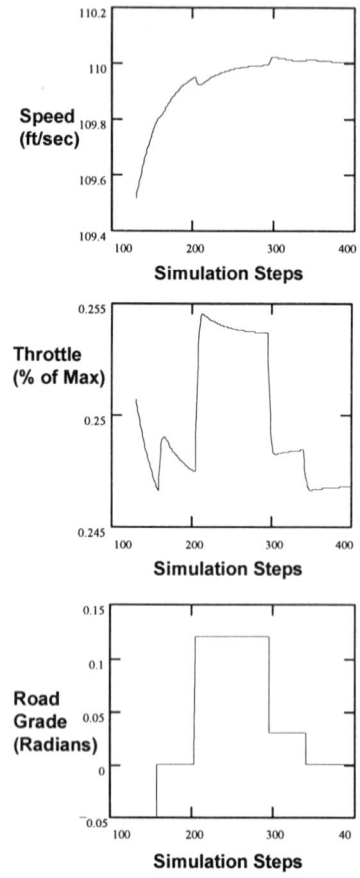

Figure 5.27: (a) Car speed, (b) throttle setting, and (c) road grade plotted over a subset of simulation steps of Figure 5.26. One simulation step = 0.1 seconds.

This ends our discussion of fuzzy controllers. We have investigated the tuning of membership fuctions and rules in two complete example controllers. We introduced the graph of the system's state-space evolution as a valuable tuning aid that yields insight over the entire operating region of the controller at a glance. Using the state space evoultion of a system, we sculpted the membership functions and rules to achieve improved controller performance. Finally, the sample control systems demostrated the non-linear response and ease of development that fuzzy controllers possess.

Decision Support System

Even from its earliest formulation by Lotfi Zadeh, the original intent of fuzzy logic was to provide a framework for the automation of modes of human reasoning that are approximate rather than exact. Sometimes referred to as rules-of-thumb or heuristics, these inexact reasoning methods give us the ability to make concrete choices given imprecise, uncertain, ambiguous, or even contradictory information. Fuzzy logic serves as a particularly useful way to capture and automate any human expertise that exists already in the form of rules-of-thumb or heuristics. A system that involves an automated form of human expertise in a specific domain can be made available as a decision aid to anyone working in that domain. Today, such a system is called an expert system. All of the control systems we have see thus far can be described as embedded expert systems.

In addition to providing embedded expert systems, fuzzy logic is useful in a lager role as a decision-aiding technology. Deciding among a variety of choices, in the presence of goals and constraints, given imprecise inputs, is a challenge we each face every day. Whether we are deciding which used car to buy, which political candidate to vote for, which route to take to work, how to spend our leisure time, or simply where to eat lunch, decision making in the real world is a complex human process. In a decision support system, fuzzy logic functions as an extension of the human decision maker by providing an intuitive, but rigorous, framework for weighing goal satisfaction in the presence of constraints and a multiplicity of choices.

First, consider a simple, every-day situation. Suppose, while driving your car, you are approaching a traffic intersection and the traffic signal changes from green (indicating *your right-of-way*) to yellow (indicating *prepare to stop*). The goals in this situation are to reach your destination safely and quickly. Your constraints are avoiding accidents and traffic tickets. You have only two choices, stop, or go through the intersection. What you do in reaction to the changing traffic signal depends on many factors including distance to intersection, current speed, actions of the vehicles in front of you, proximity of law enforcement officers, and so on. In this situation, events may happen so quickly that you have little time to consciously and deliberately weigh the pertinent factors. You must decide either to apply the brakes and stop before the intersection or to apply more gas to speed through the intersection before the light turns red. The rules-of-thumb that you use to make your decision will be a combination of crisp and fuzzy rules that are based upon both crisp and fuzzy information.

A Real Decision Support System

Now for a simple, but complete example: deciding in which region of the country to live. Supposing you have decided to relocate your family to a different region of the country, how would you decide where to go? What characteristics define your idea of a better place to live? Having identified those characteristics, how do you choose between alternatives?

Many objective measures can be used to indicate the desirability of a particular region or city. This example decision support system considers cost of living, job market, crime, health care, transportation, education, arts and culture, recreation, and climate. The purpose of this decision support system is to help you decide, given the multiple criteria listed above, which cities and regions maximize your idea of a good place to live. Consider the raw data listed in Table 5.4[63].

---

[63] All ratings in these categories for 333 U.S. metro areas appear in the *Places Rated Almanac*, 1989, Simon & Schuster Inc., New York, NY.

Chapter 5: Fuzzy Engineering II

Area	Cost of Living	Job Market	Crime	Health Care	Transit System	Education	Arts & Culture	Recreation	Climate
Atlanta, GA	227	2	285	106	4	23	14	127	24
Pittsburgh, PA	157	76	72	35	78	21	33	117	90
Dayton, OH	120	154	201	209	83	54	53	269	161
Seattle, WA	261	36	270	33	12	30	11	1	12
Boston, MA	318	29	255	5	30	4	7	57	56
New York, NY	319	71	332	2	1	1	1	15	44
Louisville, KY	89	151	137	91	60	88	36	89	60

Table 5.4: Raw data for eight US. metropolitan areas. Numbers are rankings (1 is best, 333 is worst).

How can we compute a single number, representing the overall goodness of a particular city, that appropriately considers the degree of goodness of each individual category in Table 5.4.? One answer to this question relies on the antecedent processing and rule evaluation methods of fuzzy logic. The rankings in each column can be converted to a degree of goodness ranging form 0.0 to 1.0. Then, for a particular city, the degrees of goodness in each category can be combined using an appropriate intersection operator. The resulting degree of strength reflects the overall goodness of that particular city.

Converting the category rankings into degrees of truth requires a mapping like one of the two shown in Figures 5.28(a) and (b). Considerable variation is possible in the mapping scheme, although the mapping should preserve common sense.

181

Fuzzy Logic for Real World Design

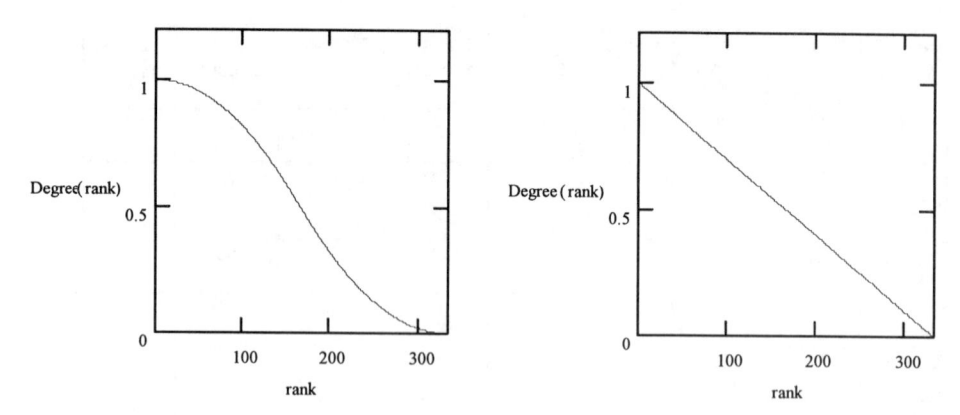

Figure 5.28: (a) S-curve mapping of city rankings to degree of goodness, and (b) Linear mapping of city rankings to degree of goodness.

The S-curve mapping shown in Figure 5.28(a) converts the rankings of Table 5.4 into the degrees listed in Table 5.5.

Area	Cost of Living	Job Market	Crime	Health Care	Transit System	Education	Arts & Culture	Recreation	Climate
Atlanta, GA	.204	1.000	.042	.800	1.000	.991	.997	.712	.990
Pittsburgh, PA	.558	.898	.909	.979	.892	.993	.981	.756	.856
Dayton, OH	.743	.575	.316	.279	.878	.949	.951	.074	.535
Seattle, WA	.094	.978	.072	.981	.998	.985	.998	1.000	.998
Boston, MA	.004	.986	.110	1.000	.985	1.000	.999	.943	.945
New York, NY	.004	.911	.000	1.000	1.000	1.000	1.000	.996	.966
Louisville, KY	.859	.592	.664	.853	.937	.863	.978	.859	.937

Table 5.5: Raw data for eight U.S. metropolitan areas. Numbers are rankings (1 is best, 333 is worst).

As the next step, we apply an intersection operator to determine the overall goodness for each city. Table 5.6 shows the results of applying the various intersection operators already presented in Chapter 3.

Chapter 5: Fuzzy Engineering II

Area	MIN	BINTER	PRODUCT	YAGER P=5.0
Atlanta, GA	.042	.000	.005	.000
Pittsburgh, PA	.558	.000	.251	.553
Dayton, OH	.279	.000	.001	.000
Seattle, WA	.072	.000	.006	.000
Boston, MA	.004	.000	.000	.000
New York, NY	.000	.000	.000	.000
Louisville, KY	.592	.000	.183	.564

Table 5.6: The result of applying a few standard intersection operators.

The interpretation of the degrees of truth listed in Table 5.6 is that the highest degree of truth indicates the best city and the lowest degree of truth indicates the worst city of the group. From the data in Table 5.6 it is clear that the choice of intersection operator can effect the result. The best city, and even the relative rankings between cities may change somewhat with the choice of intersection operator. Rather than viewing this property as a drawback however, we can use different intersection operators to more accurately reflect personal preference in the decision.

The weighted average intersection operator provides the additional flexibility needed in decision support systems. The weighted average is computed by multiplying the degree of truth for a particular input by a user-selected weight. The normalized sum of all such products yields the degree of truth for the compound statement. The equation for this method is:

$$T(\vec{x}) = \frac{\sum_i \mu_i * w_i}{\sum_i w_i}$$

$i = 1$ to number of inputs
w_i = weight applied to input i

For the city selection problem with nine inputs, a weight vector that equally weights each of the nine inputs looks like: $\vec{w} = \{1,1,1,1,1,1,1,1,1\}$.

183

Fuzzy Logic for Real World Design

We can now see how easily individual preferences are encoded in the weight vector. For example, the preferences belonging to an "urban warrior" might place more importance on Arts & Culture and Transit System while placing less emphasis on Crime and Cost of Living. A weight vector describing the preferences of the "urban warrior" could look like: $\vec{w} = \{0.2,1,0.2,1,3,1,3,1,1\}$. As another example, consider the preferences of the "nature enthusiast". The "nature enthusiast" might emphasize Climate, Recreation, and Cost of Living, while de-emphasizing Transit System and Crime. These preferences might be well represented by the weight vector: $\vec{w} = \{3,1,0.2,1,0.1,1,1,4,4\}$. Using a weighted average to arrive at an overall measure of the goodness of any geographic area allows more realistic decision making by weighting personal preferences. Table 5.7 compares the decisions reached using the "urban warrior", "nature enthusiast", and the equally weighted decision vectors.

Area	Equal Weight	Urban Warrior	Nature Enthusiast
Atlanta, GA	.748	.924	.740
Pittsburgh, PA	.869*	.912	.800
Dayton, OH	.589	.712	.495
Seattle, WA	.789	.962*	.806
Boston, MA	.775	.952	.763
New York, NY	.764	.954	.776
Louisville, KY	.838	.891	.868*

Table 5.7: Decisions produced by weighted average intersection and several weighting profiles. * - indicates best choice for given preferences.

Even with the modest goals of the current example, we begin to see the richness and utility of the fuzzy paradigm applied to the decision support realm. Fuzzy logic excels not only in the representation and manipulation of imprecise knowledge, but also, as in the city selection example, in the selection of one option from the list of many. The variety of fuzzy logic operators adds considerable flexibility to the decision making process, and also, as in the city selection example, permits the nuances of personal preference to appropriately influence the system.

Summary

Fuzzy logic can be the right problem solving paradigm in a variety of situations for a variety of reasons. The successfulness of your efforts to apply fuzzy logic to your problems hinges on two key requirements. The first requirement is your direct knowledge, or access to expert knowledge, in the problem domain. The second requirement is understanding how to apply fuzzy logic principles in the solution of your problem. While the first requirement is entirely your responsibility, the second area provides the motivation for this book.

To satisfy the first requirement you must have knowledge pertinent to the solution of your particular problem. If you are designing a machine to monitor a patient during the administration of general anesthesia you must possess expert knowledge yourself or you ought to have close consultations with practicing, certified, anesthesiologists. Keep in mind however, that meaningful knowledge is not necessarily precise. Human knowledge, which is characterized by approximations, shades of meaning, degrees of belief, rules-of-thumb, etc., is robust, insightful, and essentially imprecise.

The aim of this book is to help you address the second requirement: understanding how fuzziness applies to your problem, and the mechanics of implementing solutions with the fuzzy logic paradigm. Since the prevailing fuzzy engineering methodology leans heavily to the empirical side, we have presented tools and techniques that aid in that realm. We have introduced the concept of the state space evolution, and detailed its use in two fuzzy control examples. Mathematical simulation of the system to be controlled, while not possible for every problem, aids in the development and tuning of the fuzzy controller. The state space evolution tells, at a glance, the behavior of the whole controlled system.

Tuning of the fuzzy model is accomplished through systematic tuning of first the rules, and second, the membership functions. If satisfactory performance is not achieved immediately, the number of membership functions may be increased, or an additional input variable may be used. In the information systems applications of fuzzy modeling, the goal usually is to aid or mimic a human decision maker. In such systems the choice of mathematical operators in the fuzzy decision algorithms may be significant. Operators like the weighted average intersection permit the nuances of the human decision maker to be reflected in the fuzzy system.

We next turn our attention to the job of implementing fuzzy models in real-world systems. The next several chapters turn the abstract theory and practical advice of the previous chapters into cold, hard code that will run on microcontrollers and desktop PCs.

Chapter 6:
Embedded Systems Considerations

The invention of the microprocessor in 1969-70 resulted in two broad branches of applications. The second and most recognizable branch spawned was the personal computer. However, the earlier branch, and the one that concerns this chapter, is the embedded system.[64]

The term embedded system refers to a piece of electronic or electromechanical equipment that uses a microcomputer as an integral component. Since the microcomputer in an embedded system rarely contains the peripherals associated with today's conception of a computer, namely a video monitor, mass storage, mouse, and keyboard, the embedded microcomputer is most often referred to as a microcontroller. By simply roaming through your home with open eyes you will see several examples of these systems. Microwave ovens, VCRs, stereos, TVs, video cameras, and most other electronic consumer goods contain micro controllers to orchestrate their operations. With today's automobiles supporting features like electronic fuel injection, electronic ignition, anti-lock brakes, automatic traction control, digital

---

[64] The first recognized microprocessor, the Intel 4004, made its debut in the Busicom desk-top calculator in 1971.

climate control, cruise control, and trip computers, some automobiles contain more micro controllers than all of the appliances in your home combined.

To date, the majority of embedded fuzzy applications perform control functions. The systems typically controlled by these fuzzy models include electronic and electro-mechanical systems. The block diagram in Figure 6.1 shows the relationship between the fuzzy controller and attached system for a simple application. In the figure, the combination of the plant and the fuzzy controller make up the system. We assume that the current state of the plant can be described by the system of equations $f_1(x_1, x_2,...), f_2(x_1, x_2,...), \cdots$. The variables $x_1, x_2, \cdots$ make up the state vector. The system has at least one observable output which is represented as Y in the figure. The output of the controller, u_{FC}, along with the system of equations describing the plant determine the time evolution of the state vector of the plant. The differential equation in Figure 6.1 is just a compact way to write all of the applicable equations.

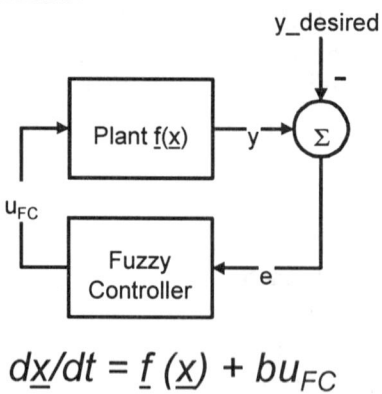

$$d\underline{x}/dt = \underline{f}(\underline{x}) + bu_{FC}$$

Figure 6.1: Representative fuzzy control diagram.

With the proliferation of successful fuzzy controller implementations one might think that fuzzy logic is just a novel control technology. Certainly this is not the case. Limiting the use of fuzzy logic to control systems would be an unjustified and artificial constraint. Rather, control is just the first well-explored area of applications of fuzzy logic.

In the role of intelligent system controller, fuzzy logic brings two main capabilities to bear. The first, and most important capability, is the automation of human expertise. The human expertise, which is expressed by means of rules and fuzzy set definitions, is most efficiently imparted to an embedded system by means of fuzzy logic. The second important capability imparted to the embedded system by fuzzy logic is the ability to express non-linear relationships between the control inputs and control outputs. In this regard, the fuzzy model performs non-linear interpolation between the peaks of the output membership functions. Certainly fuzzy logic is but one of many non-linear control techniques. However, unless you happen to be a control theorist, fuzzy logic may be the most accessible of these non-linear control techniques.

Clearly, there exists a variety of control technologies that might be applied to embedded systems. Fuzzy logic makes a compelling technology to use in a variety of situations. Those situations include:

1. The plant to be controlled contains elements that either are not or cannot be known precisely. Any system that currently uses a human in the control loop is an example of this situation.

2. The characteristics of the plant may change significantly over time. As an example of this situation, anti-lock automobile brakes must maintain effective and safe operation given considerable wear and aging of the brake system components.

3. The controller must be developed before the final plant design is known. This situation is actually being encouraged under the recent industry trend called *concurrent engineering*.

4. An embedded controller must control a diverse set of plants under a diverse set of conditions. An example of this type of situation can be seen in an off-the-shelf process controller.

With the increase in computing power of the microcontroller, digital signal processing (DSP) of analog signals has become feasible. Digital signal processing impacts embedded systems synergistically; it has made the entirety of digital music and digital video possible. Analog signals can be digitized, transformed by algorithms running on the microcontroller, and then converted back to analog signals. The combination results in tremendous flexibility as the algorithms for processing the signal exist entirely in the software running on the system. The algorithms can be adjusted and tuned as easily as any computer program. While retaining all of the advantages of general DSP techniques, fuzzy logic certainly possesses its own set of engineering issues. We examine those engineering issues in this chapter.

Goals and Constraints

Since we are now concerned with the practical aspects of designing and implementing fuzzy models, we must carefully consider the relevant factors. We will assume that a fuzzy model is at least a plausible alternative to solving the given problem. The main challenge at this point remains an engineering one. We must achieve the project goals while managing the constraints imposed on the project.

Project goals may include some or all of the following: satisfaction of performance metrics, low implementation cost, high supportability, future upward migration pathing, high reliability, low development cost, short development time, and others. The constraints imposed on the project may include any or all of the following: limited time, money, and personnel, technology learning curves, technology performance limits, and the ever-present performance-hungry, cost-sensitive customer.

The ways in which fuzzy logic addresses most of the macroscopic, project level issues in a real world design have been explored in detail in other chapters. Refer to table 6.1 for a recap of those issues and answers.

Chapter 6: Embedded Systems Considerations

ISSUES	SOLUTIONS
Technology learning curve	FL is closely aligned with our thinking so expressing solutions is naturally easier.
	Books like *Fuzzy Logic for Real World Design*.[65]
Development time and cost	FL is especially valuable approach for fast prototyping.
	Don't need to be a fuzzy expert to achieve good results.
	Commercial tools available at modest prices.[66]
Product cost	FL easily implemented on eight bit micros.
	Small RAM and ROM requirements.
Product maintenance and supportability	Fuzzy models are the ultimate in self-documenting solutions because of their reliance on natural language descriptions.
Performance hungry customers	FL is the most computationally efficient way to get intelligent non-linear control into product

Table 6.1: Summary of fuzzy logic advantages that improve product life-cycle costs.

The microscopic concerns of actually embbeding a fuzzy model can be roughly grouped into the following areas: architecture, processing speed, resolution, complexity, and specificity. We now turn our attention to first highlighting, and then answering, the engineering questions.

Architecture

Architectural constraints which limit fuzzy implementation choices may confront your design. Perhaps the project calls for integrating a fuzzy model into an existing system. The existing system already uses a

---

[65] There are many books on the mathematical aspects of fuzzy logic. There also exist stacks of magazine articles on specific applications. The motivated reader can also find other source of information including video tapes and even multimedia software. See Appendix A for books and reference materials.
[66] See Appendix A.

specific processor and memory configuration. Your implementation must accommodate these constraints.

On the other hand, the fuzzy model itself may impose demands of the embedded system that require a special architecture. Specific membership function shapes and complicated rule evaluation algorithms may require special purpose hardware for acceptable performance. Additionally, special purpose hardware typically does not support a wide range of membership function shapes or evaluation algorithms.

In practice, a spartan yet fully functioning fuzzy kernel can be written for an 8 bit microcontroller in as few as three hundred bytes of code and use as few as 20 bytes of run-time RAM space[67]. Moving away from the minimal implementation, the variety of fuzzy intersection, aggregation, and defuzzification methods imposes varying demands on system resources. When embedding simple fuzzy models, processor selection and memory availability rarely present significant implementation restrictions. However, complex or high speed fuzzy models do demand greater attention to architectural considerations.

Processing Speed

Processing speed is like a budget. You can't spend what you don't have. If the application requires real-time control your algorithms must not exceed the computational budget. The result of exceeding the budget may be catastrophic. Imagine an aircraft autopilot which crashes the airplane because of slow processing. You must decide on your computational priorities and spend the budget wisely.

Both the inherent demands of the specific application, and the properties of the fuzzy model make demands on the processing budget. Generally, the most important factor in the stability of control system applications is the speed with which the new inputs are acquired and propagated through the system to produce a new control output. Processing speed directly impacts this control loop timing. Integrating

---

[67]The fuzzy kernel refered to here is from Motorola and is written in assembly language for the 68HC11 family of microcontrollers.

a fuzzy model into the control path demands a careful look at the control loop timing. Assessing the processing budget with respect to the fuzzy model helps in determining the processing requirements.

Choices for rule evaluation, aggregation, and defuzzification algorithms impose varying demands on the processor. For example, using the MIN intersection instead of the Yager intersection in rule evaluation will save many CPU cycles. Also, choice of membership function shape affect processing. For instance, using triangular membership functions requires less processing than true gaussian shapes. Similarly, defuzzification operations on singleton output membership functions complete faster than the same defuzzification algorithm operating on other output shapes.

In addition to processing cycles consumed within the fuzzy model, the input and output signals of a fuzzy model may require pre- and post-processing outside of the fuzzy model. In this regard, integrating a fuzzy model into a system is similar to implementing traditional digital signal processing techniques. Analog signals acquired from the world are transformed into the digital domain, processed algorithmically, and transformed back to analog signals which finally interact with the world. Occasionally, the demand that the processor must complete this number crunching within a certain time period may dictate the use of special purpose hardware.

Stability

What is stability? We usually discuss stability in the context of some dynamic system. The dynamic system, being subject to certain variable inputs, responds in a deterministic way. It is usually our goal to exert control over the dynamic system to keep it in or near some desired state. Furthermore, the controlling agent should return the dynamic system to the desired state in the event of any disturbance from that state. As an illustrative example consider the aircraft autopilot. The dynamic system is the air craft, the controller is the autopilot, and the environment of the dynamic system is composed of the prevailing weather and atmospheric conditions. We presume that the desired state of the system is level flight, on a certain compass heading, at a certain speed. Stability in the context of the plane and autopilot system relates to how accurately and reliably the desired heading, altitude, and speed can be maintained. Stability also refers to the ability of the autopilot system to bring the system back to the desired state after suffering a disturbance from that desired state. Stability is a desireable property of any control system, regardless of the technology used in the controller.

Stability, as a topic of great importance to the general field of control systems, is a difficult and deeply mathematical subject. The difficulty of the mathematics depends upon the characteristics of the dynamic system and on the dynamics of the controller. As might be expected, some types of systems lend themselves to simpler mathematics and, consequently, to simpler answers. Linear control systems have the most tractable mathematics and, consequently, make up a large portion of all controllers. Linear systems have been extensively studied and widely utilized precisely because the accompanying math is easier. Linear systems work well in appropriate situations, but they certainly do not adequately service all situations.

Thus the issue of stability arises in the application of fuzzy models because of their inherent non-linear properties. While fuzzy systems have been proven general enough to produce any non-linear function, stability issues with those systems must be answered. The reason that stability questions remain somewhat open with fuzzy systems is the general difficulty of the mathematics required to prove stability in non-linear systems. We should not over-emphasize the stability questions however. To date, the authors are not aware of any real world fuzzy systems that have failed due to instability problems. In practice, the success of fuzzy controllers, particularly in non-linear systems, is due more to adequate simulation and testing than to the application of rigorous and technical mathematics.

Resolution

The resolution of a system typically refers to the number of bits used in digitizing and/or processing of the signals in the system. Number-of-bits resolution also parameterizes the fineness of the outputs from the system. The digitizing of signals is important to embedded fuzzy

Chapter 6: Embedded Systems Considerations

models for two reasons. First, the pervasive presence of microcontrollers make them an obvious and accessible implementation choice for fuzzy systems. Second, because there has not yet been a successful commercial application of an analog fuzzy processor, the only commercially available implementation choices are exclusively based on digital technologies.

The most basic digital signal processing theory dictates that the process of digitizing an analog value in a sampled data system results in an error which is, at most, equal to one half of the value represented by the least significant bit (LSB). For example, representing the universe of discourse of a fuzzy input with eight bits allows 256 distinct values for that input. If the universe of discourse of that input is a 0 to 5 volt analog signal that we digitize to eight bits, then the resolution is: $5 volts / 256 \cong 0.0195 volts / bit$. The maximum error encountered in digitizing the analog signal equals one half of one bit, or, $0.00975 volts$. In contrast, digitizing this same universe of discourse with sixteen bits results in a resolution of: $5 volts / 65536 = 0.0000763 volts / bit$, and a maximum digitizing error of $0.0000381 volts$.

In practice, the ideal maximum error of one half LSB is not possible. A whole host of other factors confound attempts to achieve the theoretical value. The system will be only as good as its least accurate or precise piece. The main constraint on resolution is that increasing it invariably costs more. The guiding principle will be to use only as much resolution as is necessary to produce the desired performance of the final product.

Another aspect of resolution in a fuzzy system is the representation of truth values. As Figure 6.2 illustrates, too little resolution in truth values results in membership functions with choppy shapes. In the figure, truth value has been quantized with only five different truth values. Inappropriate quantization introduces choppiness that hampers the ability of the fuzzy system to produce the desirable smooth output transitions that enhance control system stability.

Fuzzy Logic for Real World Design

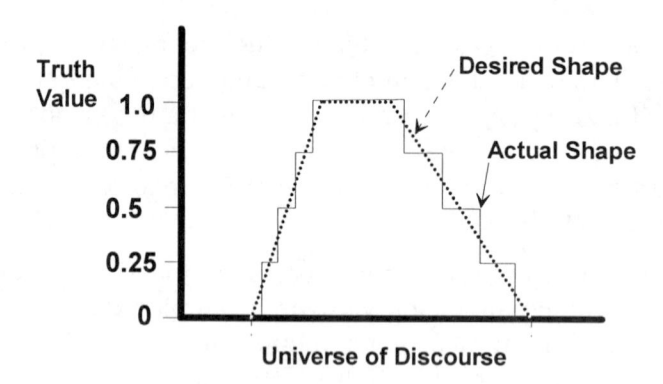

Figure 6.2: Truth values represented with too little resolution. Few quantization levels causes problems.

Complexity

Embedding the simple two input, one output, 25 rule fuzzy model is easy. A more involved application may require several fuzzy models operating in a combination of serial and/or parallel modes. Serial operation means chaining together two or more fuzzy models, where the output from one model is used as the input to another model. Parallel operation means that different fuzzy models may share the same input variables, but can be evaluated independently from each other.

A fuzzy model may not be executed all the time. Perhaps different rules will be used for different modes of operation. For example, an automatic transmission controller may support economy and performance modes. Each mode uses a different rulebase, selected based on driver input.

Specificity

Generalizing about the nature of a fuzzy implementation, we can think of the variety of implementations as existing on a continuum. At one end of the continuum lies a completely general, multi-purpose solution. At the other extreme lies a non-general, single purpose solution.

Ostensibly, a general purpose solution can be easily modified to support additional features. Or, the solution may already support many features only some of which may be used in any one specific application. A general purpose solution would be expected to support different applications with minimal modification. The multi-purpose solution would be expected to perform adequately across a number of criteria, but not exceptionally in any one area.

A single purpose fuzzy implementation can be tailored to meet the exact needs of a specific application. Freedom from reuse concerns permits optimization on one or several given criteria. Optimization tends to be a zero-sum game where an improvement in one area implies a loss in another area. For example, the MIN-MAX rule evaluation executes the fastest, but it is also the harshest because it discards information that may be contained in the truncated portion of the membership function shapes.

Solutions

The following sections present guidelines and specific answers, where appropriate, to the common design questions raised previously.

Architecture

The architecture of any embedded design involves both hardware and software aspects. Likewise, the questions that relate to implementing a fuzzy model also have hardware and software aspects. Insuring adequate attention to the hardware and software choices will lay the ground-work for a good implementation.

You must answer the following hardware questions: how will the crisp inputs enter the fuzzy model? Is analog-to-digital conversion required, or do the crisp values already exist digitally? How will the crisp output from the fuzzy system be used? Will it be used only in its digital form, or will it be applied to the external world via digital-to-analog, pulse width modulation, or other conversion techniques? As the expert in your particular application domain, only you can answer these questions.

Having clarified how the crisp inputs enter the fuzzy model and how the crisp outputs exit the fuzzy model, you can turn your attention to the computation of the fuzzy model itself. A new set of questions now appears. Will the system use a general purpose microcontroller? What type of hardware will support the membership function shapes, and rule evaluation, aggregation, and defuzzification algorithms of your particular fuzzy model? If special fuzzy processing hardware is planned, what algorithms does that hardware support? The following tables summarize the mathematical operations required by the variety of algorithms presented in Chapters 2, 3, and 4.

Membership Function Type	Mathematical Operations Used
S-, Z-, and Pi-Curves	add, subtract, multiply, divide
Trapezoid, Triangle	add, subtract, divide
Gaussian	add, subtract, multiply, divide, exponentiation
Multi-point interpolation	add, subtract, multiply, divide,

Table 6.2: Mathematical operations used during fuzzification of various fuzzy set shapes.

Intersection/Union Type	Mathematical Operations Used
MIN, MAX	compare
BINTER, BUNION	compare, add, subtract
PROD, PSUM	add, subtract, multiply
Yager Functions	compare, add, subtract, exponentiation

Table 6.3: Mathematical operations used during antecedent evaluation.

Inference Type	Mathematical Operations Used
Correlation Minimum	compare
Correlation Product	multiply

Table 6.4: Mathematical operations used during rule inference.

Aggregation Type	Mathematical Operations Used
MAX	compare
Additive	add

Table 6.5: Mathematical operations used during consequent aggregation.

Defuzzification Type	Mathematical Operations Used
Left- or Right-most maximizer	compare
Composite maximum	compare, add, divide
Centroid	add, multiply, divide

Table 6.6: Mathematical operations used for defuzzification.

Mathematical Operation	Mnemonic Instruction	CPU Clocks Consumed
Compare	CMP	1
Add	ADD	1
Subtract	SUB	1
Multiply	MUL	19
Divide	DIV	24
Exponentiation	F2XM1 and FYL2X	242+311

Table 6.7: Comparison of CPU cycles consumed on an Intel 80486 processor by various mathematical operations on 16 bit operands.

The general rule in selecting hardware is to maximize generality and flexibility. For embedded fuzzy models the usual choice is an off-the-shelf eight- or sixteen-bit microcontroller. Microcontrollers have the advantages of low cost, standard design, and highly flexible software operation. While custom hardware implementations can achieve an order of magnitude, or more, improvement in speed, they typically limit the choice of membership function shapes, and rule evaluation, aggregation, and defuzzification algorithms.

Data structures and operations upon those data structures dictate the software architecture of a fuzzy inference engine and associated fuzzy model. The two main data structures are membership functions and rules. The definition of membership functions for each input and output variable combine with rules expressing the relationship between those variables to encode the knowledge of the fuzzy model. The operations with those data structures can be roughly grouped into five areas depicted in Figure 6.3.

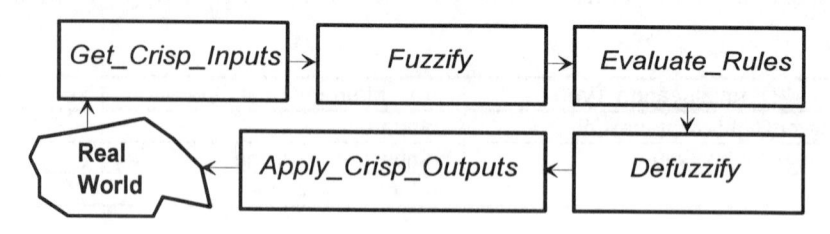

Figure 6.3: Software architecture of a fuzzy inference engine.

Get_Crisp_Inputs acts as the interface to the real world by acquiring the data and possibly transforming or scaling it for use by the fuzzy model. *Fuzzify* turns crisp inputs into degrees of membership in their corresponding membership functions. *Evaluate_Rules* uses degrees of membership and the rule structures to determine consequent truth values. *Evaluate_Rules* performs antecedent evaluation, inference and aggregation algorithms. *Defuzzify* computes a crisp output for each output variable from the aggregated consequent space produced by *Evaluate_Rules*. Finally, *Apply_Crisp_Outputs* takes the crisp output of the fuzzy model and applies it to the real world.

We will explicitly examine the software architecture for fuzzy model processing later in Chapter 7 in the section entitled *Fuzzy Kernel in C*.

Processing Speed

It is possible to make at least one incontrovertible statement about processing speed. You can never have too much processing capacity on line. It always seems possible to fill up the processor bandwidth with features that are "absolutely necessary". It also seems inevitable when you discover, at some point, that one more essential feature cannot be supported. Of course, the lesson here is that the features supported always find a way to absorb the available processing power. Fortunately, the execution of a fuzzy model requires only a modest amount of computational power.

The processing power needed by any fuzzy model is a function of many parameters. The foremost parameter affecting processing demand in the embedded system is the frequency with which the fuzzy model must be executed. If you are willing to wait all day, any processor will suffice. Of course, this is almost never the case.

Internal parameters of the fuzzy model also affect processing needs. Increasing the number of membership functions or the number of rules increases processing cycles approximately linearly. This means that doubling the number of input membership functions doubles the number of machine cycles consumed by fuzzification. Likewise, doubling the number of rules of the fuzzy model doubles the processor cycles consumed in rule evaluation. Similarly, doubling the number of output membership functions can double the processor cycles consumed in defuzzification.

Table 6.9 compares fuzzy processing capabilities across the spectrum of potential hardware platforms. Table 6.8 identifies the hardware and software configuration used for each test. Benchmarking, by its nature, seldom approaches the blind objectivity that would be most useful to engineers who must try to select the best configuration. The most often cited performance index for fuzzy systems has units of *FLIPS*, which translate to *Fuzzy Logic Inferences Per Second*. Note that several distinct operations comprise one fuzzy logic inference. The inference loop starts with acquisition of new crisp inputs, includes fuzzification, and rule evaluation, and ends with applying the defuzzified output to the attached system. FLIPS can be a misleading measure of fuzzy performance. Simply adding rules to a fuzzy model decreases the FLIPS measure of a system. What we are really interested in is the capacity of various hardware platforms to execute a range of fuzzy models. Only by using several types of performance data can we hope to compensate for disparate test conditions. In the tables below, the parameter *defuzzifications per second* is the parameter otherwise known as FLIPS.

Fuzzy Logic for Real World Design

Test Designation	Platform Configuration
Test A	Motorola 68HC11 running embedded code produced by Aptronix Corp.
Test B	Intel 8051 running embedded code produced by Inform Software Corp.
Test C	Same as Test B
Test D	Intel 80C196KD running embedded code produced by Inform Software Corp.
Test E	Same as Test D
Test F	Siemens 80C166 running embedded code produced by Inform Software Corp.
Test G	Analog Devices ADSP-2101 running embedded code produced by FDG Systems Inc.
Test H	SGS-Thomson W.A.R.P. Fuzzy Logic Processor
Test I	OKI Semiconductor MSM91U112 Fuzzy Logic Processor
Test J	VLSI Technology VY86C500 12-bit Fuzzy Computation Acceleration core
Test K	Same as Test J
Test L	NeuraLogix NLX220 Stand-Alone Fuzzy Logic Controller
Test M	Look-up table assumes 10 ns memory access time
Test N	Look-up table assumes 13.3 ns memory access time

Table 6.8: Legend describing hardware and software components of various test configurations.

Chapter 6: Embedded Systems Considerations

	Test A	Test B	Test C	Test D
Clock Frequency	8 MHz	12 MHz	12 MHz	20 MHz
Resolution	8 bits	8 bits	8 bits	8 to 16 bits
# of Inputs	2	2	2	2
Input Mbf Shape	triangular / trapezoidal	triangular	triangular	triangular
Mbfs per Input	6 for Inp#1 and 5 for Inp#2	7	7	7
# of Outputs	1	1	1	1
Output Mbf shape	singleton	singleton	singleton	singleton
Mbfs per Output	6	7	7	7
# of Rules	30	7	20	20
Algorithms	MIN/MAX	MIN/MAX	MIN/MAX	MIN/MAX
Code Size	ROM 870 bytes RAM 31 bytes	ROM 450 bytes RAM 23 bytes	ROM 540 bytes RAM 23 bytes	ROM 820 RAM 63 bytes
Inference Speed	2.2 ms per loop	1.0 ms per loop	1.4 ms per loop	0.28 ms per loop
Rules/sec	13,636	7,000	14,285	71,428
Rules/sec/MHz	1,705	583	1,190	3,571
Input Mbf Fuzzifications /sec	5,000	14,000	10,000	50,000
Input Mbf Fuzzifications /sec/MHz	625	1,167	833	2,500
Defuzzifications /sec	455	1,000	714	3,571
Defuzzifications /sec/MHz	57	83	60	179

Table 6.9(a): Performance comparisons and capabilities.

Fuzzy Logic for Real World Design

	Test E	Test F	Test G	Test H
Clock Frequency	20 MHz	40 MHz	20 MHz	40 MHz
Resolution	8 to 16 bits	8 to 16 bits	16 bits	10 bits
# of Inputs	4	2	2	5
Input Mbf Shape	triangular	triangular	triangular	piecewise linear (64 points / Mbf)
Mbfs per Input	7	7	7	8
# of Outputs	1	1	1	1
Output Mbf shape	singleton	singleton	singleton	singleton
Mbfs per Output	7	7	7	8
# of Rules	100	20	49	32
Algorithms	MIN/MAX	MIN/MAX	MIN/Centroid	Product/ Centroid
Code Size	ROM 1300 bytes RAM 75 bytes	N/A	ROM 376 bytes RAM 62 bytes	N/A
Inference Speed	1.2 ms per loop	90 us per loop	36 us per loop	1.85 us per loop
Rules/sec	83,333	222,222	1,361,111	17,297,297
Rules/sec/MHz	4,167	5,555	68,056	432,432
Input Mbf Fuzzifications /sec	23,333	155,555	388,889	21,621,622
Input Mbf Fuzzifications /sec/MHz	1,167	3,889	19,444	540,540
Defuzzifications /sec	833	11,111	27,778	540,540
Defuzzifications /sec/MHz	42	278	1,389	13,513

Table 6.9(b): Performance comparisons and capabilities.

Chapter 6: Embedded Systems Considerations

	Test I	Test J	Test K	Test L
Clock Frequency	10 MHz	20 MHz	20 MHz	10 MHz
Resolution	floating point, 8bits + 3bit exponent	12 bits	12 bits	8 bits
# of Inputs	8	8	30	2
Input Mbf Shape	triangular	piecewise linear (6 points / mbf)	piecewise linear (6 points / mbf)	triangular, trapezoidal
Mbfs per Input	8	7	7	5
# of Outputs	1	2	4	1
Output Mbf shape	singleton	singleton	singleton	singleton
Mbfs per Output	8	7	7	5
# of Rules	128	200	1000	28
Algorithms	MIN/Centroid	MIN/Centroid	MIN/Centroid	MIN/MAX
Code Size	N/A	N/A	N/A	N/A
Inference Speed	40 us per loop	370 us per loop	3.63 ms per loop	100 us per loop
Rules/sec	3,200,000	540,540	275,482	280,000
Rules/sec/MHz	320,320	27,027	13,774	28,000
Input Mbf Fuzzifications /sec	1,600,000	151,351	57,851	100,000
Input Mbf Fuzzifications /sec/MHz	160,000	7,568	2,893	10,000
Defuzzifications /sec	25,000	5,405	1,102	10,000
Defuzzifications /sec/MHz	2,500	270	55	1,000

Table 6.9(c): Performance comparisons and capabilities.

Fuzzy Logic for Real World Design

	Test M	Test N
Clock Frequency	100 MHz	75 MHz
Resolution	8 bits	10 bits
# of Inputs	2	2
Input Mbf Shape	N/A	N/A
Mbfs per Input	N/A	N/A
# of Outputs	1	1
Output Mbfshape	N/A	N/A
Mbfs per Output	N/A	N/A
# of Rules	N/A	N/A
Algorithms	N/A	N/A
Code Size	ROM 65,536 bytes	ROM 1,310,720 bytes
Inference Speed	10 ns per loop	13.3 ns per loop
Rules/sec	N/A	N/A
Rules/sec/MHz	N/A	N/A
Input Mbf Fuzzifications /sec	N/A	N/A
Input Mbf Fuzzifications /sec/MHz	N/A	N/A
Defuzzifications /sec	100,000,000	75,000,000
Defuzzifications /sec/MHz	1,000,000	1,000,000

Table 6.9(d): Performance comparisons and capabilities.

Stability

We can approach the topic of stability in fuzzy systems in two ways. The first and by far the most common approach relies on testing and the gathering of empirical evidence. The second approach, which is beyond the scope of this book, relies on detailed and deeply mathematical methods that are beyond the ken of all but the dedicated control theorist.

Empirical methods of establishing the stability of a fuzzy controller are often relied upon because of the deeply mathematical and difficult nature of stability proofs for non-linear systems. The empirical methods used usually include extensive simulation and testing of the combined fuzzy controller and dynamic system. The empirical approach to stability targets the following areas[68]:

- Testing the set of IF-THEN rules with respect to their consistencey, completeness, and continuity.

- Testing the system to be controlled with respect to controllability and observability.

- Analyzing the operating points and operating areas of the dynamic system.

If, for all possible combinations of crisp inputs, at least one rule consequent ends up with a non-zero degree of truth, that rule set can be described as *complete*. Refer to the rule coverage maps presented in Chapter 4 for a graphical depiction of completeness.

A *consistent* set of rules contains no conflicts. Two rules are said to be conflicting if they specify the same rule-antecedents but possess mutually exclusive rule-consequents. The following two rules conflict:

1. IF Pressure IS High THEN Throttle_Position IS Large_Increase
2. IF Pressure IS High THEN Throttle_Position IS Large_Decrease

Two consequents will be considered mutually exclusive if there is no region of overlap in the scope of their respective membership functions.

A fuzzy controller owes much of its ability to produce smooth output transitions to the *continuity* of its rule set. The continuity of a rule set allows the rule-consequents to blend smoothly as the crisp inputs transition through various regions of their universes of discourse. Figure 6.3 helps to illustrate the principle of continuity. Figure 6.3 depicts a matrix which encodes a fuzzy rule set. The row headings of the matrix are labeled with the membership function labels of input number one, and the column headings are labeled with the membership function labels of input number two. The labels inside the matrix belong to the membership function labels of the single output variable.

---

[68]D. Driankov et. al., *An Introduction to Fuzzy Control*,pg. 147. Springer-Verlag.

Fuzzy Logic for Real World Design

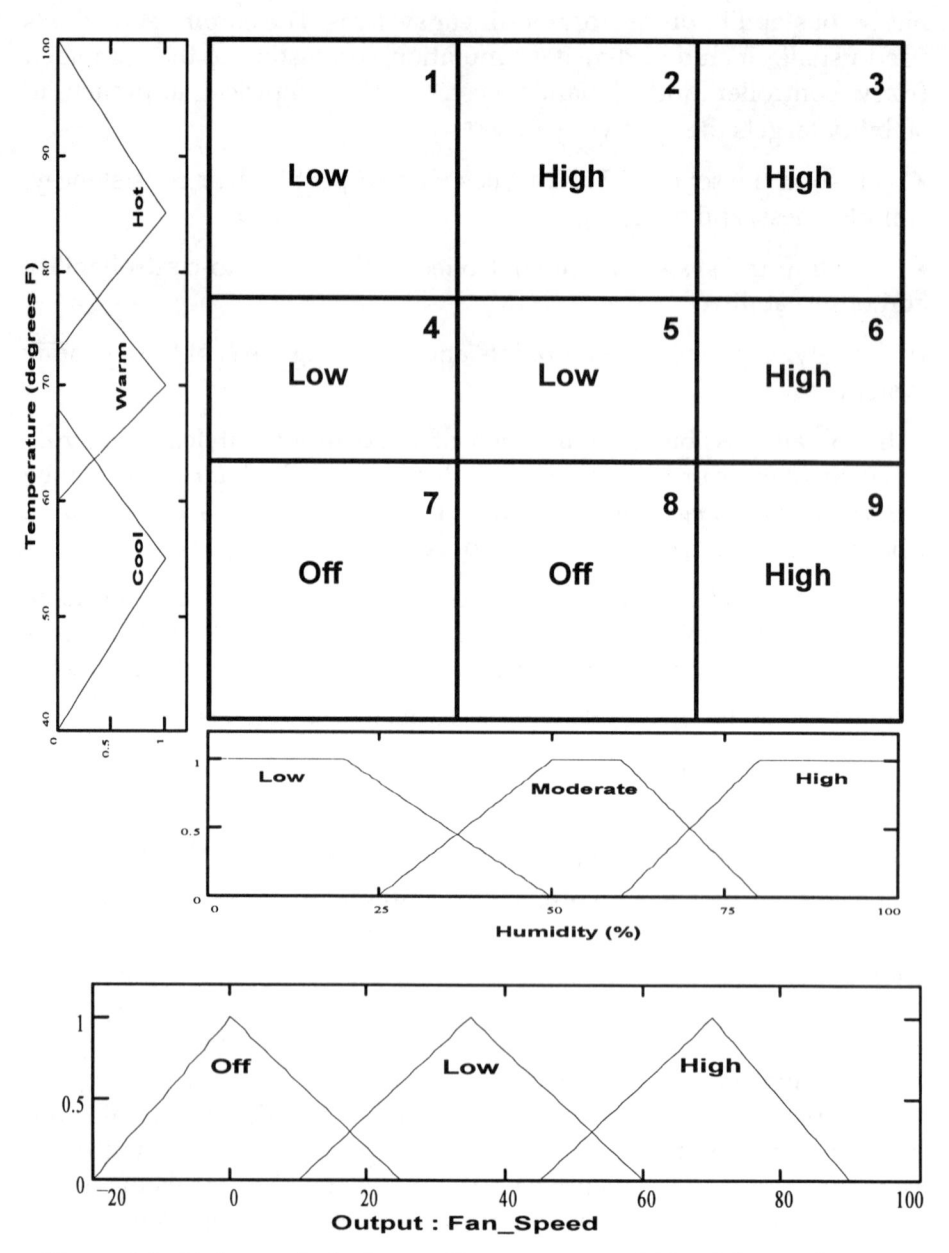

Figure 6.3: Rule matrix for inputs Temperature and Humidity, and output Fan Speed.

To clarify, for example, the cell numbered **1** expresses the rule: *IF Temperature IS Hot AND Humidity IS Low THEN Fan_Speed IS Low.*

It is now possible to define the notion of a neighboring rule. Two rules are neighboors if they occupy adjacent cells. For instance, cell number **8** has as its neighbors cells **5**, **7**, and **9**. Thus we are ready to define continuity. If a rule set has any neighboring rules whose output fuzzy sets are mutually exclusive, that set of rules is not continuous. In the example of Figure 6.3, the rule set is not continuous because the output fuzzy sets belonging to the neighboring rules in cells **8** and **9** are mutually exclusive.[69]

The mathematical approach to the stability of fuzzy systems centers on two different, but related, approaches. The first seeks to prove the stability of an existing fuzzy model. The second seeks to construct the fuzzy model in a special way which is guaranteed to result in a stable fuzzy controller[70]. While both approaches have considerable value, the second approach promises a more universal solution to the general problem of stability in fuzzy control systems. Although the details of both of these approaches are well beyond the scope of this book, several good informative texts already exist and should be consulted.

Resolution

Most embedded systems interface to the world through analog-to-digital converters. This approach is inherently integer based, with data resolutions anywhere from 6 to 20 bits. However, except for unusual applications, the most common data resolutions are 8, 10, and 12 bits. Of course, these data bit widths should not be confused with the architecture of the microcontroller that may be used to run the fuzzy algorithms. With varying ease and efficiency, any data bit width can be handled on 8 or 16 bit microcontrollers. Maintaining a consistent bit

---

[69] However, in this case continuity may be neccessary, due to the small number of Mbfs. Continuity is therefore something to be aware of but not neccessarily constrained by.

[70] Li-Xin Wang, *Adaptive Fuzzy Systems and Control*, Prentice-Hall.

width integer representation through all of the fuzzy algorithms will yield the highest throughput.

The processing time required for floating point manipulation of fuzzy variables is justifiable only where absolutely necessary. One such case might be where Yager parameterized intersection and union operators use exponents of p and $1/p$ to modify truth values. When the application needs floating point operators or algorithms, selecting a microcontroller that supports floating point operations in hardware makes a good, but more expensive, choice.

Complexity

While the complexity of fuzzy information models (like fuzzy expert systems) can be a real issue, embedded systems seldom present similar difficulties. Complexity becomes a vexing problem because as the number of fuzzy inputs increases the number of rules increases exponentially. But the typical embedded control system looks at only a few input variables. As such, the typical control system involves a completely tractable number of rules.

Specificity

The majority of fuzzy implementations rely exclusively on software. The software-only approach retains maximum flexibility while providing moderate fuzzy inference speed. Flexibility is a huge advantage of this approach. A carefully written fuzzy inference engine requires minimal modification to run on a variety of microcontroller platforms. We present just such a software implementation in the section entitled *Fuzzy Kernel in C* in the following chapter.

It is true that additional performance can be squeezed from a generic implementation by tailoring it to run a specific fuzzy model on a specific microcontroller. Due to differences in microcontroller chips, fuzzy algorithms map differently onto the machine level instructions supported by each chip. In principle, hand coding the fuzzy algorithms for execution to run on specific controller chips will both reduce code size and increase code execution speed. Since hand coding must be

Chapter 6: Embedded Systems Considerations

done at the machine instruction level, both debugging and development effort increase.

Only two situations dictate a departure from the software-only approach. In the first situation, a microcontroller is not available in the embedded system to execute fuzzy software. The best choice in this case is a stand-alone fuzzy processor[71]. The stand-alone fuzzy processor must be permanently programmed with the specifics of the fuzzy model for that specific application. In the second situation, the speed obtainable with a software-only fuzzy implementation cannot meet the speed requirements of the application. Several levels of special purpose hardware can improve inferencing speed by more than a factor of ten. A more complete picture of the state-of-the-art hardware options available is presented in Appendix A.

---

[71]One example is the NeuraLogix NLX220 stand-alone fuzzy logic controller. Once programmed with rule and fuzzy set definitions, this chip is a complete stand-alone fuzzy solution. The chip reads up to four analog inputs and produces up to four analog outputs. The chip performs 8-bit A-D conversion of the inputs, fuzzy inferencing, and D-A conversion of the outputs.

Chapter 7:
Implementation Choices

While fuzzy logic provokes interesting philosophical questions regarding the way that human expertise might be acquired, expressed, and automated, it is most valuable as a problem solving tool. Solving problems means digging into the process of implementing fuzzy models. Fuzzy logic, in contrast to most other artificial intelligence techniques, is increasingly employed by practicing engineers to solve real world problems. This chapter covers the real world implementation choices that can be made in pursuit of a working fuzzy model or fuzzy inference system.

Broadly, the implementation choices are: software, hardware, and hybrids that include elements of both. The majority of fuzzy implementations take the software approach. The advantages of software implementations include flexibility and portability. To aid your implementation efforts, we will walk through a fuzzy engine design and implementation in the section titled *Fuzzy Engine in C*. Hardware implementations involve special purpose VLSI chips that vastly accelerate fuzzy model processing. We will now explore the choices in each of these broad areas.

Software

Interest in fuzzy logic solutions has been driven, in part, by the fact that fuzzy models run well on eight bit microcontrollers. Fuzzy logic may be the only AI technique that can legitimately make such a claim. Whatever advances in embedded microcontrollers and personal computers we will see during the next decade, we can be certain that the feasibility of running more complex fuzzy models faster will only improve.

With the proliferation of general embedded microcontroller applications has come a wealth of tools for the embedded system designer. One of the most powerful recently-introduced features is support for high level programming languages[72]. High level language development yields a number of advantages. The productivity advantage is that more lines of code can be written, debugged, and documented in less time. Writing embedded software in BASIC, FORTH, or C relieves the tedium of endless lines of assembly code by providing a more powerful and compact vocabulary. Of course, people developing applications software for personal computers have been using high level languages for a long time.

Whereas prior microcontroller applications software was written in an assembly language specific to a particular brand and make of processor chip, a high level language like C is not tied to any specific chip. Because the high level code is the essence of portability, the engineering resources devoted to that code are not thrown away when migrating to the next generation of microcontroller. Optimizing compilers operate on the C code to produce compact and efficient executable code targeted for a variety of specific microcontroller chips. Hence using the same high level code on a different microcontroller only requires recompiling the C source code for a different processor.

---

[72]Rather than implying that this is a new development, it is an approach to embedded system implementation that is quickly growing in popularity.

Fuzzy Kernel in C

The designing and writing of a fuzzy inference engine is an enormously instructive project. The act of writing will solidify your understanding of the two main constructs, membership functions and rules, and the three operations on those constructs, fuzzification, rule evaluation, and defuzzification. The interaction of these constructs and operations is illustrated in Figure 7.1. Below we present our approach, with variations, to the inference engine project.

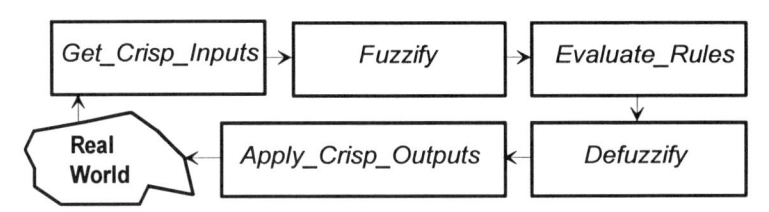

Figure 7.1: Information flow through the modules of the inference engine.

Data Structures

The first step in designing the inference engine involves creating the data structures that represent membership functions and rules. The fuzzy kernel presented in this chapter relies on two static data blocks to provide all information relevant to the fuzzy model. The first data block is called the CONFIG_BLOCK. The CONFIG_BLOCK contains all of the membership function definitions. It also contains the system parameters such as number of inputs, number of outputs, and the number of membership functions used for each fuzzy variable. The second data block is called the RULE_BLOCK. The RULE_BLOCK contains all of the rules of the fuzzy model.

The code of the fuzzy kernel requires both the CONFIG_BLOCK and RULE_BLOCK data blocks to conform to predetermined formats. The data blocks will then be continuously accessed by the fuzzification, rule evaluation, and defuzzification routines.

The computationally limited microcontroller targets that we anticipate using to execute the algorithms developed in this section

Fuzzy Logic for Real World Design

dictate limiting the fuzzy sets to simple shapes. Our example supports triangular and trapezoidal shapes for input variables and singletons for output variables. Such membership function shapes can be fully parameterized in several different ways. Figure 7.2(a) uses four points while Figure 7.2(b) parameterizes the membership function with two points and two slopes. Figure 7.2(c) shows the singleton membership function representation.

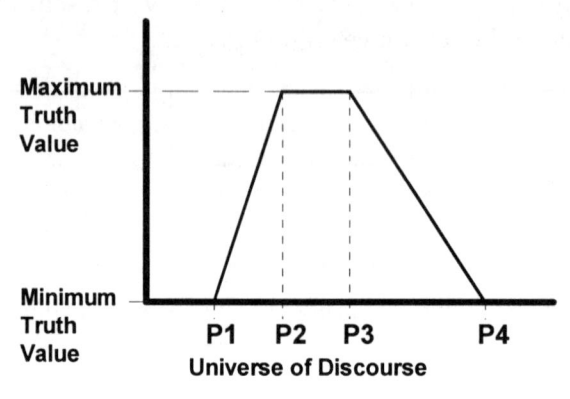

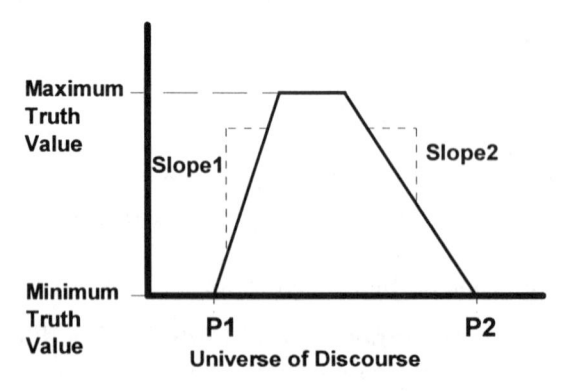

Figure 7.2(a): Four parameters describe trapezoidal or triangular membership functions.

Figure 7.2(b): Four parameters describe trapezoidal or triangular membership functions.

216

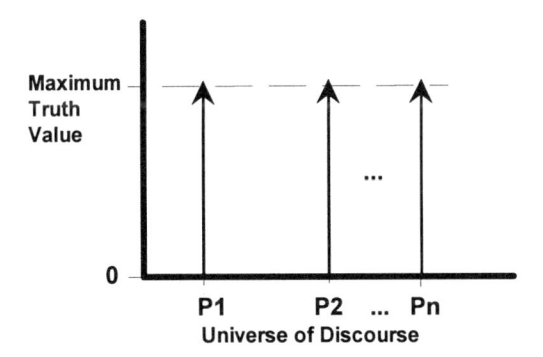

Figure 7.2(c): Singleton representation requires one parameter per output membership function.

The fuzzification algorithm takes a crisp input for each input fuzzy variable and computes the degrees of membership of those crisp inputs in the membership functions defined for each fuzzy variable. Clearly, the fuzzification algorithm is intimately tied to the exact data structure of the membership functions. The flow charts on the following two pages illustrate how the fuzzification procedure differs when the fuzzy set definitions of Figure 7.2 (a) and (b) are used.

Fuzzy Logic for Real World Design

Fuzzification for Figure 7.2(a)

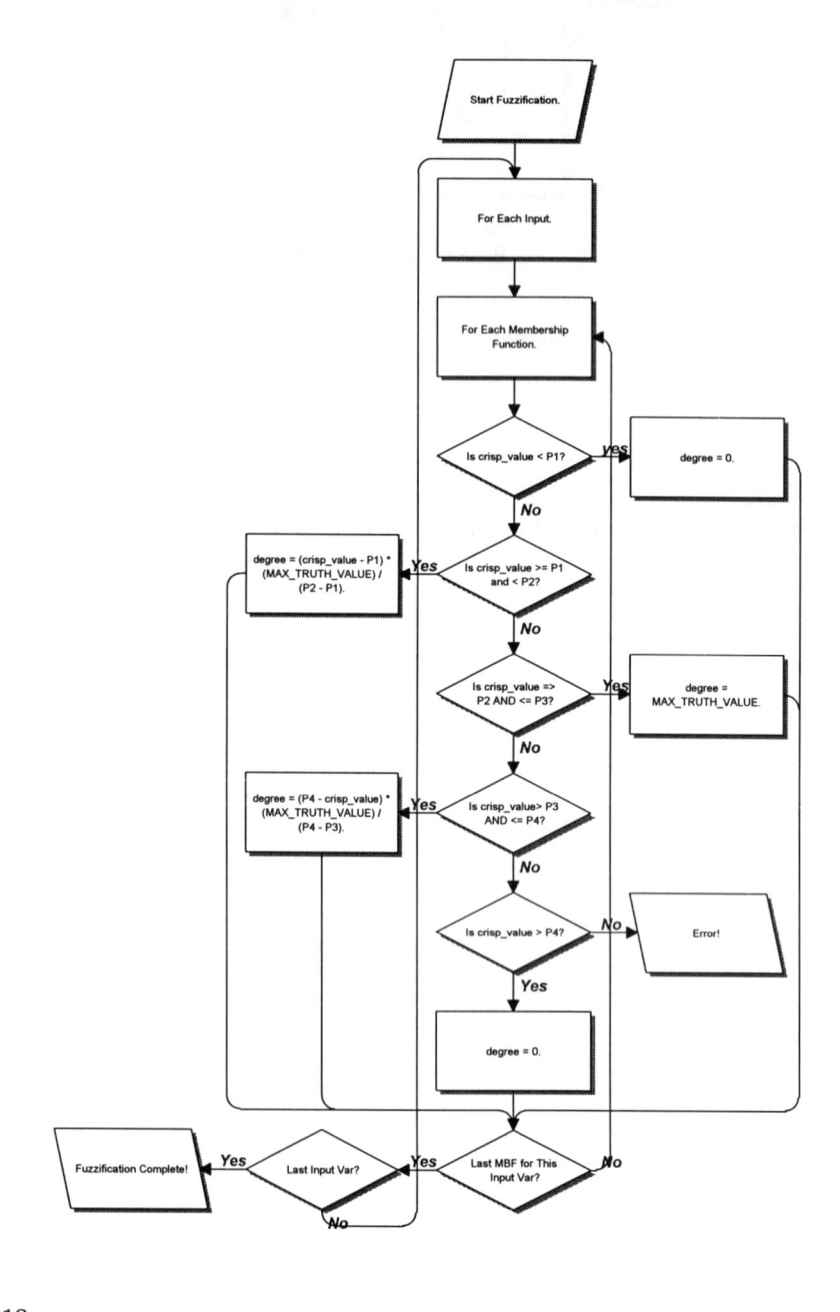

```
                    ┌──────────────────┐
                    │ Start Fuzzification. │
                    └──────────────────┘
                             │
                             ▼
                    ┌──────────────────┐
                    │   For Each Input.│
                    └──────────────────┘
                             │
                             ▼
                    ┌──────────────────┐
                    │ For Each Membership
                    │     Function.     │
                    └──────────────────┘
                             │
                             ▼
                    ╱ Is crisp_value < P1? ╲──Yes──▶ degree = 0.
                    ╲                      ╱
                             │ No
                             ▼
   degree = (crisp_value - P1) *    ╱ Is crisp_value >= P1 ╲
   (MAX_TRUTH_VALUE) /      ◀──Yes──╲    and < P2?         ╱
   (P2 - P1).                        
                             │ No
                             ▼
                    ╱ Is crisp_value =>  ╲──Yes──▶ degree =
                    ╲ P2 AND <= P3?      ╱         MAX_TRUTH_VALUE.
                             │ No
                             ▼
   degree = (P4 - crisp_value) *    ╱ Is crisp_value > P3  ╲
   (MAX_TRUTH_VALUE) /      ◀──Yes──╲    AND <= P4?        ╱
   (P4 - P3).
                             │ No
                             ▼
                    ╱ Is crisp_value > P4? ╲──No──▶ Error!
                    ╲                      ╱
                             │ Yes
                             ▼
                         degree = 0.
                             │
                             ▼
  Fuzzification Complete! ◀──Yes── Last Input Var? ◀──Yes── Last MBF for This ──No──┐
                                         │                    Input Var?            │
                                         No                                         │
                                                                                    │
```

218

Chapter 7: Implementation Choices

Fuzzification for Figure 7.2(b)

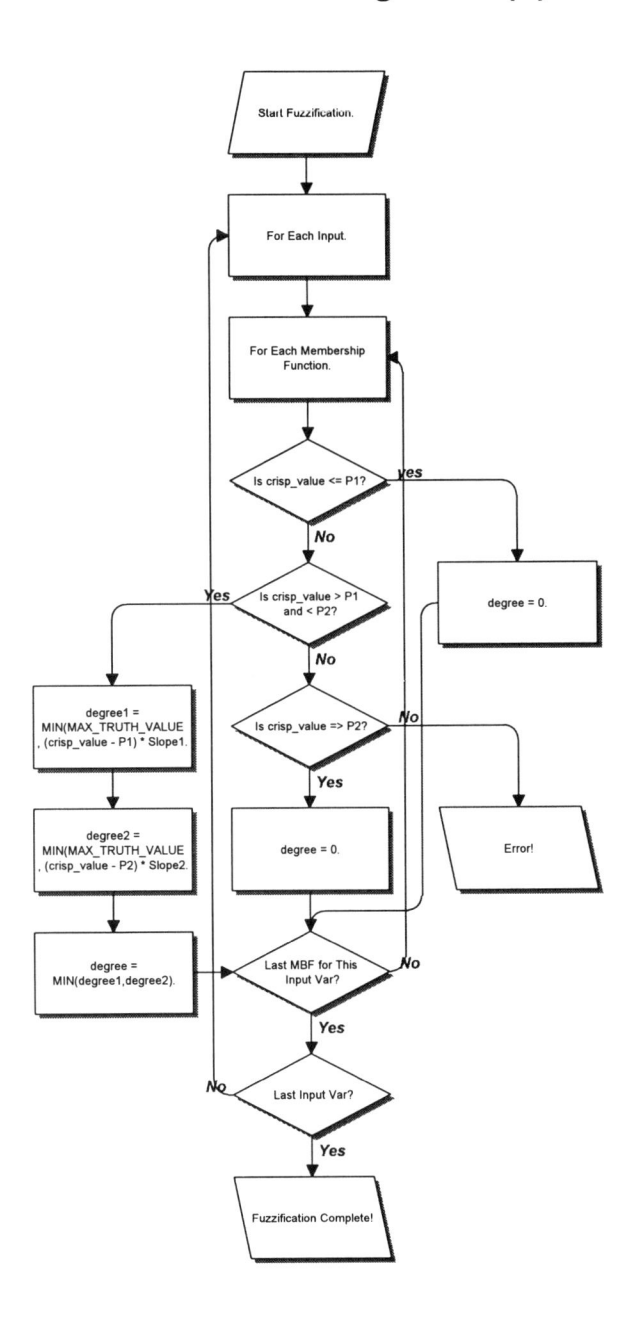

The goal here is the selection of both a fuzzy set representation and a fuzzification algorithm. We are interested in fast and compact code. Comparing the two previous flow charts highlights the basic differences and advantages of each approach. The flow chart of Figure 7.2(a) has the advantage of using a single calculation to determine the degree of membership. It also has the disadvantage that the slope required to compute the degree of membership between P1-P2 and P3-P4 must be recalculated each time through the fuzzification routine. The flow chart of Figure 7.2(b) has the advantage that any slope required to calculate a degree of membership is an intrinsic parameter. However, it has the disadvantage of requiring more calculations than the previous method.

Combining the two fuzzification methods preserves the best characteristics of each. The combined membership function representation includes four points and two slopes. The final flow chart for fuzzification is shown in Figure 7.3. The revision requires the storage of two parameters in addition to the original four. This a small penalty to pay since these parameters will normally be stored in microcontroller ROM.

Fuzzification (c)

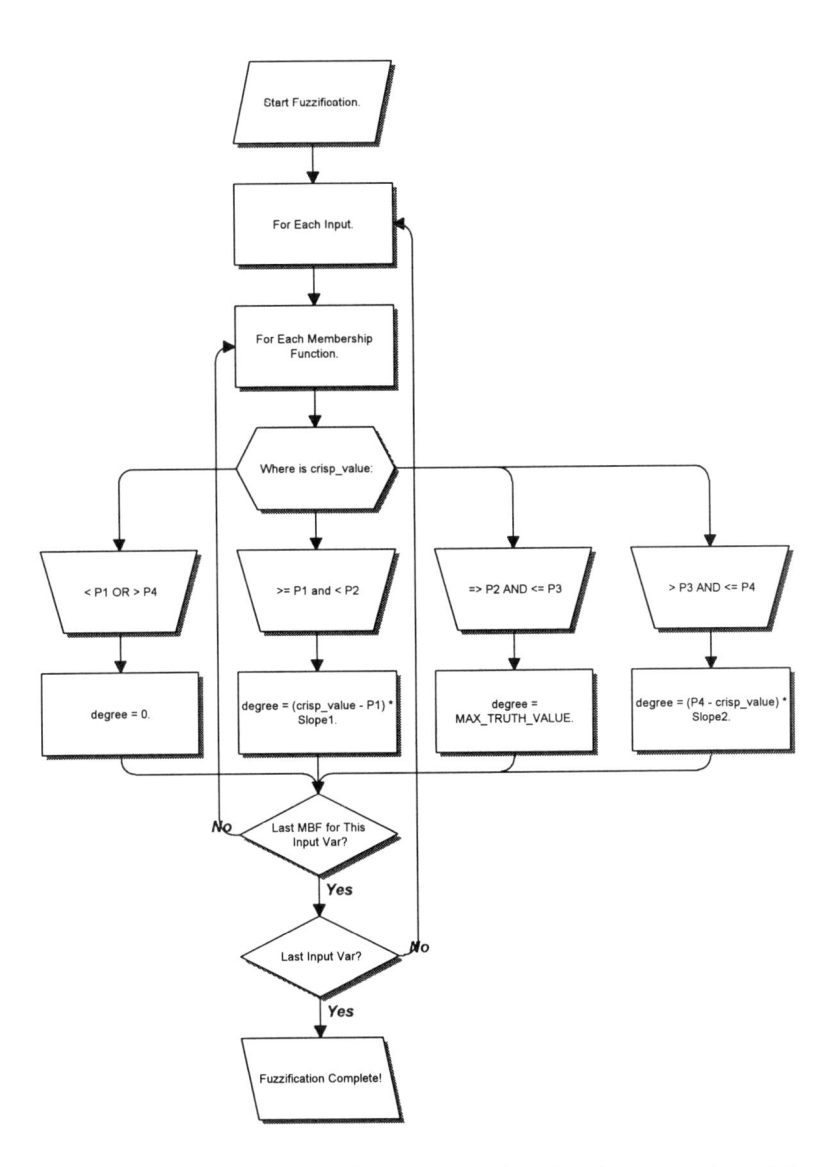

Figure 7.3: Final fuzzification flow chart using the six parameter mbf representation.

Fuzzy Logic for Real World Design

The next design item to finalize is the actual organization of the membership function parameters in the data block. Figure 7.4 shows how the parameters are formatted and stored in the CONFIG_BLOCK.

CONFIG_BLOCK_START:	NUM_OF_INPUTS
	NUM_OF_OUTPUTS
# of mbf.s for first input	NUM_OF_MBFS
first mbf	P1, P2, P3, P4, Slope1, Slope2
second mbf	P1, P2, P3, P4, Slope1, Slope2
o	o
o	o
o	o
last mbf	P1, P2, P3, P4, Slope1, Slope2
o	o
o	o
o	o
# of mbf.s for last input	NUM_OF_MBFS
first mbf	P1, P2, P3, P4, Slope1, Slope2
o	o
o	o
o	o
last mbf	P1, P2, P3, P4, Slope1, Slope2
# of mbf.s for first output	NUM_OF_MBFS
first singleton	S1
second singleton	S2
o	o
o	o
o	o
last singleton	Sn
o	o
o	o
o	o
# of mbf.s for last output	NUM_OF_MBFS
first singleton	S1
second singleton	S2
o	o
o	o
o	o
CONFIG_BLOCK_END :	Sn

Figure 7.4: Organization of the CONFIG_BLOCK containing system and fuzzy set definitions.

Moving on to the format of the RULE_BLOCK, let us consider a few simple rules.

Rule 1: IF Angle is Zero AND Speed is Small_Negative THEN Voltage is Small_Positive

Rule 2: IF Angle is Large_Negative THEN Voltage is Large_Positive

Rule 3: Voltage is Zero

As in the example of Rule 1, the most common format for a fuzzy rule contains one antecedent for each fuzzy input and one consequent for each fuzzy output. However, as discussed in previous chapters, other rule formats are possible. Rule 2 says that if *Angle is Large_Negative* the output *Voltage is Large_Positive* does not depend on *Speed*. Rule 2 can be thought of as a macro because it eliminates the need to write many individual rules which include the *Speed* variable when *Angle* is *Large_Negative*. Rule 3 is an example of an assertion. Assertions, which possess consequents but no antecedents, act to bias the output of the fuzzy system in the direction of the assertion's consequent value. Rules 2 and 3 are examples of useful rule formats that the fuzzy kernel will support.

Support for all three rule types above requires a flexible rule structure. The fuzzy kernel accomplishes this flexibility by using a special rule delimiting character between antecedents and consequents. Using the delimiter between the If-side and Then-side of a rule means that rules can have any number of antecedents. Zero antecedents corresponds to Rule 3. A number of antecedents equaling the number of fuzzy input variables corresponds the Rule 1. Some number of antecedents between zero and the maximum corresponds to Rule 2. The delimiting characters will be recognized by the rule evaluation routine and will permit all three rule types to be handled correctly.

Rules encode knowledge by relating a combination of input fuzzy sets to one or more output fuzzy sets. In order to represent rules in a compact way, we have made use of information that implicitly exists in the CONFIG_BLOCK. The membership functions are defined in the CONFIG_BLOCK in a specific sequence. That sequence allows each membership function to be numbered sequentially starting from 0. For example, if the first input has defined five membership functions they are considered to be mbf0, mbf1, mbf2, mbf3, and mbf4. If the second input also has defined five membership functions they will be considered to be mbf5, mbf6, mbf7, mbf8, and mbf9. Then an output with five membership functions will be counted as mbf10, mbf11, mbf12, mbf13, and mbf14. Using the implicit numbering of the membership functions allows the compact organization of the RULE_BLOCK as demonstrated in Figure 7.5.

Fuzzy Logic for Real World Design

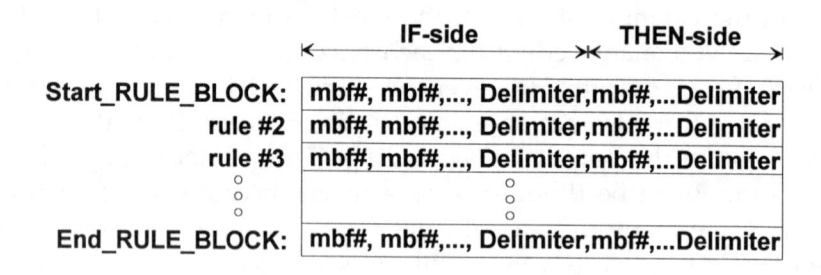

	IF-side	THEN-side
Start_RULE_BLOCK:	mbf#, mbf#,..., Delimiter, mbf#,...Delimiter
rule #2	mbf#, mbf#,..., Delimiter, mbf#,...Delimiter
rule #3	mbf#, mbf#,..., Delimiter, mbf#,...Delimiter
⋮	⋮
End_RULE_BLOCK:	mbf#, mbf#,..., Delimiter, mbf#,...Delimiter

Figure 7.5: Organization of the RULE_BLOCK containing all rules for the fuzzy model.

Approach to Rule Evaluation and Defuzzification

We will now finish the flowcharting process started in the previous section. In the same way that the variety of potential rule formats demands certain design decisions, other aspects of rule evaluation also require design decisions. Rule evaluation in the fuzzy kernel is composed of three basic elements, IF-side evaluation, inference, and aggregation. During IF-side evaluation of a particular rule, the truth value of the IF-side is computed from the truth value of its antecedents using an intersection operator. Of the many choices for the intersection operator, MIN and PRODUCT are most frequently applied to embedded systems because they are easily computed. The inference procedure takes the computed truth value of an IF-side and applies it to scale or otherwise modify the consequents of that particular rule. The two most common inference methods are correlation minimum and correlation product. When using singleton output membership functions, as we do in the fuzzy kernel, both inference methods produce identical results. The last choice to make in the design of the rule evaluation process is the type of aggregation to perform. Aggregation fuses the results of all of the rules into an overall fuzzy output set. The two methods, MAX and additive aggregation, both work well in the embedded systems domain. The flow chart in Figures 7.6 (a) and (b) summarize the control flow and decision points encountered in rule evaluation.

Chapter 7: Implementation Choices

Rule Evaluation

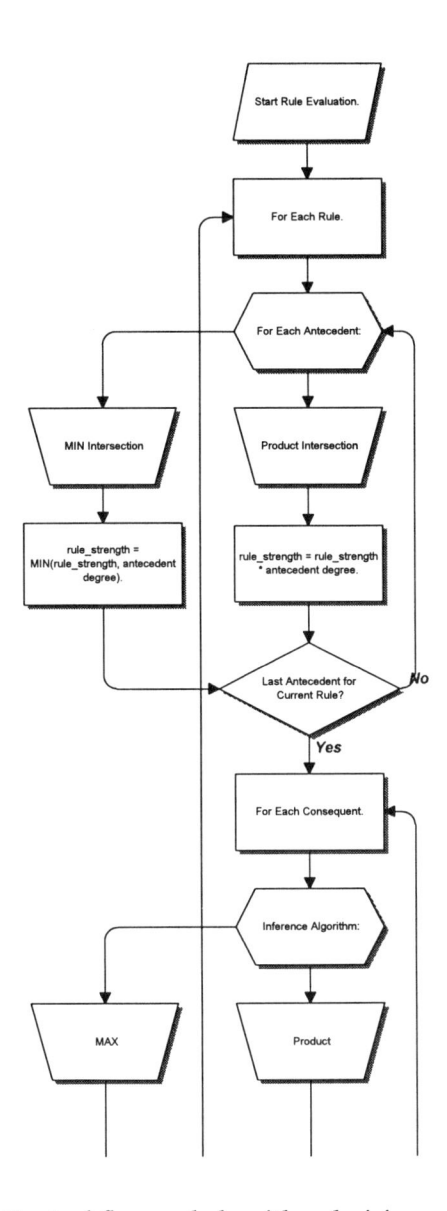

Figure 7.6(a): Control flow and algorithm decision points of the rule evaluation procedure.

Fuzzy Logic for Real World Design

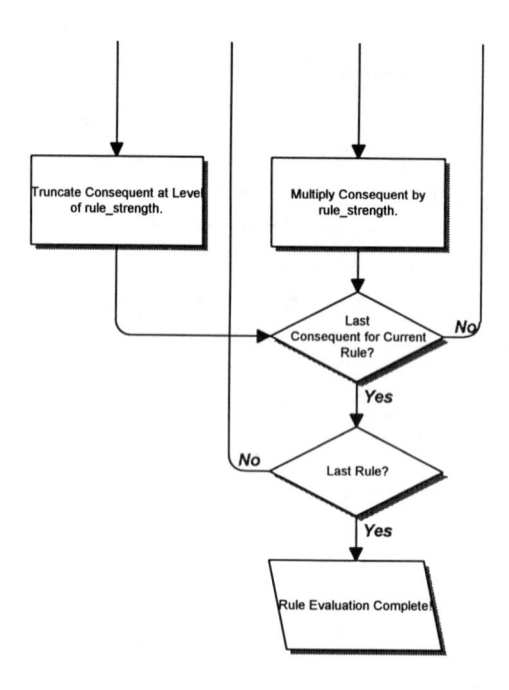

Figure 7.6(b): Control flow and algorithm decision points of the rule evaluation procedure.

The final algorithms to be designed perform aggregation and defuzzification. An aggregated fuzzy space can be created using either additive or MAX aggregation algorithms. The aggregation procedure uses the rule strengths produced in the rule evaluation procedure. Defuzzification produces one crisp value for each fuzzy output variable. Of the many possibilities, our fuzzy kernel supports MAX and Weighted Average defuzzification methods. With the singleton output membership functions of this fuzzy kernel, the weighted average can be considered equivalent to centroid defuzzification. Fuzzy singletons have no area, so the centroid algorithm technically does not apply. However, if output singletons are considered as very narrow, symetrical membership functions of identical widths, the area parameter divides out of the centroid equation. The flow chart of MAX and equally weighted singleton defuzzification is shown in Figures 7.7 (a) and (b).

Chapter 7: Implementation Choices

Defuzzification

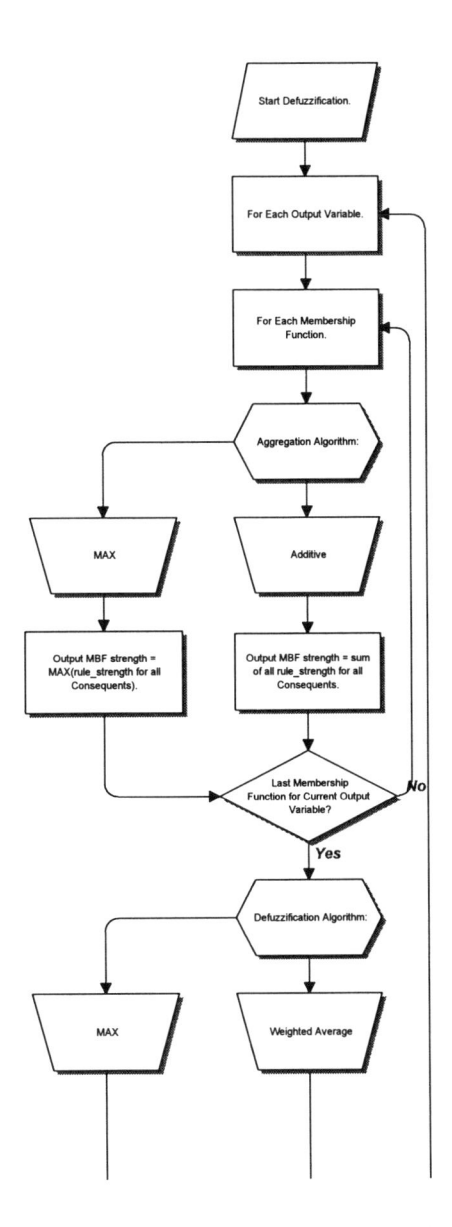

Figure 7.7(a) Control flow of defuzzification algorithm in the fuzzy kernel.

Fuzzy Logic for Real World Design

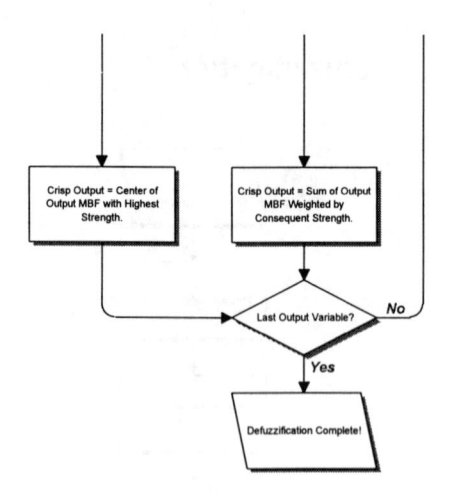

Figure 7.7(b): Control flow of defuzzification algorithm in the fuzzy kernel.

The Code

The organization of the fuzzy kernel follows the flow indicated in Figure 7.1. Several procedures are needed in addition to fuzzification, rule evaluation, and defuzzification. Our complete fuzzy inference engine requires the following components:

1. *Initialize_Fuzzy_Engine*
2. *Get_Crisp_Inputs*
3. *Fuzzify*
4. *Evaluate_Rules*
5. *Defuzzify*
6. *Apply_Crisp_Outputs*

Initialize_Fuzzy_Engine will be executed once, before any fuzzy inferencing is attempted. The routine is responsible for allocating RAM for local variables and initializing those variables to known values. The fuzzy kernel requires one variable for each input and output membership function, one variable for each crisp input, one variable for each crisp output, and several workspace variables. Thus, for a fuzzy model containing two inputs, one output, five membership functions per fuzzy variable, and twenty-five rules, eighteen local variables will be required in RAM for the crisp inputs (two inputs in this example),

Chapter 7: Implementation Choices

crisp outputs (one output in this example), and membership functions (15 membership functions in this example). Up to an additional eleven variables will be required for keeping track of calculations in progress.

Get_Crisp_Inputs is responsible for acquiring and scaling the analog or digital inputs for use by the fuzzy model. We have extensively discussed the *Fuzzify, Evaluate Rules,* and *Defuzzify* routines. *Apply_Crisp_Outputs* is responsible for converting and scaling the crisp output of the fuzzy model for use externally.

Finally, the C code for the fuzzy kernel is presented in Listing 7.8. Because our intention is to cross compile the code developed in this section to run on any microcontroller we have used only ANSI standard C syntax. Using ANSI C minimizes problems encountered when running commercial and shareware cross compilers. Note that the samples CONFIG_BLOCK and RULE_BLOCK in Listing 7.8 correspond to the inverted pendulum problem that we explored in a previous chapter. The following code is included on the companion disk as FZKERN.C.

```c
/************************************
* Module: FZKERN.C
* Fuzzy Logic Kernel written in C
* specifically for cross compilation
* on 8 and 16 bit microcontrollers
* Created: 6/2/94
* Modified: 12/19/94
* Copyright 1994 by Ted Heske
************************************/
/************************************
 * This kernel requires two static data structures: CONFIG_BLOCK, and
 * RULE_BLOCK. Those structures reflect the membership function parameters,
 * and rule data of a predefined fuzzy model. This kernel requires these
 * data to conform to a predefined format. That format is spelled out
 * in the text of the Embedded Systems Chapter.
 */

#include <stdio.h>  /* Needed only for the printf() used in testing. Remove
                       this #include and all printf() statements from the
                       code before embedding the fuzzy kernel into
                       a microcontroller application. */

#define   MAX_MBFS    255  /* Allows up to 255 mbfs. Number of mbfs can be
                              reduced to minimize RAM required, but cannot
                              be increased. Decreasing MAX_MBFS reduces the
```

Fuzzy Logic for Real World Design

```c
                                RAM required by the kernel.*/
#define   MAX_INPUTS    8    /* Allows up to 8 input fuzzy vars. */
#define   MAX_OUTPUTS   4    /* Allows up to 4 output fuzzy vars. Both
                                MAX_INPUTS and MAX_OUTPUTS can be changed
                                to suit your application. Increasing/decreasing
                                either the number of inputs or outputs
                                increases/decreases the RAM required. */
#define   MAX_TRUTH 0xffff   /* Maximum truth value for 16 bit inputs. */
                             /* Use MAX_TRUTH 0x0fff for 12 bit inputs. */
                             /* Use MAX_TRUTH 0x03ff for 10 bit inputs. */
                             /* Use MAX_TRUTH 0x00ff for 8 bit inputs. */
                             /* Minimum truth value is assumed = 0. */
#define   RULE_DELIMITER 255 /* Separator between IF-side, THEN-side, Rule. */

/* Config block (fuzzy vars and membership funtions) and
   Rules will be compiled into ROM for embedded application. */

/*
 * The sample config block included here is from the inverted
 * pendulum problem.
 */
unsigned int CONFIG_BLOCK[] =
  {
  2, 1,     /* Fuzzy model has 2 inputs and  1 output. */
  5,                        /* First input has 5 mbfs. */
  0, 0, 11468, 27852, 65535, 3,       /* mbf 0 */
  11468,27852,27852, 32767, 3, 13,    /* mbf 1 */
  27852,32767,32767, 37683, 13, 13,   /* mbf 2 */
  32767,37683,37683, 54067, 13, 3,    /* mbf 3 */
  37683, 54067, 65535, 65535, 3, 65535, /* mbf 4 */
  5,                        /* Second input has 5 mbfs. */
  0, 0, 13107, 26214, 65535, 5,       /* mbf 5 */
  13107, 26214, 26214, 32767, 5, 10,  /* mbf 6 */
  26214, 32767, 32767, 39321, 10, 9,  /* mbf 7 */
  32767, 39321, 39321, 52428, 9, 5,   /* mbf 8 */
  39321, 52428, 65535, 65535, 5, 65535, /* mbf 9 */
  5,                        /* Output has 5 singleton mbfs. */
  10194,                              /* mbf 10 */
  24029,                              /* mbf 11 */
  32768,                              /* mbf 12 */
  41506,                              /* mbf 13 */
  55341                               /* mbf 14 */
  }; /* End CONFIG_BLOCK */

/*
 * The sample rule block included here is from the inverted
 * pendulum problem.
 */
unsigned char RULE_BLOCK[] =
  {
  25, /* Rule block contains 25 rules. */
  2, 5, 255, 13, 255,        /* Rule 0 */
```

Chapter 7: Implementation Choices

```
  2, 6, 255, 13, 255,       /* Rule 1 */
  2, 7, 255, 12, 255,       /* Rule 2 */
  2, 8, 255, 11, 255,       /* Rule 3 */
  2, 9, 255, 11, 255,       /* Rule 4 */
  0, 5, 255, 14, 255,       /* Rule 5 */
  0, 6, 255, 14, 255,       /* Rule 6 */
  0, 7, 255, 14, 255,       /* Rule 7 */
  0, 8, 255, 14, 255,       /* Rule 8 */
  0, 9, 255, 13, 255,       /* Rule 9 */
  1, 5, 255, 14, 255,       /* Rule 10 */
  1, 6, 255, 14, 255,       /* Rule 11 */
  1, 7, 255, 13, 255,       /* Rule 12 */
  1, 8, 255, 12, 255,       /* Rule 13 */
  1, 9, 255, 12, 255,       /* Rule 14 */
  3, 5, 255, 12, 255,       /* Rule 15 */
  3, 6, 255, 12, 255,       /* Rule 16 */
  3, 7, 255, 11, 255,       /* Rule 17 */
  3, 8, 255, 10, 255,       /* Rule 18 */
  3, 9, 255, 10, 255,       /* Rule 19 */
  4, 5, 255, 11, 255,       /* Rule 20 */
  4, 6, 255, 10, 255,       /* Rule 21 */
  4, 7, 255, 10, 255,       /* Rule 22 */
  4, 8, 255, 10, 255,       /* Rule 23 */
  4, 9, 255, 10, 255        /* Rule 24 */
  }; /* End RULE_BLOCK */

/*
 * RAM data.
 */
unsigned int Num_of_Inputs, Num_of_Outputs;
unsigned int Mbf_Degree[MAX_MBFS];
unsigned int Crisp_Input[MAX_INPUTS];
unsigned int Crisp_Output[MAX_OUTPUTS];
unsigned int cfg_index, m_index;    /* Indexes used for stepping through
                                        the CONFIG_BLOCK and RULE_BLOCK data
                                        structures. */
unsigned char j;
unsigned char Num_of_Rules;  /* Maximum of 255 rules. */

/*
 * Initialize fuzzy kernel by reading fuzzy model parameters
 * from CONFIG_BLOCK and RULE_BLOCK structures.  Initialize
 * input, output and membership function arrays with zeros.
 */
void Init_Fuzzy_Kernel(void)
  {
  /* Find out from config and rule blocks how many inputs, outputs,
   * and rules there will be. */
  Num_of_Inputs = CONFIG_BLOCK[0];
  Num_of_Outputs = CONFIG_BLOCK[1];
  Num_of_Rules = RULE_BLOCK[0];
```

Fuzzy Logic for Real World Design

```c
    /* Zero out all input, output, and membership funtion array locations.
    */
    for (j=0; j < MAX_MBFS; j++)
        Mbf_Degree[j] = 0;
    for (j = 0; j < MAX_INPUTS; j++)
        Crisp_Input[j] = 0;
    for (j=0;j < MAX_OUTPUTS; j++)
        Crisp_Output[j] =0;
    return;
    }

/*
 * This routine, provided for testing only, loads some
 * crisp inputs into the Crisp_Inputs array and then prints the array to the
 * screen.  In an actual embedded application A-to-D conversion loads
 * digitized values into the Crisp_Input array. The A-to-D values can be
 * any precision up to 16 bits.
 */
void Get_Crisp_Inputs(void)
    {
    /* Load one crisp value for each fuzzy model input variable. */
    Crisp_Input[0] += 1000;
    Crisp_Input[1] -= 1000;
    printf("Input_1 = %u, Input_2 = %u,  ",Crisp_Input[0],Crisp_Input[1]);
    return;
    }

/*
 * Determine degree of membership of new crisp inputs in all input mbfs.
 */
void Fuzzify_Inputs(void)
    {
    unsigned char num_of_mbfs;
    unsigned int k,tmp1,tmp2;

    cfg_index = 2;      /* Index to step through the CONFIG_BLOCK. */
    m_index = 0;        /* Index to point into the Mbf_Degree array. */
    for (j = 0; j < Num_of_Inputs; j++)   /* Calculate one input var at a
    time.*/
        {
        num_of_mbfs = CONFIG_BLOCK[cfg_index];
        for (k = 0; k < num_of_mbfs; k++)
            {   /* Calc degree here. */
                /* CONFIG_BLOCK[cfg_index+1] contains P1,
                   CONFIG_BLOCK[cfg_index+2] contains P2,
                   CONFIG_BLOCK[cfg_index+3] contains P3,
                   CONFIG_BLOCK[cfg_index+4] contains P4,
                   CONFIG_BLOCK[cfg_index+5] contains Slope1,
                   CONFIG_BLOCK[cfg_index+6] contains Slope2.  */

            if (    (Crisp_Input[j]<CONFIG_BLOCK[cfg_index+1])
                 || (Crisp_Input[j]>CONFIG_BLOCK[cfg_index+4]) )
```

Chapter 7: Implementation Choices

```c
                Mbf_Degree[m_index++] = 0;

        else if (    (Crisp_Input[j]<=CONFIG_BLOCK[cfg_index+3])
                  && (Crisp_Input[j]>=CONFIG_BLOCK[cfg_index+2]) )
              Mbf_Degree[m_index++] = MAX_TRUTH;

        else if ( Crisp_Input[j]<CONFIG_BLOCK[cfg_index+2] )
              Mbf_Degree[m_index++] = ((Crisp_Input[j]
                         - CONFIG_BLOCK[cfg_index+1])
                         * CONFIG_BLOCK[cfg_index+5]);

        else
              Mbf_Degree[m_index++] = ((CONFIG_BLOCK[cfg_index+4]
                         - Crisp_Input[j]) * CONFIG_BLOCK[cfg_index+6]);
    cfg_index += 6; /* Look at value just after mbf Slope2. */
    } /* End for k */
cfg_index++;
} /* End for j */
 tmp1 = cfg_index;  /* Save current index for use later. */
 tmp2 = m_index;    /* Save current index for use later. */

 /* Next clear output mbf degree. */
 for (j = 0; j < Num_of_Outputs; j++)
    {
    num_of_mbfs = CONFIG_BLOCK[cfg_index];
    for (k = 0; k < num_of_mbfs; k++)
       {
       Mbf_Degree[m_index++] = 0; /* Clear degree here. */
       cfg_index++;
       } /* End for k */
    } /* End for j */
 cfg_index = tmp1; /* Restore index. */
 m_index = tmp2;   /* Restore index. */
 return;
 }

/*
 * Use fuzzified inputs to calculate strength of each rule in rulebase.
 */
void Eval_Rulebase(void)
   {
   unsigned long rule_strength;
   unsigned char index;
   unsigned int rule_index;

   rule_index = 1;
   for (j = 0; j < Num_of_Rules; j++)
      {
      rule_strength = MAX_TRUTH;       /* Each rule starts at MAX_TRUTH. */
      /* Process IF-side of rule. */
      while ((index = RULE_BLOCK[rule_index++]) != RULE_DELIMITER) /* IF-Side*/
         { /* Min Intersection: */
```

233

```c
                rule_strength = ((rule_strength > Mbf_Degree[index]) ?
                                (Mbf_Degree[index]) : rule_strength);
    /* Product Intersection: To use this intersection method
     * uncomment the following statement and comment out the
     * previous rule_strength statement used for MIN Intersection. */
    /*          rule_strength = rule_strength * Mbf_Degree[index] / MAX_TRUTH; */

            } /* End while */
         /* Update THEN-side mbfs with rule_strength. */
         while((index=RULE_BLOCK[rule_index++])!=RULE_DELIMITER)/* THEN-side */
            {   /* MAX Aggregation: */
            if (Mbf_Degree[index]<rule_strength)
                Mbf_Degree[index] = rule_strength;
    /*
     * Additive Aggregation: To use this aggregation method
     * uncomment the following statement and comment out the
     * previous if() statement used for MAX aggregation.
     */
    /*          Mbf_Degree[index] += rule_strength; */
            } /* End while */
        } /* End for j */
      return;
      }

    /*
     * Calculate 16 bit crisp outputs.
     */
    void Defuzzify_Rulebase (void)
      {
      unsigned int k,num_mbfs;
      unsigned long temp1,temp2,degree;

      for (j = 0; j < Num_of_Outputs; j++)
            {
            temp1 = 0;
            temp2 = 0;
            num_mbfs = CONFIG_BLOCK[cfg_index++];

         /* Weighted average defuzzification. You can replace this
          * with MAX defuzzification by commenting it out and
          * uncommenting the MAX defuzzificaiton code, shown in comments below.
          */
            for (k = 0; k < num_mbfs; k++)
                {
                degree = Mbf_Degree[m_index++];
                temp1 += degree;
                temp2 += (CONFIG_BLOCK[cfg_index++]) * degree;
                } /* End for k */
            Crisp_Output[j] = (temp2/temp1);

         /*
```

Chapter 7: Implementation Choices

```
    * MAX defuzzification.  Can be uncommented and used as an alternative
    * to weighted-average defuzzification.
    */
   /* for (k = 0; k < num_mbfs; k++)
        {
        temp1 = Mbf_Degree[m_index++];
        if (temp1 > temp2)
            {
            temp2 = temp1;
            Crisp_Output[j] = CONFIG_BLOCK[cfg_index++];
            }
        else
        cfg_index++;
        }    */
     }  /* End for j */
  return;
  }

/*
 * Print crisp output to screen.  In an actual embedded application
 * D-to-A conversion is applied to system.  This routine is provided
 * so you can see some results on the screen for testing.
 */
void Apply_Crisp_Outputs(void)
  {
  printf ("Crisp Output is %u \n",Crisp_Output[0]);
  return;
  }

/*
 * Initialize fuzzy kernel and then enter loop to
 * get crisp inputs, process them through the fuzzy kernel,
 * and apply crisp outputs to the system.
 */
int main()
  {
  int i;

  Init_Fuzzy_Kernel();
  for (i = 0; i < 60; i++)   /* Do 60 fuzzy inference loops. */
      {
      Get_Crisp_Inputs();
      Fuzzify_Inputs();
      Eval_Rulebase();
      Defuzzify_Rulebase();
      Apply_Crisp_Outputs();
      } /* End for i */
  return(0);
  }
```

Listing 7.8: Complete ANSI C code listing for the fuzzy kernel: FZKERN.C.

235

Fuzzy Logic for Real World Design

Kernel Performance

Due to the flexibility of the kernel there are a number of different combinations of inference algorithms that may be applied. In order to compare the different combinations, we compiled the kernel to run on an industry standard (Intel Corp x86 compatible CPU) PC. The fuzzy model obtained the following performance parameters:

Target: Intel 80386DX 32-bit microprocessor;

> clock frequency: 33 MHz
> resolution: 16 bits
> inputs: 2
> input mbf shape: triangular, trapezoidal
> mbfs per input: 5
> outputs: 1
> output mbf shape: singleton
> mbfs per output: 5
> rules: 25
> rule evaluation/defuzzification algorithm: MIN/MAX
> code size: ROM 800 bytes, RAM 46 bytes
> speed: 333us per loop
> rules per second: 75,075
> rules per second per MHz: 2,275
> input mbf fuzzifications per second: 30,030
> input mbf fuzzifications per second: per MHz: 910
> defuzzifications per second: 3003
> defuzzifications per second per MHz: 91

The following table summarizes the inference loop timings for various combinations of algorithms in the kernel.

Intersection	Aggregation	Defuzzification	Loop Timing
MIN	MAX	MAX	333us
MIN	MAX	Weighted Average	348us
MIN	Additive	MAX	328us
MIN	Additive	Weighted Average	342us
Product	MAX	MAX	390us
Product	MAX	Weighted Average	403us
Product	Additive	MAX	380us
Product	Additive	Weighted Average	393us

Table 7.1: Loop timing of different fuzzy kernel configurations running the inverted pendulum fuzzy model on a 33MHz 80386DX equipped PC.

Cross Compiling the Kernel

Cross compilation, while an advantage in many ways, still demands a unique software tool for each different microcontroller target. Luckily, an ANSI C standard compiler already exists for almost every 8 and 16-bit microcontroller that you might consider using in your application. While cross compiler tools are fairly easy to use, they do vary somewhat in the steps required for a successful compilation. Some tools take the FZKERN.C file directly and churn out assembly code customized for that specific microcontroller. Other tools may require a predefined memory map and register allocation table. To complete the embedded application the assembly code produced by cross compiling the fuzzy kernel must then be linked together with all of the remaining application code. However, anyone working on embedded systems is already familiar with these steps.

To demonstrate the ease with which the fuzzy kernel can be retargeted to a variety of microcontrollers we have provided several examples. Appendix B contains the code for the Motorola 68HC11 8-bit microcontroller which was cross compiled using a shareware non-ANSI C compliant compiler, the cross compiled code for the Intel 80C196 16-bit microcontroller using a commercially available ANSI C compiler, and the code for the Motorola 68HC16 16-bit microcontroller which was cross compiled using another commercially available ANSI C compiler. The Appendix includes details specific to each of the three examples including information relevant to the software tools used.

Hardware

On a fast enough processor, for a small fuzzy model, fuzzy inferencing speeds can approach tens of microseconds per inference loop. This realm of performance easily enables the real-time control required for automotive applications like anti-lock braking or automatic traction control. This level of fuzzy inferencing speed can also handle the control requirements of most industrial processes. However, certain applications make higher demands on inferencing speed. Real time pattern recognition is one such example. For example, in order to perform general voice or handwriting recognition, anywhere from dozens to thousands of patterns must be compared to find a best match. For a fuzzy model with many inputs, outputs, and rules, a dedicated, high speed architecture may be the only viable approach.

Special purpose fuzzy processing hardware provides a factor of ten or more improvement in inferencing speed. Certain limitations accompany these speed improvements, however. Where a primarily software-based solution allows for tremendous flexibility in the use of a variety of inference algorithms, the typical dedicated hardware solution relies on a single set of algorithms which cannot be modified. Also, dedicated fuzzy hardware will not be capable of executing anything other than the fuzzy model. So the standard microcontroller is still a necessary component in the embedded solution.

Below we present two approaches to accelerating fuzzy inference. The first approach concentrates on digital implementation. Currently, digital technology provides the only commercially available fuzzy processors. The second approach concentrates on analog implementation. Although not commercially available, analog fuzzy processing offers exceptional speed potential on the order of one hundred nanoseconds per inference loop. We include both approaches for a balanced perspective.

Digital Architectures for Acceleration

The architecture of a typical dedicated fuzzy processing chip is a combination of three main elements: parameter memory, arithmetic processing units, and data path controls. The parameter memory contains information specific to your particular fuzzy model including the definition of input and output fuzzy sets and the rules governing your model. The arithmetic processing units handle the algorithmic functions of the chip including fuzzification, rule evaluation, and defuzzification. Finally, the data path controls inside the chip orchestrate the movement of partial results flowing through the arithmetic units. In a pipelined fashion, the results of fuzzification flow to the arithmetic units comprising rule evaluation, the results of which flow, in turn, to the arithmetic units comprising the defuzzification

Chapter 7: Implementation Choices

algorithm, and so on. The block diagram of Figure 7.9 illustrates the general functional layout of a dedicated fuzzy processor.

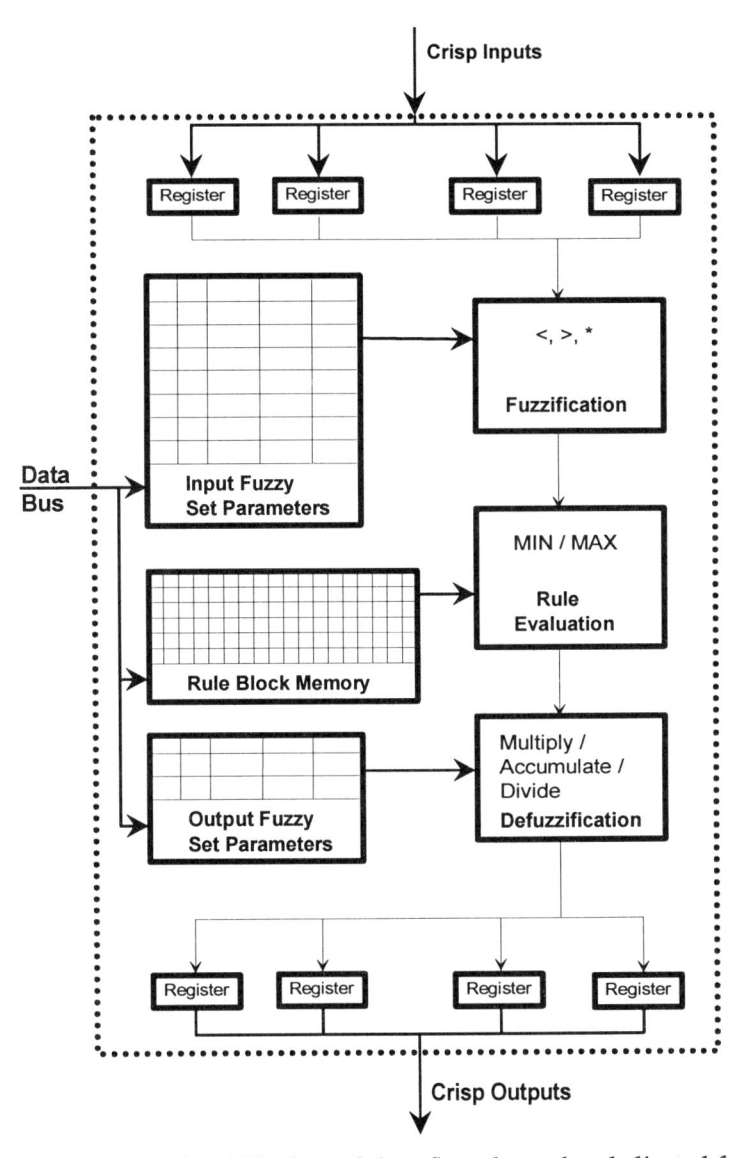

Figure 7.9: Functional blocks and data flow through a dedicated fuzzy processor.

At present, several manufacturers offer dedicated fuzzy processors. However, each architecture differs to the extent that simple comparisons between the various choices are not terribly meaningful. For further information on specific suppliers refer to Appendix A.

Analog Architectures for Acceleration

Fuzzy logic is a technology and methodology that embraces and fully uses the inherently analog nature of the world. It therefore seems ironic that, to date, the sole vehicle for the delivery of commercial fuzzy systems has been the digital computer. Although it is not commonly done, analog fuzzy processing is viable. As an example, consider the MAX circuit of Figure 7.10 and the MIN circuit of Figure 7.11.[73]

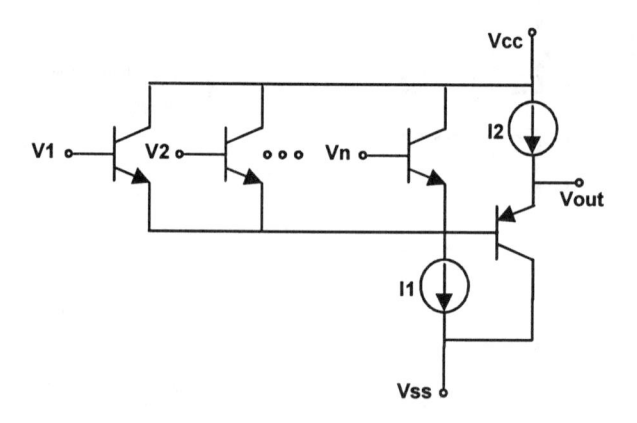

Figure 7.10: MAX processing of analog inputs.

---

[73]Toshiro Terano et al, *Applied Fuzzy Systems*, pp 273-281. AP Professional publishers.

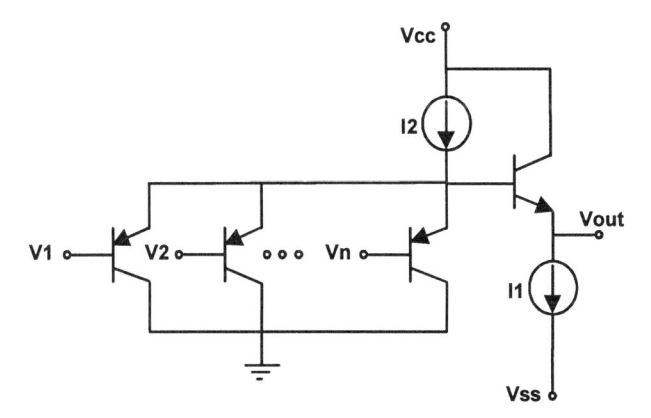

Figure 7.11: MIN processing of analog inputs.

These two voltage mode circuits share a number of elements. NPN and PNP bipolar junction transistors are used throughout. The two current sources depicted, *I1* and *I2*, must each source equivalent currents. Truth values on the continuum 0 to 1 are encoded as analog voltages 0 to +5 volts. *Vcc* must be maintained above approximately 6 volts, and *Vss* must be maintained below -1 volt. Both circuits are examples of emitter-coupled fuzzy inference logic gates.

The MAX circuit of Figure 7.10 compares the input voltages *V1*, *V2*, *etc.*, and the maximum of the various input voltages appears at the circuit's output *Vout*. The MIN circuit of Figure 7.11 compares the input voltages *V1*, *V2*, etc., and the minimum of the various input voltages at the circuit's output *Vout*.

The response time for either of the circuits can be on the order of 10 nanoseconds. Additional advantages of these circuits is their built-in temperature compensation (assuming all transistors are fabricated on the same substrate), and their ability to work under a very wide range of supply voltages.

The main drawback to an analog implementation of fuzzy logic is the same drawback that has prompted tremendous growth in the digital signal processing arena. Analog systems are designed and constructed with a specific problem in mind. Once constructed, they are well-suited to solving that one problem. Analog solutions lack the flexibility of a general purpose microcontroller. Changing the behavior of an analog circuit requires rewiring and changing analog components whereas a new program can be downloaded with great ease to change the microcontroller's behavior.

Hybrid

A hybrid approach to accelerating the fuzzy inference process involves a mix of hardware and software elements. Hardware may be used to speed up only a portion of the inference process while software running on a nearby microcontroller handles the balance of the computations.

Since there are no common practices in designing hybrid fuzzy architectures, they must be viewed as the engineer's last choice. The hybrid approach may work where both cost and performance in the end product become issues of extreme importance. The disadvantages of the hybrid approach include both increased development time and increased development cost due to the unavailability of appropriate design tools.

Chapter 8:
Object-Oriented Design of a Fuzzy Application

In recent years the object-oriented paradigm has enjoyed increasing popularity. Its popularity is due primarily to its partitioning of a system into self-contained components called classes, which can be enhanced or reused with minimal effort. We have found this technique to be extremely effective in the building and management of large systems, particularly simulations and information systems, and so we have decided to present our design and an implementation (in the following chapter) in some detail.

This chapter presents an overview of the object-oriented design of a fuzzy logic application. Primary emphasis will be placed on the design of a fuzzy engine class and its components. We'll begin by showing a static view of the system; that is, fuzzy logic classes and their relationships to each other. We'll then describe the dynamic design; how messages flow through the class hierarchies. The design reflects as closely as possible the fuzzy logic concepts presented in earlier chapters. Because this is a design, and not an analysis, we do make accommodations for the subsequent implementation (for example, some fuzzy logic methods are combined for efficiency). However, the design does not reflect any language-specific details. Constructors and overloaded operator methods, for instance, are a necessity in almost any C++ implementation but they are not shown in our class diagrams because they are not meaningful to the generic fuzzy engine design.

In the following chapter we'll illustrate a complete C++ implementation of our design using a very simple example, a thermostatic control problem. This chapter assumes some familiarity with object-oriented analysis and design concepts, while the following chapter assumes knowledge of C++.

The Static Design

This section defines our fuzzy logic classes and their relationships to each other. To describe our class relationships we will use a subset of notation from the Object Modeling Technique (OMT), as defined by Rumbaugh, et. al.[74] Figure 8.1 shows some of the notation that we will be using for classes and relationships between classes. Additional notation will be introduced as needed.

---

[74] James Rumbaugh, et. al., Object-Oriented Modeling and Design, 1991, Prentice-Hall.

Chapter 8: Object-Oriented Design of a Fuzzy Application

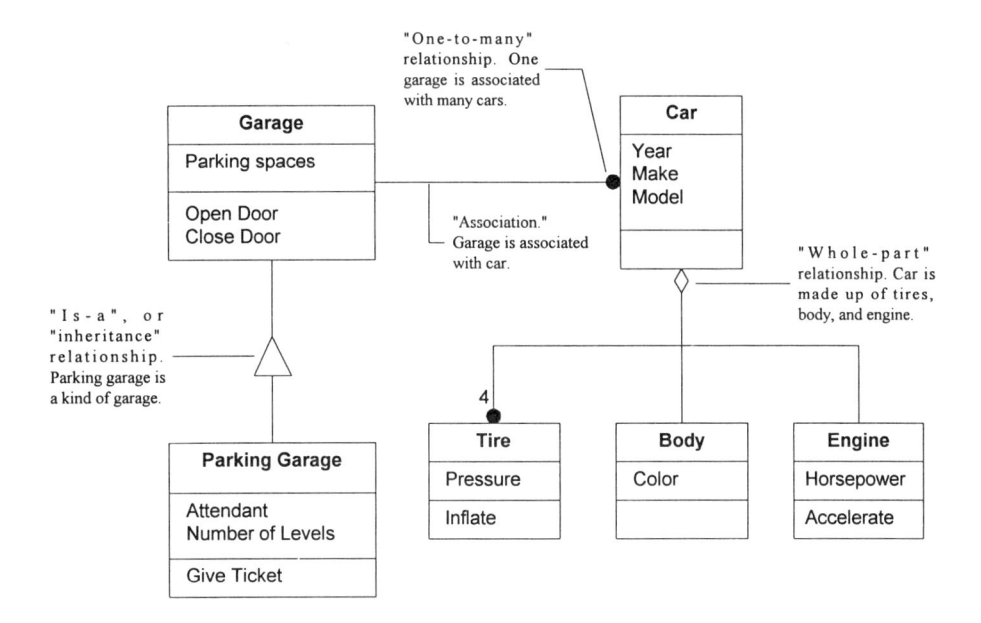

Figure 8.1: Object-Oriented class notation.

Each box in the diagram represents one class. At the top of each box is the class's name. Below the name are the class's data members (attributes). Below the attributes are the class's functions (methods). The diagram also shows several types of relationships:

Inheritance: ParkingGarage *is a kind of* Garage.

Association: Garage *is associated with*, or *uses*-a Car. The black dot at the right of the Garage-Car association line indicates *multiplicity*. A Garage can have multiple Car objects.

Whole-Part Car *has* an Engine, a Body, and Tire objects. We could also say that Car *is assembled from* Engine, Body, and Tire objects.[75] This differs from association in that cars and garages are

---

[75]The whole-part relationship is also known as "aggregation," a term which we will avoid in this context in order to prevent confusion with "fuzzy rule aggregation."

245

completely separate entities. Neither is dependent upon the other for existence. A car, on the other hand, is really not a car without its integral parts: tires, body, and engine.

Fuzzy Logic Classes

Chapters 2 and 3 discussed in detail the fundamental components of a fuzzy system: *fuzzy variables* and *rules*. Each of these entities has its own distinct characteristics, or *attributes*, and its own behaviors, or *methods*. Not surprisingly, these building blocks of fuzzy logic provide the foundation for our object-oriented design. From these building blocks we will create the classes that comprise our fuzzy engine.

Fuzzy Rules

Fuzzy rules are composed of two primary components: a set of antecedents and a set of consequents. Our Rule class has three attributes: a list of antecedents, or IfSide; a list of consequents, or ThenSide; and a Strength attribute, which will hold the result of the rule evaluation process. Figure 8.2 below shows the class relationship between Rule and its subcomponents.

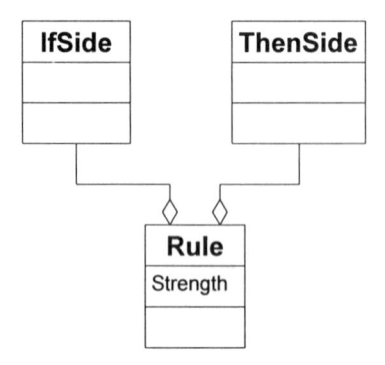

Figure 8.2: Fuzzy Rule *class and its relationships.*

Rule actually has a whole-part relationship with its subcomponents, IfSide and ThenSide. This is a somewhat stronger relationship than

Chapter 8: Object-Oriented Design of a Fuzzy Application

simple association. It indicates that `IfSide` and `ThenSide` are integral parts of `Rule`. They do not normally exist as independent entities.

Antecedents and consequents are fuzzy propositions. Recall from Chapter 3 that a fuzzy proposition consists of a single fuzzy variable and a qualifying membership function. Fuzzy propositions are the building blocks of fuzzy rules. The diagram below points out the fuzzy propositions for one rule.

IF Temp IS Hot   AND   Humidity IS High   THEN   Fanspeed IS High

Fuzzy Propositions

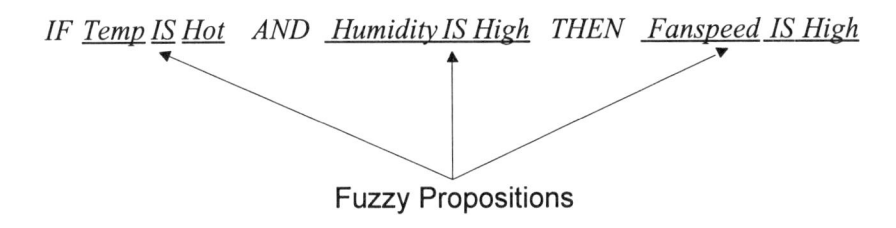

Figure 8.3: A rule and its fuzzy propositions.

Antecedent and consequent fuzzy propositions appear to be identical. They each contain a fuzzy variable which is qualified by a membership function. However, they do differ in the operations that are performed on them. For example, each antecedent will be evaluated during rule evaluation, and each consequent's membership function will be scaled during inference and aggregation. We can abstract the similarities between antecedents and consequents into one class, called `FuzzyProposition`, and then derive two classes `Antecedent` and `Consequent` from the parent class as follows:

Fuzzy Logic for Real World Design

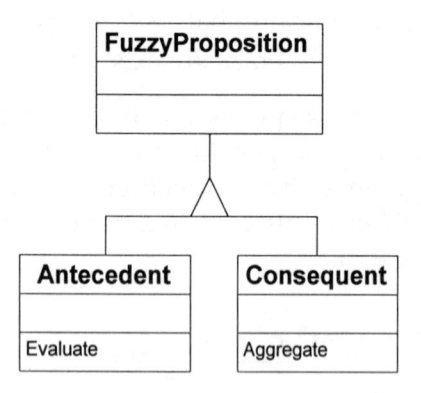

Figure 8.4: `FuzzyProposition` *class and derived classes* `Antecedent` *and* `Consequent`.

We can now build `Rule`'s `IfSide` and `ThenSide` from `Antecedent` and `Consequent` primitives:

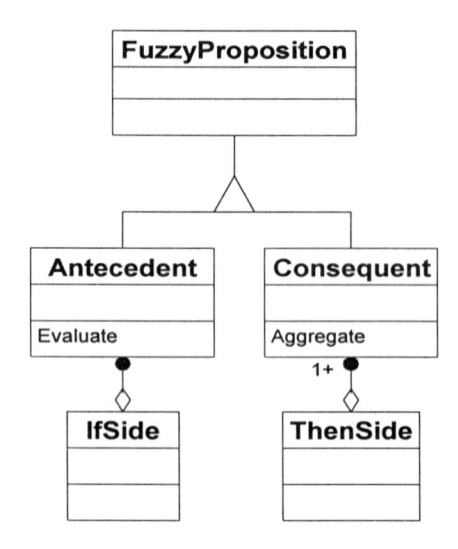

Figure 8.5: IfSide and ThenSide classes.

Chapter 8: Object-Oriented Design of a Fuzzy Application

There is a multiplicity relationship between `IfSide` and `Antecedent`, and between `ThenSide` and `Consequent`, as indicated by the black dots next to the association lines. That is, `IfSide` and `ThenSide` can have many fuzzy propositions, while each fuzzy proposition is owned by only one `Antecedent` or `Consequent`. The `1+` next to `ThenSide` indicates that `ThenSide` must have at least one fuzzy proposition, ie., that rules must each have at least one consequent. Notice that there is no `1+` next to `IfSide`, indicating that it does not have to have at least one fuzzy proposition, ie., that our design supports assertions (rules without any antecedents).

Our discussion above centers on the structure of a single rule. The intelligence of a fuzzy system, however, resides not in one rule, but in a collection of rules, or *Rulebase*, shown below:

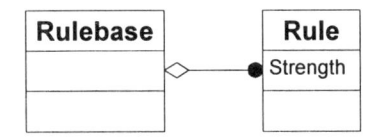

Rulebase	Rule
	Strength

Figure 8.6: Rulebase and its relationship to Rule.

`Rulebase` is nothing more than a "collection manager." It contains a set of `Rules` (as indicated by the black dot to the left of `Rule`), and provides all of the methods needed to manage the collection, such as adding and deleting `Rules`, and searching for `Rules`. The collection-manager can iterate through the collection applying one of `Rule`'s methods, like `Evaluate` or `Aggregate`, to each `Rule` object. We don't show collection-specific methods on our class diagrams because they are inherent to any collection.

The value of collection managers will become more apparent as our discussion continues. When we create our `FuzzyEngine` class below, we'll see that it is easier and more intuitive to include an object which already has the intelligence to manage `Rule` objects, rather than have the engine provide that intelligence itself. Notice above that `IfSide`

249

and `ThenSide` are also collection managers, freeing the `Rule` object from the necessity of containing the methods to manage collections of `FuzzyProposition` objects.

Fuzzy Variables

As we learned in Chapter 2, a fuzzy variable has a universe of discourse, as defined by a minimum crisp value and a maximum crisp value, and a set of membership functions. Fuzzy variables can be fuzzified and defuzzified. Each membership function has a label, a shape (which you'll probably implement as an array of points), and, with respect to a particular crisp value, a degree of truth. A membership function's methods include a fuzzify procedure for calculating the degree of truth of a particular crisp value. We can also create groups of fuzzy variables via a collection manager, which we'll call `FuzzyVarList`.

Figure 8.7 below shows `FuzzyVar`, `Mbf` (short for Membership Function, and `FuzzyVarList` classes. A "/" next to a class's attribute means that the attribute is calculated from the class's other attributes. For example, we can use `Mbf`'s shape (which is simply a collection of points) to calculate its `Area` and its `Centroid`. Methods for calculating these so-called "derived attributes" are implied and so we do not explicitly list them on the class diagrams.

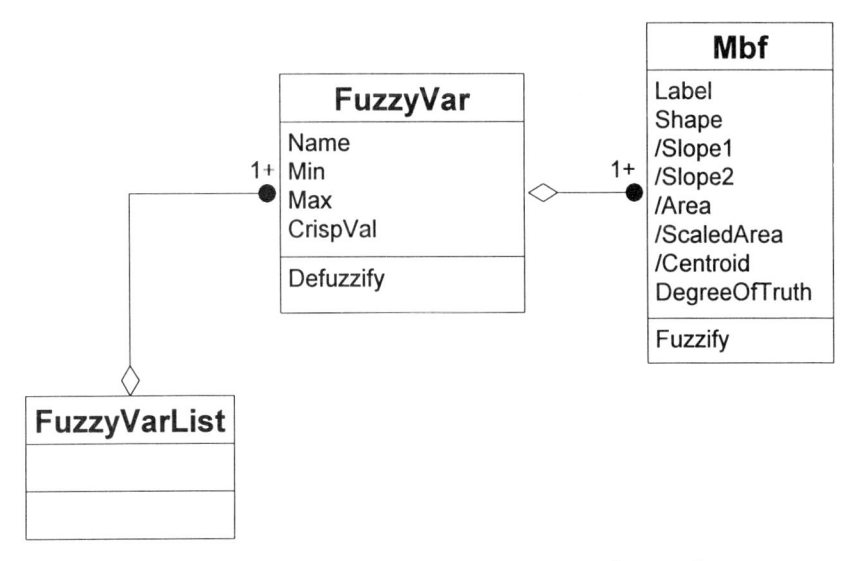

Chapter 8: Object-Oriented Design of a Fuzzy Application

```
                          ┌─────────────────┐        ┌─────────────────┐
                          │    FuzzyVar     │        │      Mbf        │
                          ├─────────────────┤        ├─────────────────┤
                          │ Name            │        │ Label           │
                       1+ │ Min             │    1+  │ Shape           │
               ┌──────◆───│ Max             │◇──────●│ /Slope1         │
               │          │ CrispVal        │        │ /Slope2         │
               │          ├─────────────────┤        │ /Area           │
               │          │ Defuzzify       │        │ /ScaledArea     │
               │          └─────────────────┘        │ /Centroid       │
               │                                     │ DegreeOfTruth   │
               ◇                                     ├─────────────────┤
        ┌──────────────┐                             │ Fuzzify         │
        │ FuzzyVarList │                             └─────────────────┘
        ├──────────────┤
        ├──────────────┤
        └──────────────┘
```

Figure 8.7: `FuzzyVarList`, `FuzzyVar`, *and* `Mbf` *classes.*

Instead of designing one `FuzzyVar` class, we could have created two separate classes, `InputFuzzyVar`, and `OutputFuzzyVar`. However, as we have designed more complex systems using these basic components we have found that input fuzzy variables and output fuzzy variables are more similar than different, particularly in applications that use outputs from one fuzzy system as inputs into another fuzzy system.

The Relationship between Rules and Fuzzy Variables

While in reality a fuzzy proposition contains references to both a fuzzy variable and a membership function, in our design we provide an association only between `FuzzyVar` and `Mbf`. This is because a membership function does not exist outside the scope of the fuzzy variable that defined it. An antecedent, for example, contains not just any membership function, but specifically a membership function defined by its fuzzy variable component. Figure 8.8 highlights this relationship.

251

Fuzzy Logic for Real World Design

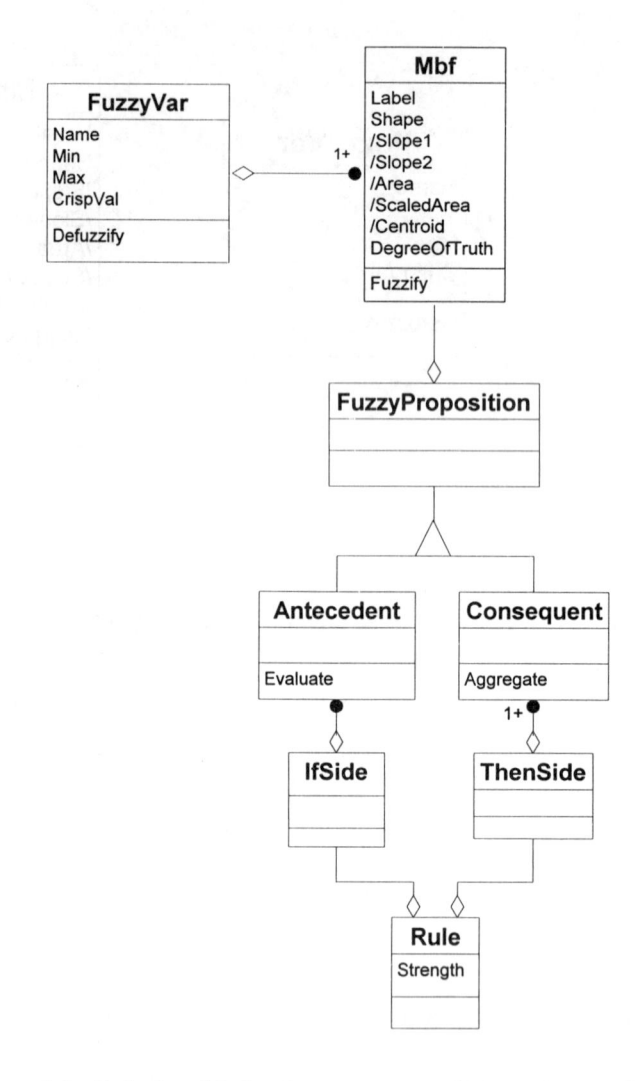

Figure 8.8: Relationship between `Rule` *and* `FuzzyVar` *classes.*

Notice that `FuzzyProposition` is not directly associated with `FuzzyVar`. This prevents us from creating a fuzzy proposition whose membership function is not defined in its fuzzy variable. We could not, for instance, create fuzzy proposition *Temperature IS Fast* since *Fast* would not be one of *Temperature*'s membership functions.

Chapter 8: Object-Oriented Design of a Fuzzy Application

By providing an association with a Mbf, we are implicitly providing an association with the appropriate FuzzyVariable. It would probably make sense to impement Mbf with methods to access information about its FuzzyVar. That way if a FuzzyProposition has to print itself out, for example, it could ask Mbf to retrieve the name of its FuzzyVariable.

The Fuzzy Engine Class

A fuzzy engine's job is to take an input data tuple, process it using fuzzy logic, and produce an output tuple.[76] Accordingly, our FuzzyEngine class interfaces directly to the highest-level components of a fuzzy system: FuzzyVarList, and Rulebase.

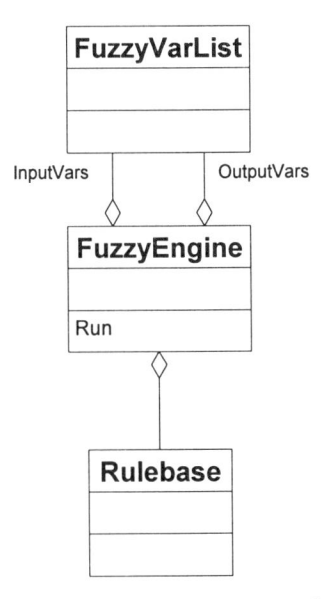

Figure 8.9: FuzzyEngine class.

---

[76] A tuple is an ordered set of values. An input tuple contains one crisp input for each input fuzzy variable. After fuzzy engine processing, an output tuple will contain one crisp output for each output fuzzy variable.

253

`FuzzyEngine` has two distinct associations with `FuzzyVarList`. This emphasizes the fact that a fuzzy engine will contain one list of *input* fuzzy variables and a separate list of *output* fuzzy variables. `FuzzyEngine` contains one method, "Run", which controls the fuzzy logic processing of the data tuple.

Our `FuzzyEngine` design is completely generic. The `Rules` and `FuzzyVars` (and the `FuzzyVars'` `Mbfs`) can describe any system at all. There is no inherent restriction on the numbers of `Rule`, `FuzzyVar`, or `Mbf` objects. Where the description of the system comes from will depend upon the implementation that you choose. `Rulebase`'s design, for example, specifies nothing about how to construct a `Rulebase`. The specifics of `Rulebase` construction depend upon your implementation. You may decide to read fuzzy rules from a file, as we do in our implementation in Chapter 9, or you may decide to let the user build them interactively, as we do in **FuzzyLab**'s Rules Editor.[77] These are implementation details which are not the responsibility of, and therefore are not restricted by, our design.

The Application Class

A fuzzy application gets input tuples from a source, either the user, a file, a database, or a hardware device, passes them to a fuzzy engine for processing, and reports, stores, or applies the crisp outputs. Figure 8.10 shows the `FuzzyAppl` class diagram.

FuzzyAppl	FuzzyEngine
GetCrispInputs ControlLoop ApplyCrispOutputs	Run

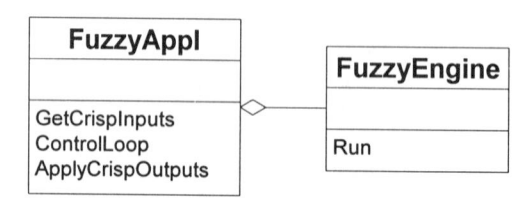

Figure 8.10: `FuzzyAppl` *class.*

---

[77]**FuzzyLab**, a Window's-based fuzzy logic instructional program, can be found on your companion disk.

Chapter 8: Object-Oriented Design of a Fuzzy Application

Once again, the class design is not concerned with the specific origin or destination of inputs and outputs. It merely requires that inputs be obtained and processed, and outputs reported, and leaves the I/O details to the implementor and the processing to the fuzzy engine.

Putting the Design Together

In describing the various class hierarchies of a fuzzy system we have thus far ignored the bigger picture of exactly how all of these objects fit together into one design. Figure 8.11 compiles all of the classes described above into one comprehensive design diagram. Pay particular attention to the links between classes:

1. *Inheritance*, as in `Antecedent` "is a kind of" `FuzzyProposition`.

2. *Simple Association*, as in `FuzzyAppl` "uses-a" `FuzzyEngine`

3. *Whole-part (special case of association)*, as in `Rulebase` "is composed of" `Rules`.

The Dynamic Design

To this point we have described only the static relationships between classes. We have seen which classes derive from, comprise, or contain instances of, other classes. Now we'll look more closely at class' methods and how control flows through the class hierarchy via messages and data. Rather than present a rigorous dynamic design, with complex state diagrams and pseudo code, we'll keep things simple by restricting our discussion to how messages and data flow between classes.

There are two primary areas of control flow in our fuzzy application design: *the fuzzy application's main control loop,* and *fuzzy engine processing of a data tuple.* We will describe each of these areas in detail below. First, however, we must say a few words about how objects communicate with each other, as this mechanism will be necessary to your understanding of the second area of control flow, data tuple processing.

Fuzzy Logic for Real World Design

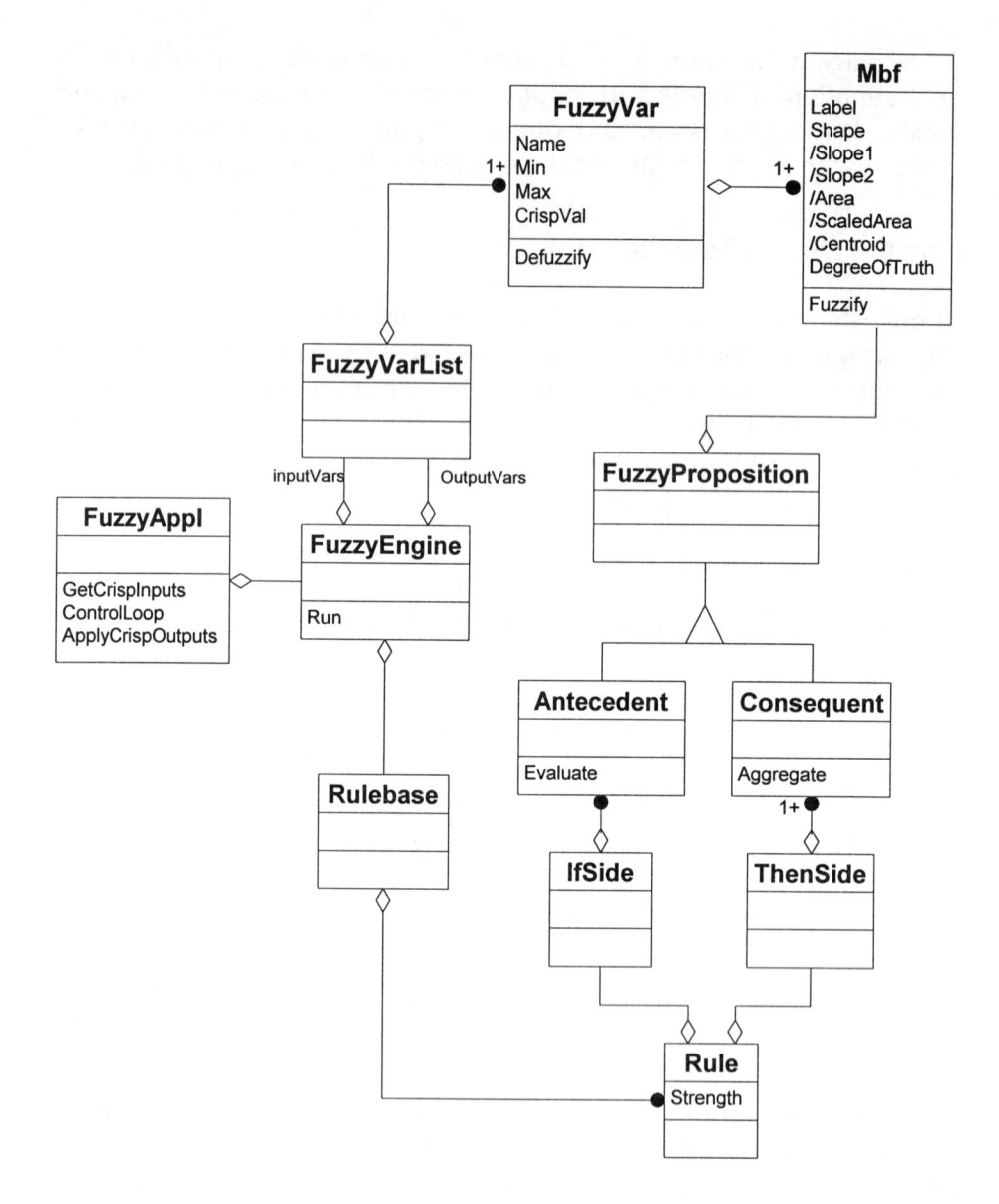

Figure 8.11: Complete fuzzy logic application class hierarchy.

Chapter 8: Object-Oriented Design of a Fuzzy Application

Message Passing

In an object-oriented design information passes between objects via messages, which may contain data. You can think of a message as a function call and the data as parameters. The sending object sends a message, invoking one of the receiving objects' methods. This means that a class's design must provide a method for every type of message that an object can receive. Often the method simply provides a means to propagate the message to another part of the class hierarchy.

To demonstrate how messages propagate through the class hierarchy, let's say that we want to provide the ability to make a copy of a `Rulebase` object. Figure 8.12 shows how the copy message flows through the object hierarchy. We use the word "object" here instead of "class" because we are referring to instances of classes. Each object is labeled with its name, and the name of its class is in parentheses.

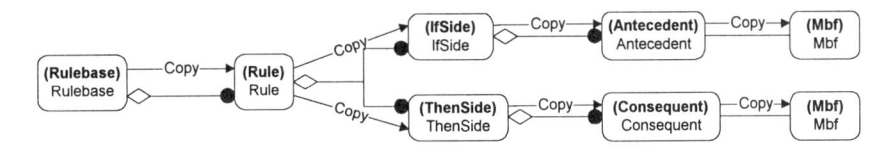

Figure 8.12: How copy message propagates through object hierarchy.

To copy `Rulebase`, we'll have to copy each of its `Rule` objects. Actually, it is more realistic to say that `Rulebase` merely initiates the copying of each `Rule`. As is typical in an object-oriented design, `Rulebase` sends a copy command, or "message" to its `Rules`, each of which then proceeds to copy itself by sending the copy message to its `IfSide` and its `ThenSide`, etc.. This distinction is important in that it illustrates that each object is responsible for copying itself. The details of duplication are completely "encapsulated" in each class. A `Rulebase` need not know the specifics of copying each `Rule`. `Rule`, in turn, relies on `IfSide` and `ThenSide` to copy themselves, etc. You'll see the power and importance of encapsulation later, when we show

257

you how to construct your own fuzzy application by simply including an instance of `FuzzyEngine` into your `FuzzyAppl` class.

The Control Loop

`FuzzyAppl`'s control loop collects data tuples from the user, or from some mechanical device, passes them to the `FuzzyEngine` for processing, and applies the crisp outputs generated by the `FuzzyEngine`. It will continue collecting and processing tuples until there are no more to process. Figure 8.12 shows a high-level flow chart of the control loop.

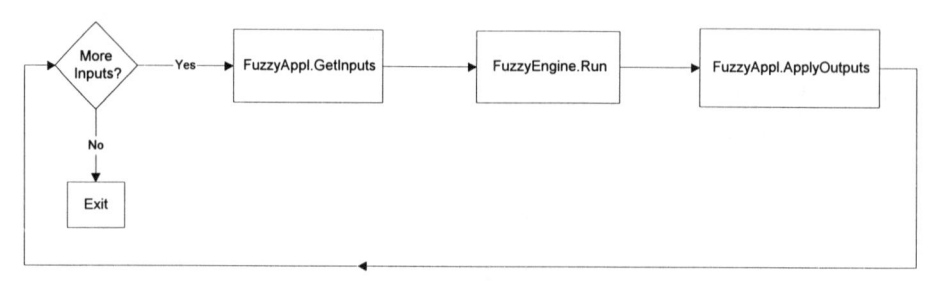

Figure 8.13: High-level flow of `FuzzyAppl`'s control loop.

Data Tuple Processing

This is where most of the work in a fuzzy application takes place, and it is, accordingly, the more complex of the control flow areas that we will discuss. `FuzzyAppl` initiates the process by passing a data tuple to its `FuzzyEngine`'s `Run` method. `Run` itself does very little work. Instead, it simply passes messages and data on to `Rulebase`, `InputVars`, and `OutputVars` so that they can do all the work.

Recall from Chapter 3 that there are five basic methods involved in processing one data tuple: *input fuzzification, rule evaluation, rule inference, rule aggregation, and output defuzzification*. For efficiency (more specifics about this in Chapter 9), we have combined inference and

Chapter 8: Object-Oriented Design of a Fuzzy Application

aggregation into one method, which we have called Aggregation. FuzzyEngine's run method contains four steps, which the engine executes in the following sequence, described in detail below:

1. InputVars.Fuzzify
2. Rulebase.Evaluate
3. Rulebase.Aggregate (*combines aggregation and inference*)
4. OutputVars.Defuzzify

1. Fuzzify Inputs

Fuzzification involves determining the degree of truth of a crisp input in a fuzzy variable's membership functions. Recall from our static design that class Mbf has an attribute called DegreeOfTruth. Mbf, therefore, will ultimately be responsible for receiving the fuzzify message and a crisp input and calculating the value of its own DegreeOfTruth attribute. Since the fuzzify message originates from FuzzyEngine's Run method, the message will have to propagate through the class hierarchy until it reaches Mbf.

Finally, each Mbf object uses its Fuzzify method to determine the DegreeOfTruth of the crisp input. The flow diagram below shows how the fuzzify message and the data tuple flow through the class hierarchy.

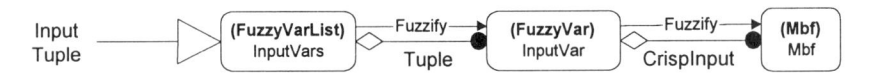

Figure 8.14: Input variable fuzzification

2. Evaluate Rulebase

Recall from Chapter 3 that rulebase evaluation involves evaluation of each rule's antecedents. Evaluation is a means of combining the truth values of all of the antecedents's fuzzy propositions into one value, called the strength. The strength is passed back and stored as an attribute of `Rule`.

This process works similarly to `Fuzzify`, but in this case `FuzzyEngine` passes the evaluate message to `Rulebase`. `Rulebase`'s `Evaluate` method sends the evaluate message to each `Rule` object in the collection. `Rule` has its own `Evaluate` method, which passes the evaluate message to `IfSide`. `IfSide` then applies the selected fuzzy rule evaluation method and returns the result to `Rule`. `Rule` stores the returned value in its `Strength` attribute.

```
(Rulebase)  -Evaluate→  (Rule)  -Evaluate→  (IfSide)  -Evaluate→  (Antecedent)
Rulebase                Rule                IfSide                Antecedent
```

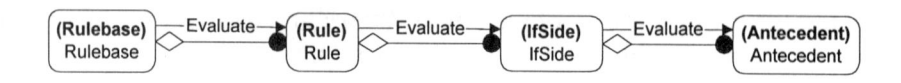

Figure 8.15: Rule evaluation.

3. Aggregate Rulebase

For computational efficiency, we have designed our classes to allow this method to combine inference with aggregation. You'll see the implementation details in Chapter 9.

```
(Rulebase) -Aggregate→ (Rule) -Aggregate→ (ThenSide) -Aggregate→ (Consequent) -Aggregate→ (Mbf)
Rulebase               Rule               ThenSide               Consequent               Mbf
```

Figure 8.16: Rulebase aggregation.

Inference is a means of applying a rule's strength (the result of `IfSide` evaluation) to each of the `Rule`'s consequent membership functions (see Chapter 3). We designed our engine for Product Inference[78], which scales (multiplies) a consequent's membership

---

[78] An object-oriented design may accomodate a particular implementation by providing class with specific attributes. We knew that our subsequent

260

Chapter 8: Object-Oriented Design of a Fuzzy Application

function by the rule's strength, resulting in a "scaled area." If more than one rule contains an antecedent which points to a particular membership function, that membership function will have multiple scaled areas.

Aggregation adds up a membership function's scaled areas into one composite scaled area, represented by the `ScaledArea` attribute in the `Mbf` class. This process will be discussed in detail in Chapter 9. For now it is sufficient to say that our `Aggregate` method takes as its input the `Rule` object's `Strength`, and stores the results in one `ScaledArea` for each output `Mbf`.

4. Defuzzify Outputs

Defuzzification produces one crisp output value for each output fuzzy variable from the composite fuzzy sets produced by aggregation (see Chapter 3).

Figure 8.17: Output variable defuzzification.

`FuzzyVarList` compiles its `FuzzyVars`' crisp outputs into one data tuple, which it returns to `FuzzyEngine`. The final result will be an output data tuple, containing the crisp outputs to be returned to `FuzzyAppl`.

---

implementation would use product inference, so in our design we provided a `ScaledArea` attribute to the `Mbf` class. This points out one difference between object-oriented analysis and and object-oriented design. In the earlier and higher-level analysis phase, `Mbf` would have been more generic. `Mbf` would have had an inference method because all fuzzy engines use some sort of inference. `Mbf` would not, however, have had the `ScaledArea` attribute since not all fuzzy engines use *product* inference.

Reusing Our Design

You might wonder why we discuss reuse in terms of a design, as opposed to implementation. Code reuse, after all, is the most widely appreciated benefit of object-orientation. An implementation, however, is merely an encoding of a design. It is in the design phase (and, of course, the analysis phase) that organization and problem-solving take place and most opportunities for class-reuse can be found. Class reuse generally translates directly into code reuse. In this section we'll discuss some possibilities for how you might reuse the classes presented in this chapter.

Our primary motivations in creating an object-oriented design were flexibility and reusability. Based on this design and the implementation presented in Chapter 9 we have constructed a variety of DOS and Windows-based products. We have also found reuses for these classes in more advanced designs and applications which are beyond the scope of this book, such as adaptive fuzzy control systems and model-free estimation. Below is a sampling of some of the ways that you might reuse our class hierarchies to build and tune your own application designs.

Different-shaped fuzzy sets

We have generally found trapezoidal membership functions to be sufficient for our purposes and so we constructed our design with these in mind. It is conceivable, however, that your designs will require other fuzzy set shapes. The diagram below shows one way that you could design a new membership function class, in this case a piecewise linear membership function (see Chapter 2).

Chapter 8: Object-Oriented Design of a Fuzzy Application

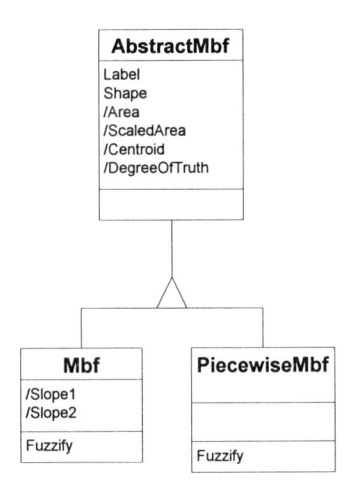

```
         AbstractMbf
    ─────────────────
    Label
    Shape
    /Area
    /ScaledArea
    /Centroid
    /DegreeOfTruth
    ─────────────────
```

```
    Mbf              PiecewiseMbf
  ─────────        ──────────────
  /Slope1
  /Slope2
  ─────────        ──────────────
  Fuzzify          Fuzzify
```

Figure 8.18: Class design for multiple membership function shapes.

We have abstracted the similarities between the trapezoidal and the piecewise linear membership function into one `AbstractMbf` class, and then derived the two shapes from the parent. The two classes inherit their common attributes from the parent class, `AbstractMbf`, but the actual algorithms for calculating these attributes will depend upon the shape (ie., the class) of a given fuzzy set. `Slope1` and `Slope2` are particular to trapezoidal membership functions, so we leave these two attributes with our original `Mbf` class.

You may recall from Chapter 2 that a trapezoidal membership function is really just a special case of the piecewise linear fuzzy set. Theoretically, we could have used `PiecewiseMbf` to implement all of our membership functions. However, the general-purpose methods provided by `PiecewiseMbf` would carry too great a computational cost, especially since many applications require only the more streamlined trapezoidal membership function design. If you do find that your applications call for more than occasional use of piecewise linear membership functions, then the overall cost of the general-purpose design might be small enough to warrant its use.

Rewriting fuzzy logic methods

We designed our fuzzy engine with some specific methods in mind. For example, in Chapter 9 we'll provide an implementation that uses the product evaluation method to evaluate rules. In fact, there is nothing in our design that indicates which algorithm you must use, only that you must provide some method to evaluate rules. You might, however, like to offer not one, but a variety of rule evaluation methods (perhaps you're writing a general-purpose simulation in which the user gets to choose which type of rule evaluation to use).

Once again, you could use the approach taken above: create an abstract base class to encapsulate the methods and attributes that are common to all types of rules. The base class would probably provide a default `Evaluate` method. From the base class you would derive another class, which would inherit all of the base class's methods and attributes, but would provide its own `Evaluate` method, overriding the default method in the base class.

Adding Support for Hedges

Let's say that you decide that you want your design to support hedges (see Chapter 2). A hedge operates on a rule's fuzzy proposition components. It further qualifies an antecedent or a consequent, usually by scaling, translating, or otherwise transforming its membership function during rule evaluation (if your implementation will support antecedent hedges) and/or aggregation (if your implementation will support hedges in consequents).

A hedge is an attribute of a fuzzy proposition. To support hedges, you can enhance our existing design by adding a `Hedge` attribute to class `FuzzyProposition` as shown below. At implementation time, then, you would have to account for a hedge's effect on a fuzzy proposition during rule evaluation if you're supporting input hedges, and during aggregation if output hedges are supported.

Chapter 8: Object-Oriented Design of a Fuzzy Application

FuzzyProposition
Hedge

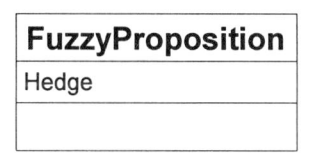

Figure 8.19: Fuzzy proposition with hedge support.

If you decide you would like to preserve the hedge-free FuzzyProposition, you can use the approach taken in Figure 8.18. That is, you can create abstract class AbstractFuzzyProposition and derive from it FuzzyProposition and FuzzyPropWithHedge.

Adding a user-interface

This area has provided us the greatest opportunity for reuse of our classes. Simply by adding various interfaces to FuzzyEngine and its components, we have designed applications ranging from **Tempcntl**, the simple DOS-based thermostatic control simulation presented in Chapter 9, to the more sophisticated **FuzzyLab**, the Windows-based instructional toy which you can find on your companion disk.

Our fuzzy engine is not bound to any particular interface. We provided the FuzzyAppl interface class to show you how the fuzzy logic "black box" fits into the framework of an application. We have found FuzzyAppl to be quite useful for basic applications, but in designing more sophisticated projects it has been necessary to replace it. In the case of **FuzzyLab**, for example, we exchanged the simple FuzzyAppl interface class used by **Tempcntl** for Borland's Windows-aware TApplication and TWindow interface classes, into which we simply integrated (included as an attribute) our own FuzzyEngine. We also created an interface class, or "view," as such classes are often called, for Rule, in order to create FuzzyLab's Rules Editor.

Summary

In this chapter we have presented a design for a simple fuzzy application, the most important component of which is a fuzzy engine. While we constructed our design with some implementation details in mind, there is nothing which ties this design to any particular programming language. In Chapter 9 we will present one possible implementation in C++.

There are, of course, many different ways to design fuzzy engines and applications. We believe that you will find this one most useful for simulations, information systems, and decision support systems, where the varieties of interfaces needed warrant a modular and flexible design. For embedded control, or for any other system in which flexibility and component reusability must be traded for size and speed, we recommend the more streamlined, less object-oriented approach presented in Chapter 7.

Chapter 9:
Object-Oriented Implementation of a Fuzzy Application

The implementation presented in this chapter is, with only a few differences, a C++ encoding of the design described in Chapter 8. Our sample application, thermostatic control, is a simple one, and is designed to be free of any details that could distract from explanations of the C++ implementation of fuzzy logic. We present the source code in its entirety so that you can see how the user interface (the FuzzyAppl class) and fuzzy engine objects (the FuzzyEngine class hierarchy) work together to form a complete solution. We have kept the user interface simple so that you can concentrate on the inner workings of the fuzzy engine.

Before reading this chapter we recommend that you experiment with the sample program, TEMPCNTL.EXE, provided on the companion disk.

Fuzzy Logic for Real World Design

Getting the Most from this Chapter

This chapter serves two purposes. The first is to present and explain in some detail a C++ implementation of an application that uses fuzzy logic. The second is to provide you with some guidance in reusing the code for your own applications. While the application consists of a great deal of code, almost all of the code comprises the inner workings of the fuzzy engine, which you do not have to change or even understand in detail to create and customize your own application. If you are interested only in how to embed a fuzzy engine in your application, and you are not concerned with the inner workings of our code, then we suggest you skim most of this chapter, concentrating on the section titled *How to Build Your Own Fuzzy Application Using Our Code*. That section tells you how to use the code presented in this chapter, which you can also find on your companion disk.

TEMPCNTL.EXE

TEMPCNTL.EXE simulates an air-conditioner controller. The controller keeps your house at the ideal comfort-level by running the air-conditioner when it is too hot, or when the humidity level makes it seem hot. The A/C runs for some percentage of every "cycle." A cycle is a fixed amount of time, say 10 minutes. The percentage of a given cycle that the fan will run is called a "duty cycle." The duration of each duty cycle will depend upon how hot it is in comparison to your ideal temperature. Humidity is also a factor since high humidity makes it seem hotter than it is.

TEMPCNTL has a simple DOS interface. When you start up TEMPCNTL.EXE, you will see the message:

```
Enter Temperature Setpoint:
```

Chapter 9: Object-Oriented Implementation of a Fuzzy Application

The set-point is the temperature that you think is ideal for your home, assuming average humidity. You enter this number only once, at the beginning of program execution. The application then enters a loop. Each pass through the loop represents one control loop. At the beginning of the loop you will be asked to enter the current temperature and humidity as follows:

```
Enter Current Temperature (degrees) [0 to exit]:
Enter Current Humidity:
```

These are the crisp inputs for the current control loop. The application passes your crisp inputs to the fuzzy engine, which produces a crisp output and displays it:

```
The Fan will be on N Percent of this Cycle
```

where N is the duty-cycle that will be needed to bring the comfort level of your home closer to the set-point.

To exit TEMPCNTL, enter a 0 when asked to enter the temperature.

TEMPCNTL's Fuzzy Model

Our fuzzy temperature controller has two inputs, *TempDiff* and *Humidity*, and one output, *DutyCycle*. *TempDiff* represents not absolute temperature, but the difference between the current temperature, input by the user at the beginning of each control loop, and the temperature set-point, which is specified by the user at the start of the application.

Fuzzy Variables and Membership Functions

We partition each fuzzy variable into three membership functions. Figure 9.1 shows the fuzzy variables and membership functions for our controller.

Fuzzy Logic for Real World Design

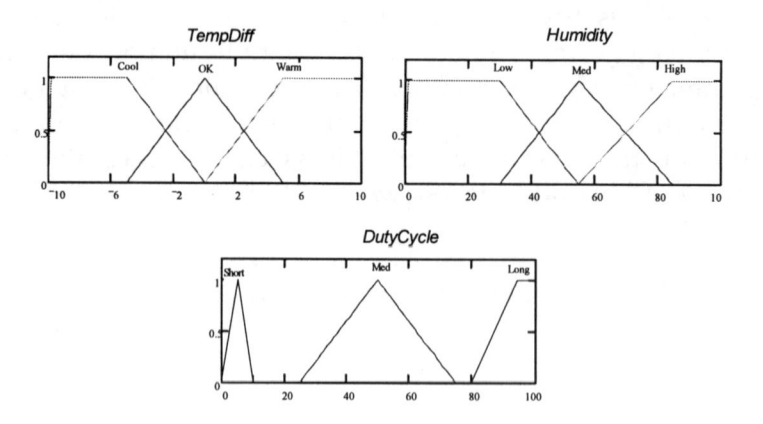

Figure 9.1: TEMPCNTL*'s fuzzy variables and membership functions.*

Our implementation will read the fuzzy variables and their membership functions from a text file, called a configuration file. TEMPCNTL.CFG contains the definitions for the fuzzy variables and membership functions shown above. It is shown in Figure 9.2.

```
INPUT TempDiff degrees 3 -10 10
Cool   -10  -10   -5    0
OK      -5    0    0    5
Warm     0    5   10   10
INPUT Humidity percent 3 0 100
Low      0    0   30   55
Med     30   55   55   85
High    55   85  100  100
OUTPUT DutyCycle percent 3 0 100
Short    0    5    5   10
Med     25   50   50   75
Long    80   95  100  100
```

Figure 9.2: TEMPCNTL.CFG *contains text descriptions of* TEMPCNTL*'s input and output fuzzy variables and the fuzzy variables' membership functions.*

Chapter 9: Object-Oriented Implementation of a Fuzzy Application

`TEMPCNTL.CFG` contains text descriptions for input and output fuzzy variables and membership functions. There is one line for each fuzzy variable in the format:

`INPUT|OUTPUT FuzzyVarName Units NumOfMbfs Min Max`

Units is the unit of measure (degrees, minutes, etc.), and **Min** and **Max** are the boundaries of the universe of discourse. **NumOfMbfs** is the number of membership functions into which the fuzzy variable is partitioned. Directly following the fuzzy variable there are **NumOfMbfs** lines, one for each membership function in the format: Membership function description format consists of four X values.

`MbfLabel P1 P2 P3 P4`

We assume that the Y values are consistent with a trapezoidal membership function. Figure 9.3 illustrates the format of TEMPCNTL.CFG.

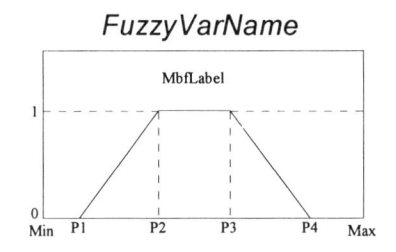

Figure 9.3: A fuzzy variable and one of its membership functions. Only four X values, P1 - P4, are specified in the config file. Our implementation assumes Y values to be 0, 1, 1, and 0 respectively.

Fuzzy Rules

TEMPCNTL's rules are described in text file `TEMPCNTL.RB`, shown in Figure 9.4.

Fuzzy Logic for Real World Design

```
IF TempDiff IS Cool AND Humidity IS Low  THEN DutyCycle IS Short
IF TempDiff IS Cool AND Humidity IS Med  THEN DutyCycle IS Short
IF TempDiff IS Cool AND Humidity IS High THEN DutyCycle IS Med
IF TempDiff IS OK   AND Humidity IS Low  THEN DutyCycle IS Short
IF TempDiff IS OK   AND Humidity IS Med  THEN DutyCycle IS Med
IF TempDiff IS OK   AND Humidity IS High THEN DutyCycle IS Med
IF TempDiff IS Warm AND Humidity IS Low  THEN DutyCycle IS Med
IF TempDiff IS Warm AND Humidity IS Med  THEN DutyCycle IS Med
IF TempDiff IS Warm AND Humidity IS High THEN DutyCycle IS Long
```

Figure 9.4: `TEMPCNTL.RB` contains `TEMPCNTL`'s rulebase.

C++ Source Code

The code presented in this section is *not* compiler independent. It was developed and tested in `Borland C++ 4.02`. We selected the `Borland` environment primarily because of its excellent implementation of template-based container classes. We discuss how we use the container classes in our implementation following the source listings.

The source code makes limited use of *Hungarian notation*, a variable naming convention which adds prefixes to variables to indicate their type. For example, the variable `fStrength` is of type `float`, thus the "f" prefix. There is no standard for how or when to use Hungarian notation. We simply use it where we feel it adds to code readability. We have found it especially helpful in `MS Windows` applications, as the `Windows` display environment is entirely integer-based and we often have to do `int-to-float` conversions and vice-versa. The prefixes we use most often are:

n	integer
f	float
i	loop iterator
sz	NULL-terminated string
fc	fuzzy class

Chapter 9: Object-Oriented Implementation of a Fuzzy Application

```
/***************************************************************
 * FUZZAPPL.H:  Header file for fuzzy application class         *
 *              fcFuzzyAppl                                     *
 ***************************************************************/
#ifndef FUZZAPPL_H          // Prevent multiple copies of this
                            //     include file
#define FUZZAPPL_H
#include <fuzzengn.h>

const short MAXTUPLESIZE = 100;    // Maximum length of a data
                                   //     tuple

/***************************************************************
 * Fuzzy Class fcFuzzyAppl (fuzzy application) class definition.
 *
 * Contains a fuzzy engine. Responsible for accepting inputs
 * from the user (or from  sensors in a real-life system),
 * running the fuzzy model, and reporting outputs to the user
 *(or applying outputs to the cooling fan in a real-life
 *system).
 ***************************************************************/
class fcFuzzyAppl
        {
        public:
          // Constructors and destructor
          fcFuzzyAppl(char *szCfgFileName, char *szRbFileName);
          ~fcFuzzyAppl();
          // "Primary" methods, to retrieve/change class's
          //     attributes
          BOOL GetCrispInputs();
          // Class's methods
          void ControlLoop();
          void ApplyOutputs(istrstream& OutputTuple);
        private:
          fcFuzzyEngine *FuzzEngn;    // Fuzzy engine
          char szTuple[MAXTUPLESIZE]; // Tuple to pass to
                                      //     fuzzy engine
          float TempSetpoint;         // Temperature set point
        };
#endif   // FUZZAPPL_H
```

```
/***************************************************************
 * FUZZAPPL.CPP: Implementation file for fuzzy application      *
 * class FuzzyAppl                                              *
 ***************************************************************/
//
// Simple air-conditioner control problem to demonstrate the
// mechanics of fuzzy logic.  Controller keeps the user's home
```

273

Fuzzy Logic for Real World Design

```c
// at the "ideal" temperature by running a cooling fan N percent
// of each cycle.  A cycle is simply given amount of time, for
// example, 10 minutes. The percentage of the cycle that the fan
// will run is called a "duty cycle." At start of application
// user specifies their ideal home temperature.  Application
// then enters the cycle loop.  At the beginning of each cycle
// the user is asked for the current temperature and humidity.
// The application runs the fuzzy model on the current
// conditions and then reports the results, (the duty cycle).
//
#include <conio.h>
#include <fuzzappl.h>


// fcFuzzyAppl Constructor.  Get TempSetpoint from user.
fcFuzzyAppl::fcFuzzyAppl(char *szCfgFileName,char *szRbFileName)
        {
        FuzzEngn = new fcFuzzyEngine (szCfgFileName,szRbFileName);
        cout << "Enter temperature setpoint [Degrees]: ";
        cin >> TempSetpoint;
        }

//  Destructor. Delete the fuzzy engine.
fcFuzzyAppl::~fcFuzzyAppl()
        {
        if (FuzzEngn)
           delete FuzzEngn;
        }

// Ask user for current crisp inputs, temperature and humidity.
// Calculate difference between current temperature and user's
// ideal temperature. This difference value will be an input to
// the fuzzy system.
BOOL fcFuzzyAppl::GetCrispInputs()
          {
          float  Temp, Humidity;
          cout << "\n";
          cout << "\nEnter current temperature [Degrees] (0 to Exit): ";
          cin >> Temp;
          if (Temp == 0)
            return (FALSE);
          float TempDiff = Temp - TempSetpoint;
          cout << "Enter current humidity level [Percent]: ";
          cin >> Humidity;
          sprintf (szTuple,"%f %f",TempDiff,Humidity);
          return (TRUE);
          }

// Get ideal temperature from user. Enter loop: ask user for
// current conditions (temp & humidity), run fuzzy model, and
// display results (how long fan will run this cycle).
void fcFuzzyAppl::ControlLoop()
```

Chapter 9: Object-Oriented Implementation of a Fuzzy Application

```cpp
    {
    while (TRUE)
      {
      if (!GetCrispInputs())    // If user quits
            return;
      istrstream OutputTuple = FuzzEngn->Run (istrstream (szTuple));
      ApplyOutputs(OutputTuple);
      }
    }

// Display the results.  Use the output var's name (specified in
// TEMPCNTL.CFG) to get a pointer to the fuzzy output var object
// so the object's crisp value can be accessed.
void fcFuzzyAppl::ApplyOutputs(istrstream& OutputTuple)
    {
    float DutyCycle;

    OutputTuple >> DutyCycle;
    cout << "Fan will be on " << DutyCycle << "% of the time.";
    getch();
    }

// Construct/initialize fuzzy engine and enter its ControlLoop.
void main ()
    {
    fcFuzzyAppl FuzzyAppl ("TEMPCNTL.CFG","TEMPCNTL.RB");
    FuzzyAppl.ControlLoop();
    }


/***********************************************************
 * FUZZENGN.H: Header file for fuzzy engine class fcFuzzyEngine*
 ***********************************************************/
#ifndef FUZZENGN_H              // Prevent multiple copies of this
                                //    include file
#define FUZZENGN_H
#include <stdio.h>
#include <rule.h>

/***********************************************************
 * Fuzzy Class fcFuzzyEngine (fuzzy engine) class definition
 *
 * Contains list of input fuzzy vars, list of output fuzzy vars
 * and a rulebase.  Responsible for running one input tuple
 * through the fuzzy logic process and returning one output
 * tuple.
 ***********************************************************/
class fcFuzzyEngine
    {
    public:
```

```
        // Constructors and destructor
        fcFuzzyEngine();
        fcFuzzyEngine(char *CfgFileName, char *RBFileName);
        ~fcFuzzyEngine();
        // Class's methods
        istrstream Run (istrstream& Tuple);
    private:
        fcFuzzyVarList *InputVars;
        fcFuzzyVarList *OutputVars;
        fcRulebase *Rulebase;
    };
#endif   // FUZZENGN_H
```

```
/******************************************************************
 *  FUZZENGN.CPP: Implementation file for fuzzy engine class      *
 *                  fcFuzzyEngine.                                 *
 ******************************************************************/
#include <fuzzengn.h>

/******************************************************************
 * Fuzzy Engine class member function definitions.
 ******************************************************************/

// Default constructor.
fcFuzzyEngine::fcFuzzyEngine()
    {
    InputVars = OutputVars = 0;
    Rulebase = 0;
    }

// Construct a fuzzy engine from a config file and a rulebase
//     file.
fcFuzzyEngine::fcFuzzyEngine(char *CfgFileName,char *RBFileName)
    {
    InputVars = new fcFuzzyVarList (CfgFileName,nINPUTS);
    OutputVars = new fcFuzzyVarList (CfgFileName,nOUTPUTS);
    Rulebase = new fcRulebase (RBFileName,*InputVars,*OutputVars);
    }

// Destructor.  Delete InputVars, OutputVars, and Rulebase.
fcFuzzyEngine::~fcFuzzyEngine()
    {
    if (InputVars)
       delete InputVars;
    if (OutputVars)
       delete OutputVars;
    if (Rulebase)
       delete Rulebase;
    }
```

Chapter 9: Object-Oriented Implementation of a Fuzzy Application

```
// Run fuzzy engine on one data tuple.
istrstream fcFuzzyEngine::Run(istrstream& Tuple)
      {
      InputVars->Fuzzify(Tuple);
      Rulebase->Evaluate();
      Rulebase->Aggregate();
      return (OutputVars->Defuzzify());
      }
```

```
/*******************************************************************
 * RULE.H: Header file for fuzzy rule class fcRule and fuzzy       *
 *         rulebase class fcRulebase.                              *
 *******************************************************************/
#ifndef RULE_H            // Prevent multiple copies of this include
                          //      file
#define RULE_H
#include <fuzzyvar.h>
#include <ifthen.h>

/*******************************************************************
 * Fuzzy Class fcRule (fuzzy rules) class definition
 *
 * Each rule has an IfSide, a ThenSide and a Strength (result of
 * evaluation).
 *******************************************************************/
class fcRule
      {
      public:
        // Constructors and destructor
        fcRule();
        fcRule(istrstream& RuleStream, const fcFuzzyVarList& InputVars,
                              const fcFuzzyVarList& OutputVars);
        ~fcRule();
        // Overloaded operators
        BOOL operator== (const fcRule& Rule2) const;
        // Class's methods
        void Aggregate();
        void Evaluate();
        // "Primary" methods, used to change/retrieve class's
        //     attributes
        float GetStrength() const
                    {return fStrength;}
      private:
        fcIfSide *IfSide;      // List of antecedents
        fcThenSide *ThenSide;  // List of consequents
        float fStrength;       // Result of rule evaluation
      };
```

277

```
/*****************************************************************
 * Fuzzy Class fcRulebase (list of Rules) class definition
 *
 * Contains a list of Fuzzy Rules, which is an instance of
 * BORLAND's double linked list Template class.  Each member of
 * the list is a pointer to a fcRule object.
 *****************************************************************/
class fcRulebase
      {
      friend class fcRuleListIter;
      public:
        // Constructors and destructor
        fcRulebase() {};
        fcRulebase(char *szRulebaseFileName,
                   const fcFuzzyVarList& InputVars,
                   const fcFuzzyVarList& OutputVars);
        ~fcRulebase();
        // Class's methods
        void Aggregate();
        void Evaluate();
        // TIDoubleListImp Collection class wrapper functions.
        int AddAtTail (fcRule *Rule)
              {return Rules.AddAtTail(Rule);}
        void Flush (TShouldDelete::DeleteType DelType)
              {Rules.Flush(DelType);}
        unsigned GetItemsInContainer() const
              {return Rules.GetItemsInContainer();}
        int IsEmpty() const
              {return Rules.IsEmpty();}
      private:
        TIDoubleListImp<fcRule> Rules;       // List of rules
                                             //    in the rulebase
      };

// Iterator for fcRulebase class.
class fcRuleListIter : public TIDoubleListIteratorImp<fcRule>
      {
      public:
           fcRuleListIter(const fcRulebase& Rulebase)
             : TIDoubleListIteratorImp<fcRule>(Rulebase.Rules) {}
      };
#endif    //RULE_H




/*****************************************************************
 * RULE.CPP: Implementation file for fuzzy rule class fcRule     *
 *           and fuzzyrulebase class fcRulebase.                 *
 *****************************************************************/
#include <rule.h>
```

Chapter 9: Object-Oriented Implementation of a Fuzzy Application

```cpp
/*****************************************************************
 * fcRule class's methods
 *****************************************************************/

// Default constructor.
fcRule::fcRule()
    {
    IfSide = 0;
    ThenSide = 0;
    fStrength = 0;
    }

// Construct from a line of file input which represents a rule.
// This is used when we read the rule from the file put it into
// memory.
fcRule::fcRule(istrstream& RuleStream, const fcFuzzyVarList& InputVars,
               const fcFuzzyVarList& OutputVars)
    {
    IfSide = new fcIfSide(RuleStream, InputVars);
    ThenSide = new fcThenSide(RuleStream, OutputVars);
    fStrength = 0;
    }

// Destructor.  Call destructors for IfSide and ThenSide.
fcRule::~fcRule()
    {
    if (IfSide)
      delete IfSide;
    if (ThenSide)
      delete ThenSide;
    }

// Overloaded equality test operator.  No need to compare fStrengths.
BOOL fcRule::operator== (const fcRule& Rule2) const
    {
    if ( (*IfSide == *Rule2.IfSide) && (*ThenSide == *Rule2.ThenSide) )
      return TRUE;
    return FALSE;
    }

// EVALUATE a rule.  The result is the rule's strength.
inline void fcRule::Evaluate()
    {
    fStrength = IfSide->Evaluate();
    }

// Do Inference/Aggregation to rule's ThenSide.
inline void fcRule::Aggregate()
    {
    ThenSide->Aggregate (fStrength);
    }
```

279

```
/******************************************************************
 * fcRulebase class's methods
 ******************************************************************/

// Construct the rulebase from a file.  Each line in the file
// represents one rule.  Insert one rule at a time into the
// rulebase.  If the file is not found the rulebase is still
// constructed but there won't be any rules in it.
fcRulebase::fcRulebase(char* szRulebaseFileName,
                  const fcFuzzyVarList& InputVars,
                  const fcFuzzyVarList& OutputVars)
   {
   ifstream RulebaseInFile (szRulebaseFileName,ios::in);
   char szInputLine[MAXLINE];

   while (RulebaseInFile.getline(szInputLine,MAXLINE))
      {
      istrstream RuleStream (szInputLine,strlen(szInputLine));
      AddAtTail (new fcRule (RuleStream,InputVars,OutputVars));
      }
   RulebaseInFile.close();
   }

// Destructor.  Releases all memory associated with the rulebase.
fcRulebase::~fcRulebase()
   {
   Flush(TShouldDelete::Delete);
   }

// Aggregate rulebase one rule at a time.  If rule's strength is 0 then
// don't bother aggregating it.  A strength of 0 means that the rule
// didn't "fire" during evaluation.
void fcRulebase::Aggregate()
   {
   fcRuleListIter iRulebase(*this);

   for (iRulebase.Restart(); iRulebase; iRulebase++)
       {
       if (iRulebase.Current()->GetStrength())
         {
         iRulebase.Current()->Aggregate();
         }
       }
   }

// Have each rule in the rulebase evaluate its own IfSide.
void fcRulebase::Evaluate()
   {
   fcRuleListIter iRulebase(*this);

   for (iRulebase.Restart(); iRulebase; iRulebase++)
```

Chapter 9: Object-Oriented Implementation of a Fuzzy Application

```
        iRulebase.Current()->Evaluate();
    }




/*******************************************************************
 *   IFTHEN.H: Header file for if/then class fcFuzzyProposition     *
 *             if/then classes fcIf and fcThen, ifside              *
 *             class fcIfSide, and thenside class fcThenSide.       *
 *******************************************************************/
#ifndef IFTHEN_H              // Prevent multiple copies of this
                              //    include file
#define IFTHEN_H
#include <fuzzyvar.h>

/*******************************************************************
 * Fuzzy Class fcFuzzyProposition class definition
 *
 * This class describes generic fuzzy propositions.
 * Each fuzzy proposition points to one membership function.
 *******************************************************************/
class fcFuzzyProposition
        {
        public:
          // Constructors and destructor
          fcFuzzyProposition(fcMbf *NewMbf = 0);
          fcFuzzyProposition(istrstream& Rule,
                  const fcFuzzyVarList& FuzzyVars);
          fcFuzzyProposition(const fcFuzzyProposition& FromFuzzyProp);
          // Overloaded operators
          BOOL operator== (const fcFuzzyProposition& FuzzyProp) const;
          BOOL operator!= (const fcFuzzyProposition& FuzzyProp) const;
          // "Primary" methods, to retrieve/change class's attributes.
          fcMbf *GetMbf() const
                      {return (Mbf);}
          void  PutMbf (fcMbf *NewMbf)
                      {Mbf = NewMbf;}
        protected:
          fcMbf *Mbf;
          fcFuzzyVar *FuzzyVar;
        };

/*********************************************************************
 * Fuzzy Class fcIf class definition
 *
 * This class describes antecedents (IFs).   fcIf is derived publically
 * from fcFuzzyProposition, and  provides an evaluate method.
 *********************************************************************/
class fcIf : public fcFuzzyProposition
    {
        public:
```

281

Fuzzy Logic for Real World Design

```cpp
        // Constructors and destructor
        fcIf(fcMbf *NewMbf = 0)
                    : fcFuzzyProposition (NewMbf) {};
        fcIf(istrstream& Rule, const fcFuzzyVarList& InputVars)
                    : fcFuzzyProposition (Rule, InputVars) {};
        fcIf(const fcIf& FromIf)
                    : fcFuzzyProposition (FromIf) {};
        // Class's Methods
        float Evaluate();
    };

/************************************************************************
 * Fuzzy Class fcThen class definition
 *
 * This class describes fuzzy consequents (THENs).  fcThen is derived
 * publically from fcFuzzyProposition and has an Aggregate method.
 ************************************************************************/
class fcThen : public fcFuzzyProposition
    {
    public:
        // Constructors and destructor
        fcThen(fcMbf *NewMbf)
                        : fcFuzzyProposition (NewMbf) {};
        fcThen(istrstream& Rule, const fcFuzzyVarList& OutputVars)
                    : fcFuzzyProposition (Rule, OutputVars) {};
        fcThen(const fcThen& FromThen)
                    : fcFuzzyProposition (FromThen) {};
        // Class's methods
        void Aggregate(const float fStrength);
    };

/************************************************************************
 * Fuzzy Class fcIfSide (list of antecedents) class definition.
 *
 * Contains a list of antecedents (fcIfs). The list is an instance of
 * BORLAND's double linked list Template class. Each member of the list
 * is a pointer to one fcIf object.
 ************************************************************************/
class fcIfSide
    {
    friend class fcIfListIter;
    public:
        // Constructors and destructor
        fcIfSide() {};
        fcIfSide(istrstream& RuleStream, const fcFuzzyVarList&
InputVars);
        ~fcIfSide();
        // Overloaded operators
        fcIfSide& operator= (const fcIfSide& FromIfSide);
        int operator== (const fcIfSide& IfSide2) const;
        float Evaluate();
        // TIDoubleListImp Collection class wrapper functions
```

282

Chapter 9: Object-Oriented Implementation of a Fuzzy Application

```cpp
            int AddAtTail (fcIf *IfArg)
                        {return IfList.AddAtTail(IfArg);}
            void Flush (TShouldDelete::DeleteType DelType)
                        {IfList.Flush(DelType);}
            unsigned GetItemsInContainer() const
                        {return IfList.GetItemsInContainer();}
            int IsEmpty() const
                        {return IfList.IsEmpty();}
        private:
            TIDoubleListImp<fcIf> IfList;
        };

// fcIfSide list iterator
class fcIfListIter : public TIDoubleListIteratorImp<fcIf>
        {
        public:
            fcIfListIter(const fcIfSide& IfSide)
                    : TIDoubleListIteratorImp<fcIf>(IfSide.IfList) {}
        };

/************************************************************************
 * Fuzzy Class fcThenSide (list of consequents) class definition.
 *
 * Contains a list of consequents (fcThens). The list is an instance of
 * BORLAND's double linked list Template class. Each member of the list
 * is a pointer to one fcThen object.
 ************************************************************************/
class fcThenSide
        {
        friend class fcThenListIter;
        public:
            // Constructors and destructor
            fcThenSide() {};
            fcThenSide(istrstream& RuleStream, const fcFuzzyVarList&
OutputVars);
            ~fcThenSide();
            // Overloaded operators
            virtual fcThenSide& operator= (const fcThenSide& FromThenSide);
            int operator== (const fcThenSide& ThenSide2) const;
            void Aggregate ();
            void Aggregate(const float fStrength);
            // TIDoubleListImp Collection class wrapper functions.
            int AddAtTail (fcThen *ThenArg)
                        {return ThenList.AddAtTail(ThenArg);}
            void Flush (TShouldDelete::DeleteType DelType)
                        {ThenList.Flush(DelType);}
            unsigned GetItemsInContainer() const
                        {return ThenList.GetItemsInContainer();}
            int IsEmpty() const
                        {return ThenList.IsEmpty();}
        private:
            TIDoubleListImp<fcThen> ThenList;
```

283

```cpp
            };

    // fcThenSide list iterator
    class fcThenListIter : public TIDoubleListIteratorImp<fcThen>
            {
            public:
                fcThenListIter(const fcThenSide& ThenSide)
                        : TIDoubleListIteratorImp<fcThen>(ThenSide.ThenList) {}
            };
#endif   // IFTHEN_H
```

```cpp
/***********************************************************************
 *   IFTHEN.CPP: Implementation file for fuzzy proposition             *
 *               class fcFuzzyProposition, fuzzy antecedent-list       *
 *               class, fcIfSide, and fuzzy consequent-list            *
 *               class, fcThenSide.                                    *
 ***********************************************************************/
#include <ifthen.h>

/***********************************************************************
 * fcFuzzyProposition class's methods
 ***********************************************************************/

// Default constructor.  Mbf is a POINTER to an Mbf.
fcFuzzyProposition::fcFuzzyProposition (fcMbf *NewMbf)
        {
        Mbf = NewMbf;
        }

// Parse one IfThen component of a rule.  Throw away "IS" separator.
fcFuzzyProposition::fcFuzzyProposition (istrstream& Rule,
                                        const fcFuzzyVarList&
                                            FuzzyVars)
        {
        char szIS[10],   // "IS" string (throwaway)
                szMbfLabel[MAXLABELSIZE],
                szFuzzyVarName[MAXLABELSIZE];

        Rule >> szFuzzyVarName >> szIS >> szMbfLabel;  // FuzzyVarName
"IS" MbfLabel
        FuzzyVar = FuzzyVars.FindFuzzyVar(szFuzzyVarName);   // Find var
in list
        Mbf = FuzzyVar->FindMbf(szMbfLabel);  // Point to correct Mbf
        }

// Copy constructor.
fcFuzzyProposition::fcFuzzyProposition(const fcFuzzyProposition&
FromFuzzyProp)
        {
```

Chapter 9: Object-Oriented Implementation of a Fuzzy Application

```
            Mbf = FromFuzzyProp.Mbf;
        }

// Overloaded equality test operator.
BOOL fcFuzzyProposition::operator== (const fcFuzzyProposition& FuzzyProp)
const
        {
        if ( *Mbf == *(FuzzyProp.Mbf) )
           return TRUE;
        return FALSE;
        }

// Overloaded inequality test operator.
int fcFuzzyProposition::operator!= (const fcFuzzyProposition& FuzzyProp)
const
        {
        return !((*this) == FuzzyProp);
        }
/***********************************************************************
 * fcIf class's methods
 ***********************************************************************/

// Return degree of truth of crisp val for this antecedent.
float fcIf::Evaluate()
        {
        return Mbf->GetDegree();
        }
/***********************************************************************
 * fcThen class's methods
 ***********************************************************************/

// Apply inference/aggregation methods to consequent. fStrength is the
// rule's strength, which will be used for the PRODUCT-CORRELATION
// INFERENCE, which multiplies the consequent's Mbf's fArea by fStrengh.
void fcThen::Aggregate(const float fStrength)
        {
        Mbf->Aggregate(fStrength);
        }

/***********************************************************************
 * fcIfSide class's methods
 ***********************************************************************/

// Constructor function.  Constructs an IfSide or a ThenSide from a
// rule string.  All the rules exist already and are being read in
// from a file.  This routine will be called twice by an fcRule
// constructor, once for the IfSide, and again for the ThenSide (the
// second half of the rule).
fcIfSide::fcIfSide(istrstream& RuleStream,
                     const fcFuzzyVarList& InputVars)
```

285

Fuzzy Logic for Real World Design

```
            {
            char szSeparator[5];    // Separators "IF", "THEN", and "AND"
                                    //      (throwaways)
            RuleStream >> szSeparator;

            while (stricmp (szSeparator,"THEN"))
                    {
                    AddAtTail (new fcIf (RuleStream,InputVars));
                    RuleStream >> szSeparator;
                    }
    }

// Destructor.  Flush/delete list of antecedents/consequents.
fcIfSide::~fcIfSide()
        {
        Flush(TShouldDelete::Delete);
        }

// Overloaded assignment operator.  Can't simply do an element-by-element
// copy since the IfSides might be different lengths.  Instead,
// delete "this" IfSide and reconstruct it. WATCH OUT!! This releases
// all dynamic memory associated with "this" IfThen before assigning
// the new IfThen objects to it.
fcIfSide& fcIfSide::operator= (const fcIfSide& FromIfSide)
        {
        fcIfListIter iIfSide(FromIfSide);

        // Prevent assignment of object to itself
        if (this == &FromIfSide)
          return (*this);

        if (!IsEmpty ())
          Flush (TShouldDelete::Delete);
        for (iIfSide.Restart(); iIfSide; iIfSide++)
             AddAtTail (new fcIf (*iIfSide.Current()));
        return (*this);
        }

// Overloaded equality test operator. Member-by-member comparison. fcRule
// calls this from its overloaded equality operator function.
// Overloaded equality test operator.
int fcIfSide::operator== (const fcIfSide& IfSide2) const
        {
        if (GetItemsInContainer() != IfSide2.GetItemsInContainer())
          return (FALSE);

        fcIfListIter iIfSide1(*this), iIfSide2(IfSide2);

        for (iIfSide1.Restart(),iIfSide2.Restart(); iIfSide1,iIfSide2;
              iIfSide1++,iIfSide2++)
             {
             if (*(iIfSide1.Current()) != *(iIfSide2.Current()))
```

Chapter 9: Object-Oriented Implementation of a Fuzzy Application

```
                        {
                        return FALSE;
                        }
                  }
            return TRUE;
            }

// Evaluate an IfSide by multiplying together the degrees of truth
// of its Antecedents.  If one degree of truth is 0 then set
// strength to 0 & quit.  No need to multiply anything else by 0.
float fcIfSide::Evaluate()
            {
            float fDegree, fStrength;
            fcIfListIter iIfSide(*this);

            //Initialize strength to 1 so 1st multiplication is * 1
            fStrength = 1;
            for (iIfSide.Restart(); iIfSide; iIfSide++)
                  {
                  fDegree = iIfSide.Current()->GetMbf()->GetDegree();
                  if (fDegree == 0)
                     return (0);
                  fStrength *= fDegree;
                  }
            return (fStrength);
            }

/**********************************************************************
 * fcThenSide class's methods
 **********************************************************************/

// Constructor function.  Constructs an ThenSide or a ThenSide from a
// rule string.  All the rules exist already and are being read in
// from a file.  This routine will be called twice by an fcRule
// constructor, once for the ThenSide, and again for the ThenSide (the
// second half of the rule).
fcThenSide::fcThenSide(istrstream& RuleStream,
                  const fcFuzzyVarList& OutputVars)
            {
            char szSeparator[5];    // Keywords IS, and AND (throwaways)

            do
             {
             AddAtTail (new fcThen (RuleStream,OutputVars));
             RuleStream >> szSeparator;
             } while (strlen(szSeparator));  // While stream is not empty
            }

// Destructor.  Flush/delete list of antecedents/consequents.
fcThenSide::~fcThenSide()
            {
            Flush(TShouldDelete::Delete);
```

```cpp
        }

// Overloaded assignment operator.  Can't simply do an element-by-element
// copy since the ThenSides might be different lengths.  Instead,
// delete "this" ThenSide and reconstruct it. WATCH OUT!! This releases
// all dynamic memory associated with "this" object before assigning
// the new Then objects to it.
fcThenSide& fcThenSide::operator= (const fcThenSide& FromThenSide)
        {
        fcThenListIter iThenSide(FromThenSide);

        // Prevent assignment of object to itself
        if (this == &FromThenSide)
           return (*this);

        if (!IsEmpty ())
           Flush (TShouldDelete::Delete);
        for (iThenSide.Restart(); iThenSide; iThenSide++)
               AddAtTail (new fcThen (*iThenSide.Current()));
        return (*this);
        }

// Overloaded equality test operator.  Member-by-member comparison.
// fcRule calls this from its overloaded equality operator function.
int fcThenSide::operator== (const fcThenSide& ThenSide2) const
        {
        if (GetItemsInContainer() != ThenSide2.GetItemsInContainer())
           return (FALSE);

        fcThenListIter iThenSide1(*this), iThenSide2(ThenSide2);

        for (iThenSide1.Restart(),iThenSide2.Restart();
iThenSide1,iThenSide2;
                iThenSide1++,iThenSide2++)
                {
                if (*(iThenSide1.Current()) != *(iThenSide2.Current()))
                   {
                   return FALSE;
                   }
                }
        return TRUE;
        }

// Combined INFERENCE/AGGREGATION method.
void fcThenSide::Aggregate(const float fStrength)
        {
        fcThenListIter iThenSide(*this);

        for (iThenSide.Restart(); iThenSide; iThenSide++)
                iThenSide.Current()->Aggregate(fStrength);
        }
```

Chapter 9: Object-Oriented Implementation of a Fuzzy Application

```
/**********************************************************
 * FUZZYVAR.H: Header file for fuzzy variable class fcFuzzyVar *
 *             and fuzzy variable-list class fcFuzzyVarList.   *
 **********************************************************/
#ifndef FUZZYVAR_H       // Prevent multiple copies of this include file
#define FUZZYVAR_H
#include <mbf.h>

enum IOTYPE {nINPUTS,nOUTPUTS};    // Distinguish input/output fuzzyvars

/***************************************************************
 * Fuzzy Class fcFuzzyVar (fuzzy input or output var) class definition
 *
 * Contains domain, truth value axis, membership functions
 ***************************************************************/
class fcFuzzyVar
        {
        friend class fcMbfListIter;
        public:
          // Constructors and destructor
          fcFuzzyVar(const char *szNewName = "", const char *Units = "",
                  const float fNewMin = 0, const float fNewMax = 0);
          fcFuzzyVar(istrstream& InStream, ifstream& CfgFile);
          ~fcFuzzyVar();
          // Overloaded operators
          int operator== (const fcFuzzyVar& FuzzyVar) const;
          // "Primary" methods, to retrieve/change class's attributes
          char* GetName()
                     {return (szName);}
          float GetCrispVal() const
                     {return (fCrispVal);}
          // Class's methods
          virtual void Defuzzify();
          fcMbf *FindMbf(char *szMbfLabel);
          virtual void Fuzzify (float fCrispInput);
          // TIDoubleListImp Collection class wrapper functions
          int AddAtTail (fcMbf *Mbf)
                     {return MbfList.AddAtTail(Mbf);}
          void Flush (TShouldDelete::DeleteType DelType)
                     {MbfList.Flush(DelType);}
          unsigned GetItemsInContainer() const
                     {return MbfList.GetItemsInContainer();}
          int IsEmpty() const
                     {return MbfList.IsEmpty();}
        private:
          char szName[MAXLABELSIZE];   // Fuzzy var's name
          char szUnits[MAXLABELSIZE    // Unit of measure (degrees,
                                       //     inches, etc.)
          float fMin, fMax;            // Bounds of universe of discourse
```

289

```cpp
            TIDoubleListImp<fcMbf> MbfList;  // List of membership functions
            int NumOfMbfs;                   // How many Mbfs
            float fCrispVal;                 // Current crisp input/output
    };

// Iterator for fcFuzzyVars list of Mbfs, MbfList
class fcMbfListIter : public TIDoubleListIteratorImp<fcMbf>
    {
    public:
            fcMbfListIter(const fcFuzzyVar& FuzzyVar)
                    : TIDoubleListIteratorImp<fcMbf>(FuzzyVar.MbfList) {}
    };

/************************************************************************
 * Fuzzy Class fcFuzzyVarList (list of fuzzy input vars or list of fuzzy
 * output vars) class definition.
 *
 * Contains a list of Fuzzy propositions, which is an instance of
 * BORLAND's double linked list Template class. Each member of the list
 * is a pointer to a fcFuzzyVar object.
 ************************************************************************/
class fcFuzzyVarList
        {
            friend class fcFuzzyVarListIter;
        public:
            // Constructors and destructor
            fcFuzzyVarList() {};
            fcFuzzyVarList(char *szCfgFileName, IOTYPE nIOTYPE);
            ~fcFuzzyVarList();
            // Class's methods
            virtual istrstream Defuzzify();
            fcFuzzyVar *FindFuzzyVar(char *szFuzzyVarName) const;
            virtual void Fuzzify (istrstream& Tuple);
            // TIDoubleListImp Collection class wrapper functions
            int AddAtTail (fcFuzzyVar* FuzzyVar)
                                {return FuzzyVars.AddAtTail(FuzzyVar);}
            void Flush (TShouldDelete::DeleteType DelType)
                                {FuzzyVars.Flush(DelType);}
            unsigned GetItemsInContainer() const
                                {return FuzzyVars.GetItemsInContainer();}
        private:
            TIDoubleListImp<fcFuzzyVar> FuzzyVars;
        };

// Iterator class for fcFuzzyVarList
class fcFuzzyVarListIter : public TIDoubleListIteratorImp<fcFuzzyVar>
        {
        public:
            fcFuzzyVarListIter(const fcFuzzyVarList& FuzzyVarList)
                :TIDoubleListIteratorImp<fcFuzzyVar>(FuzzyVarList.FuzzyVars) {}
        };
#endif  // FUZZYVAR_H
```

Chapter 9: Object-Oriented Implementation of a Fuzzy Application

```cpp
/****************************************************************************
 * FUZZYVAR.CPP: Implementation file for fuzzy variable class               *
 *               fcFuzzyVar and fuzzy variable-list class fcFuzzyVarList.   *
 ****************************************************************************/
#include <fuzzyvar.h>

/**************************************
 * fcFuzzyVar class's methods
 **************************************/

// Default constructor.
fcFuzzyVar::fcFuzzyVar(const char* szNewName, const char* szNewUnits,
               const float fNewMin, const float fNewMax)
       {
       strcpy (szName,szNewName);
       strcpy (szUnits,szNewUnits);
       fMin = fNewMin;
       fMax = fNewMax;
       NumOfMbfs = 0;
       fCrispVal = 0;
       }

// Construct by reading fuzzy var's attributes from an input stream.
fcFuzzyVar::fcFuzzyVar (istrstream& InStream,ifstream& CfgFile)
       {
       InStream >> szName >> szUnits >> NumOfMbfs >> fMin >> fMax;
       for (int i = 0; i < NumOfMbfs; i++)
              AddAtTail (new fcMbf (CfgFile));
       fCrispVal = 0;
       }

// Destructor. Delete the MbfList.
fcFuzzyVar::~fcFuzzyVar()
       {
       MbfList.Flush(TShouldDelete::Delete);
       }

// Overloaded equality test operator.
int fcFuzzyVar::operator== (const fcFuzzyVar& FuzzyVar) const
       {
       // Lists don't even have same # of Mbfs so they can't be equal
       if (GetItemsInContainer() != FuzzyVar.GetItemsInContainer())
         return (FALSE);

       fcMbfListIter iMbfList1 (*this),
                     iMbfList2 (FuzzyVar);

       for (iMbfList1.Restart(),iMbfList2.Restart();
```

Fuzzy Logic for Real World Design

```
                iMbfList1,iMbfList2;
                iMbfList1++, iMbfList2++)
            {
            if (*(iMbfList1.Current()) != *(iMbfList2.Current()))
                {
                return (FALSE);
                }
            }
        if ( (strcmp (szName,FuzzyVar.szName))
                || (NumOfMbfs != FuzzyVar.NumOfMbfs)
                || (fMin != FuzzyVar.fMin)
                || (fMax != FuzzyVar.fMax) )
          {
          return FALSE;
          }
        return TRUE;
        }

// Search through list of Mbfs and return the one whose Label matches
// szMbfLabel.  Return 0 if no match.  This method is used by
// fcFuzzyProposition constructor to associate a fcFuzzyPropostion object
// with a membership function.
fcMbf *fcFuzzyVar::FindMbf(char *szMbfLabel)
        {
        fcMbfListIter iMbfList (*this);

        for (iMbfList.Restart(); iMbfList; iMbfList++)
            {
            if (! stricmp (iMbfList.Current()->GetLabel(),szMbfLabel))
                {
                return (iMbfList.Current());
                }
            }
        return (0);
        }

// Fuzzify a crisp input into this fuzzy variable.  This is done by
// determining the crisp value's degree of membership in each of the
// fuzzy var's membership functions.
void fcFuzzyVar::Fuzzify (float fCrispInput)
        {
        fcMbfListIter iMbfList (*this);

        // Is crisp input in universe of discourse?  If not, then put it
        //    there.
        if (fCrispInput < fMin)
          fCrispInput = fMin;
        if (fCrispInput > fMax)
          fCrispInput = fMax;
        for (iMbfList.Restart(); iMbfList; iMbfList++)
                iMbfList.Current()->Fuzzify(fCrispInput);
        }
```

Chapter 9: Object-Oriented Implementation of a Fuzzy Application

```
// Compute this fuzzy output variable's the final crisp output.
// CrispOutput equals SumOfProduct divided by SumOfAreas, as long as
// SumOfAreas is not 0.  If SumOfAreas is 0 then crisp output will just
// stay the same as it was after the previous execution of the fuzzy
// engine.  (SumOfAreas will be 0 if all rules in the rulebase had 0
// strength, that is, there no rules fired during rule evaluation).
void fcFuzzyVar::Defuzzify()
      {
      fcMbfListIter iMbfList (*this);
      float fSumOfAreas = 0,
            fSumOfProducts = 0;

      for (iMbfList.Restart(); iMbfList; iMbfList++)
             {
             fSumOfAreas += iMbfList.Current()->GetScaledArea();
             fSumOfProducts += iMbfList.Current()->GetScaledArea() *
                          iMbfList.Current()->GetCentroid();
             iMbfList.Current()->PutScaledArea (0);   // Reset Mbf's
                                                     //   scaled area
             }
      if (fSumOfAreas != 0)
        fCrispVal = fSumOfProducts / fSumOfAreas;
   }

/**********************************************************************
 * fcFuzzyVarList class's methods
 **********************************************************************/

// Construct a list of fuzzy vars.  Whether the list contains INPUT vars
// or OUTPUT vars depends on the value of nIOTYPE.  A list will never mix
// INPUT vars with OUTPUT vars.
fcFuzzyVarList::fcFuzzyVarList(char *szCfgFileName, IOTYPE nIOTYPE )
    {
         char szInputLine[MAXLINE], szIOType[10];     // Fuzzy var type-
                                                       //   Input or Output
         ifstream CfgFile(szCfgFileName,ios::in);

         while (CfgFile.getline(szInputLine,MAXLINE))
           {
           //Make an input stream (istrstream) from InputLine so we can use
           // '>>' to extract its fields one by one
           istrstream InputStream(szInputLine,strlen(szInputLine));
           InputStream >> szIOType;
           if (  ( !stricmp(szIOType,"input")  && (nIOTYPE == nINPUTS) )
                  ||( !stricmp (szIOType,"output") && (nIOTYPE ==
nOUTPUTS)) )
                  AddAtTail (new fcFuzzyVar (InputStream,CfgFile));
           }
         CfgFile.close();
         }
```

293

Fuzzy Logic for Real World Design

```
// Destructor.  Delete every fuzzy var object from the list.
fcFuzzyVarList::~fcFuzzyVarList()
      {
      Flush(TShouldDelete::Delete);
      }
// Have each fuzzy var in the list fuzzify its own crisp input. A "Tuple"
// is a set of crisp inputs. There is one crisp input for each fuzzy
// input variable.  For example, the tuple might contain the current
// temperature and humidity readings.
void fcFuzzyVarList::Fuzzify(istrstream& Tuple)
      {
      float fCrispVal;
      fcFuzzyVarListIter iInputVars(*this);

      for (iInputVars.Restart(); iInputVars; iInputVars++)
            {
            Tuple >> fCrispVal;
            iInputVars.Current()->Fuzzify(fCrispVal);
            }
      }

// Have each output fuzzy var compute its own crisp output.
istrstream fcFuzzyVarList::Defuzzify()
      {
      fcFuzzyVarListIter iOutputVars(*this);
      char Buffer[MAXLINE];
      ostrstream OutputTuple (Buffer,MAXLINE);

      for (iOutputVars.Restart(); iOutputVars; iOutputVars++)
                  {
                  iOutputVars.Current()->Defuzzify();
                  OutputTuple << iOutputVars.Current()->GetCrispVal();
                  }
      return (istrstream (Buffer));
      }

// Look through the list of fuzzy vars for the one whose name matches
// szVarName.  Return 0 if there is no match.
fcFuzzyVar *fcFuzzyVarList::FindFuzzyVar(char *szFuzzyVarName) const
      {
      fcFuzzyVarListIter iFzVarList(*this);

      for (iFzVarList.Restart(); iFzVarList; iFzVarList++)
                  {
                  if (!stricmp (iFzVarList.Current()-
>GetName(),szFuzzyVarName))
                        {
                        return (iFzVarList.Current());
                        }
                  }
      return (0);
      }
```

294

Chapter 9: Object-Oriented Implementation of a Fuzzy Application

```
/************************************************************
 * MBF.H: Header file for fuzzy membership function class    *
 *        fcMbf.                                             *
 ************************************************************/
#ifndef MBF_H            // prevent multiple copies of this include file
#define MBF_H

#include <values.h>      // Defines MAXFLOAT
#include <fstream.h>     // file streams classes
#include <deques.h>      // Container classes

const int MAXLINE      = 300;  // Max length of file input line
const int MAXLABELSIZE =  30;  // Max length of mbf's label
typedef int BOOL;

#ifndef TRUE
#define TRUE 1
#endif
#ifndef FALSE
#define FALSE 0
#endif

/*************************************************************************
 * Fuzzy Class fcMbf (membership function) class definition.
 *
 * Each membership function has a label, a shape (as defined by points
 * fP1 to fP4), slopes of its left and right sides, an area, a centroid,
 * and a degree of truth .
 *************************************************************************/
class fcMbf
      {
      public:
        // Constructors
        fcMbf(const char *szNewLabel = "", const float fNewPt1 = 0,
                    const float fNewPt2 = 0, const float fNewPt3 = 0,
                    const float fNewPt4 = 0);
        fcMbf(ifstream& CfgFile);
        // Overloaded operators
        fcMbf& operator= (const fcMbf& FromMbf);
        BOOL operator== (const fcMbf& Mbf2) const;
        BOOL operator!= (const fcMbf& Mbf2) const;
        // Class's methods
        void  Aggregate(const float fStrength);
        float CalcArea();
        float CalcCentroid();
        void  CalcSlopes();
        float Fuzzify(const float fCrispVal);
```

295

```
    // "Primary" methods, to retrieve/change class's attributes
        float GetCentroid() const {return (fCentroid);}
        float GetDegree() const
                {return (fDegree);}
        char *GetLabel()
                {return (szLabel);}
        float GetScaledArea() const
                {return fScaledArea;}
        void PutScaledArea (const float fNewScaledArea)
                {fScaledArea = fNewScaledArea;}
    private:
        char  szLabel[MAXLABELSIZE];  // Mbf's label
        float fP1, fP2, fP3, fP4;     // Points that comprise shape of
                                      //     Mbf
        float fSlope1, fSlope2;       // Slopes of sides of Mbf
        float fArea;                  // Area of Mbf
        float fScaledArea;            // Scaled area of Mbf,
                                      //     accumulated
                                      // during inference
        float fCentroid;              // Mbf's centroid
        float fDegree;                // Degree to which crisp input is
                                      //     member of Mbf
    };
#endif  // MBF_H
```

```
/************************************************************
 *  MBF.CPP: Implementation file for fuzzy membership function *
 *           class fcMbf.                                      *
 ************************************************************/
#include <mbf.h>

const float fMAXTRUTH = 1.0;

/************************************************************
 * fcMbf class's methods
 ************************************************************/

// Default Constructor function to compute shape of a membership
// function. Assigns szNewMbfName to MbfName.  Assigns fNewPt1 - fNewPt4
// to fPt1 - fPt4. Assigns nNewMbfNum to nMbfNum.   Initializes fDegree
// and computes centroid, area, and slopes.
fcMbf::fcMbf(const char* szNewLabel, const float fNewPt1, const float fNewPt2,
                const float fNewPt3, const float fNewPt4)
        {
        strcpy (szLabel, szNewLabel);
        fP1 = fNewPt1;
        fP2 = fNewPt2;
        fP3 = fNewPt3;
        fP4 = fNewPt4;
```

Chapter 9: Object-Oriented Implementation of a Fuzzy Application

```
      CalcSlopes();
      CalcCentroid();
      CalcArea();
      fDegree = 0;
      fScaledArea = 0;
      }

// Construct Mbf from a line in a config file.  Compute centroid, area,
// and slopes.
fcMbf::fcMbf(ifstream& CfgFile)
   {
   char szInputLine[MAXLINE];

   CfgFile.getline(szInputLine,MAXLINE);
   istrstream MbfStream(szInputLine,strlen(szInputLine));

   MbfStream >> szLabel >> fP1 >> fP2 >> fP3 >> fP4;
   CalcSlopes();
   CalcCentroid();
   CalcArea();
   fDegree = 0;
   fScaledArea = 0;
   }

// Overloaded assignment operator.
fcMbf& fcMbf::operator= (const fcMbf& FromMbf)
      {
      // Prevent assignment of object to itself
      if (this == &FromMbf)
         return (*this);

      fArea = FromMbf.fArea;
      fDegree = FromMbf.fDegree;
      fCentroid = FromMbf.fCentroid;
      strcpy (szLabel,FromMbf.szLabel);
      fP1 = FromMbf.fP1;
      fP2 = FromMbf.fP2;
      fP3 = FromMbf.fP3;
      fP4 = FromMbf.fP4;
      fSlope1 = FromMbf.fSlope1;
      fSlope2 = FromMbf.fSlope2;
      fScaledArea = FromMbf.fScaledArea;
      return (*this);
      }

// Overloaded equality test operator.
BOOL fcMbf::operator== (const fcMbf& Mbf2) const
   {
   if (  (!strcmp (szLabel,Mbf2.szLabel))
              &&(fP1 == Mbf2.fP1)
              &&(fP2 == Mbf2.fP2)
              &&(fP3 == Mbf2.fP3)
```

297

```
                    &&(fP4 == Mbf2.fP4) )
            return TRUE;
    return FALSE;
    }

// Overloaded inequality test operator.
BOOL fcMbf::operator!= (const fcMbf& Mbf2) const
   {
   return !((*this) == Mbf2);
   }

// Compute the area of a trapezoidal membership function. This
// calculation is done once, during Mbf construction and then retrieved
// during AGGREGATION. Area of a trapezoid = (top + base)/2) * height.
//(Height = 1). The trapezoid is really a triangle if fP2 = fP3.
float fcMbf::CalcArea()
        {
        float fBase = fP4 - fP1;
        float fTop = fP3 - fP2;
        fArea = (fTop + fBase)/2;
        return (fArea);
        }

// Calculate the Centroid of a trapezoidal Mbf.
// The centroid is calculated once, during Mbf construction,
// and then retrieved during ThenSide AGGREGATION.
float fcMbf::CalcCentroid()
        {
        fCentroid = ((fP4*fP4 + fP3*fP3 + fP4*fP3)
                                    - (fP2*fP2 + fP1*fP1 + fP2*fP1))
                                / ( 3 * (fP4 + fP3 - fP2 - fP1));
        return (fCentroid);
        }

// Calculate slopes from points P1-P4.  MAXFLOAT is used to
// approximate an infinite slope.
void fcMbf::CalcSlopes()
   {
   if (fP1 == fP2)
           fSlope1 = MAXFLOAT;
   else
           fSlope1 = fMAXTRUTH/(fP2-fP1);
   if (fP3 == fP4)
           fSlope2 = MAXFLOAT;
   else
           fSlope2 = fMAXTRUTH/(fP4-fP3);
   }

// Combined inference/aggregation.  Multiply Mbf's area by rule's
// strength (inference), and then accumulate the result into Mbf's scaled
// area (aggregation).
void fcMbf::Aggregate(const float fStrength)
```

Chapter 9: Object-Oriented Implementation of a Fuzzy Application

```
{
fScaledArea += fArea * fStrength;
}

// Calculate a crisp input's degree of membership in this membership
// function. This will be called by FuzzyVar.Fuzzify method during
// FUZZIFICATION.
float fcMbf::Fuzzify(const float fCrispVal)
        {
        float fDelta1 = fCrispVal - fP1;
        float fDelta2 = fP4 - fCrispVal;

        if ((fCrispVal < fP1) || (fCrispVal > fP4))
                fDegree = 0;
        else if ((fCrispVal <= fP3) && (fCrispVal >= fP2))
                fDegree = fMAXTRUTH;
        else if (fCrispVal < fP2)
                fDegree = fSlope1 * fDelta1;
        else if (fCrispVal > fP3)
                fDegree = fSlope2 * fDelta2;
        return (fDegree);
        }
```

Container Classes

Container classes have nothing in particular to do with fuzzy logic, but we discuss them here because they provide much of the framework for our implementation. We also don't want you to have to wonder about all those AddAtTail() and Flush() calls, and all the loops with strange iterators. Coverage of the full power and functionality offered by Borland's container class libraries is far beyond the scope of this chapter. We'll just try to give you enough of a taste to understand our code. To learn more you should refer to the Borland documentation and the books listed in the bibliography.

BIDS, or Borland International Data Structures, are Borland's container class libraries. The container classes are implemented using templates, a relatively new and extremely powerful addition to C++. Containers are objects that implement common data structures, like linked lists, binary trees, vectors, etc. They provide all of the member functions for adding and accessing a container's data elements, while hiding all of the details from the user. A container can hold a collection of any type of object, including complex, user-defined objects. In short, we can build data structures to contain any type of data that we want, without having to write a single line of data structure code.

Container classes are specifically designed to implement the multiplicity relationships that we discussed in Chapter 8. They are the "collection managers" that appear so frequently in our design. Recall that a collection manager is really just a data structure. The "collection" is the group of objects, like rules, or membership function. The "manager" is all the routines to build and traverse the collection.

We use containers to implement all of our collections. For example, recall from the design in Chapter 8 that a `Rulebase` has a collection of `Rule` objects. We implement a rulebase with the following class definition, found in RULE.H.

```
class fcRulebase
    {
    private:
        TIDoubleListImp<fcRule> Rules;
    };
```

`fcRulebase` contains the list of rules. `Rules` is a doubly-linked list of pointers to instances (objects) of `class fcRule`. The BIDS library offers a wide variety of data structures, including several varieties of linked lists. We selected a doubly-linked list because it gives us the flexibility to add items at both the head and tail of the list and we can traverse the list in either direction. `TIDoubleListImp` is Borland's indirect doubly-linked list container class. "Indirect" means that the list holds *pointers* to objects as opposed to containing the actual objects.

The syntax `<fcRule>` indicates that the pointers point to `fcRule` objects. Now we can use all of the doubly-linked list container class's list manipulation methods. We can do things like

```
AddAtTail (Rule)
```

to add a rule pointer object to the end of the rulebase, and

```
Flush(TShouldDelete::Delete)
```

to delete all the rules in the rulebase. `TShouldDelete::Delete` is the boolean value 1, which, when sent to the `Flush()` routine, permits `Flush()` to delete not only all the pointers in the container, but also the objects that they point to.

Collection Iterators

Container classes also provides us with a simple way to traverse the items in our collection. There is an iterator class designed especially for `TIDoubleListImp`, called `TIDoubleListIteratorImp`. To traverse an `fcRulebase` object we have derived an `fcRulebase`-specific iterator class, called `fcRuleListIter`, from `TIDoubleListIteratorImp`. We'll discuss the implementation of `fcRuleListIter` in more detail later in the chapter.

Whenever we want to traverse an `fcRulebase` we must create a variable of type `fcRuleListIter` and specify which rulebase we want to iterate. For example, the statement

```
fcRuleListIter iRulebase (*Rulebase)
```

creates iterator object `iRulebase` to traverse the rulebase pointed to by `Rulebase`. We can now call `iRulebase.Restart()` to start from the beginning of `Rulebase`. `iRulebase++` points us to the next item in the rulebase, and `iRulebase.Current()` returns a pointer to the current rule in the list.

Mapping Classes from the OO Design to C++

C++ provides substantial support for mapping an OO design to an implementation. To map a class to C++ we must do two things: we must translate the class structure to a corresponding C++ data structure, and we must provide source code for all of the class's methods. To this end we have provided two C++ source files for each class pictured in Figure 8.11 (see Chapter 8): a header file, which contains the class's C++ data structure, and an implementation file, which contains the code for the class's methods.

To translate a class from the design to a C++ class, each attribute in the class's design diagram maps to a private data member of the class, each method a public function prototype. Additionally, we provide the class with design-independent, C++-specific methods, namely constructors, a destructor, copy semantics, and any overloaded operators that it needs. The table in Figure 9.5 lists all the classes from Figure 8.11, their corresponding C++ class names, and their header and implementation file names.

CLASS	C++ CLASS	HEADER FILE	IMPLEMENTATION FILE
FuzzyApplication	fcFuzzyApplication	FUZZAPPL.H	FUZZAPPL.CPP
FuzzyEngine	fcFuzzyEngine	FUZZENGN.H	FUZZENGN.CPP
Rule	fcRule	RULE.H	RULE.CPP
Rulebase	fcRulebase	RULE.H	RULE.CPP
Membership Function	fcMbf	MBF.H	MBF.CPP
Fuzzy Variable	fcFuzzyVar	FUZZYVAR.H	FUZZYVAR.CPP
FuzzyVarList	fcFuzzyVar	FUZZYVAR.H	FUZZYVAR.CPP
FuzzyProposition	fcFuzzyProposition	IFTHEN.H	IFTHEN.CPP
If	fcIf	IFTHEN.H	IFTHEN.CPP
IfSide	fcIfSide	IFTHEN.H	IFTHEN.CPP
Then	fcThen	IFTHEN.H	IFTHEN.CPP
ThenSide	fcThenSide	IFTHEN.H	IFTHEN.CPP

Figure 9.5: Classes (see Figure 8.11) and their C++ source files.

Implementing Collection Managers

Where the design calls for a collection of objects of a particular class, we create the class's collection manager, actually a separate class, in the same header file. For example, you'll find class fcFuzzyProposition in IFTHEN.H. The header file also contains two collection managers fcIfSide and fcThenSide, each of which contains a collection of fcFuzzyPropostion objects.

Chapter 9: Object-Oriented Implementation of a Fuzzy Application

As specified in the design, each collection manager contains a list of objects. We implement this by including a container object in the collection manager's private data as follows:

```
class fcIfSide
  {
  .
  .
  private:
    TIDoubleListImp<fcIf> IfList;
  .
  .
  };
```

`fcIfSide` is the collection manager, and it contains a doubly-linked list of pointers to `fcIf` objects. The application must use the collection manager's public methods to access the container, a private data member. The collection manager provides access functions such as `AddAtTail()`, `Flush()`, and `PeekHead()`, which are merely public wrappers, for the container object's member functions of the same name.

Implementing Iterators

For each collection manager we must create an iterator class, which will allow us to traverse the objects in the collection manager's container. `fcIfSide`'s iterator class looks like the following:

```
class fcIfListIter : public
                     TIDoubleListIteratorImp<fcIf>
  {
  public:
    fcIfListIter(const fcIfSide& IfSide)
      :TIDoubleListIteratorImp<fcIf>(IfSide.IfList)
      {}
  };
```

This iterator class declaration is somewhat tricky and warrants some explanation. Remember that the purpose of an iterator is to allow us to traverse a list of objects. In this case we want to traverse `IfList`, `fcIfSide`'s list of `fcIf` objects so we have to create an iterator class specifically for `fcIfSide`. Ideally, we would like to declare the iterator class as follows:

```
fcIfListIter iIfList (IfSide.IfList)
```

`IfList`, however, is not publicly accessible to the application since it is a private data member of `fcIfSide`. So to give the application access to `IfList` we declare the iterator class as follows:

```
fcIfListIter iIfList (IfSide)
```

and rely on `fcIfListIter` to manage the details. Because of the way we define `fcIfListIter`, it can do this, since its constructor takes an `fcIfSide` object as its parameter. Notice, however, that when `fcIfListIter`'s constructor calls the parent constructor, it is really constructing an iterator for `IfSide`'s private `IfList` container:

```
TIDoubleListIteratorImp<fcIf>(IfSide.IfList) {}
```

In its class declaration, `fcIfSide` declares `fcIfListIter` as a `friend` to allow it access to `IfList`.

Construction of Objects

The previous few sections detailed how we mapped our design to C++ data structures. In the following sections we'll discuss the source code in more detail, paying particular attention to algorithms and C++-specific details like object construction.

In the Dynamic Design section of Chapter 8 we described how messages propagate through an object hierarchy. We used the example of copying a rulebase (refer to Figure 8.12). A rulebase doesn't explicitly copy each of its subcomponents. Rather, it passes the copy message to its subcomponents so they can copy themselves by passing the message to their subcomponents, etc. This is exactly how our fuzzy application object hierarchy is created. Construction is initiated in `main()` with the statement:

```
fcFuzzyAppl FuzzyAppl
       ("TEMPCNTL.CFG","TEMPCNTL.RB");
```

Chapter 9: Object-Oriented Implementation of a Fuzzy Application

and propagates through the entire hierarchy via C++ constructors.

Fuzzy Application and Fuzzy Engine Construction

Main() creates a fuzzy application object by calling constructor fcFuzzyAppl() with the config and rulebase files as parameters. fcFuzzyAppl() calls fcFuzzyEngine() with the same parameters to construct the fuzzy controller. fcFuzzyEngine() in turn calls constructors to create the lists of input and output fuzzy variables and the rulebase. The next few sections detail how fuzzy variables and rules are constructed.

Fuzzy Variables Construction

Earlier in the chapter we described the configuration file, TEMPCNTL.CFG, which contains the text for our fuzzy system's input and output fuzzy variables and membership function. As specified in the design, fcFuzzyEngine() will have to create instances of fcFuzzyVarList, one for input fuzzy variables and one for output variables.

```
InputVars = new
        fcFuzzyVarList("TEMPCNTL.CFG",INPUTS);
OutputVars = new
        fcFuzzyVarList("TEMPCNTL.CFG",OUTPUTS);
```

fcFuzzyEngine() passes the config file's name as a parameter to fcFuzzyVarList's constructor. The second parameter, an enumerated type, tells fcFuzzyVarList() which type of list to construct.

FuzzyVarList Construction

fcFuzzyVarList's constructor processes the entire config file each time it is called, constructing only inputs or only outputs, depending upon the value of the second parameter, nIOTYPE, an enumerated type whose value can be either INPUTS or OUTPUTS. fcFuzzyVarList() processes the file, looking for fuzzy variables of type nIOTYPE. It reads lines sequentially from the file into a character buffer, and converts the buffer to an input stream:

```
istrstream
    InputStream(szInputLine,strlen(szInputLine));
```

so its contents can be extracted with the input operator ">>". `fcFuzzyVarList()` extracts the first field from the stream:

```
InputStream >> szIOType;
```

and determines whether it matches `nIOTYPE`. If it does then a fuzzy variable of the correct type has been found, and the constructor calls `fcFuzzyVar()` to construct a fuzzy variable object, and adds it to the collection. Notice how we combine construction of the fuzzy variable and addition of its pointer to the collection into one statement:

```
AddAtTail (new fcFuzzyVar (InputStream, CfgFile));
```

If `szIOTYPE` does not correspond to `nIOTYPE`, the constructor will ignore the line and continue to the next one. It will continue reading lines until it finds one whose first field corresponds with `nIOTYPE`. This way, `fcFuzzyVarList()` effectively ignores fuzzy variables of the other type.

FuzzyVar Construction

`fcFuzzyVar()` receives two parameters: `InputStream` contains the information about the fuzzy variable, and `CfgFile` is a handle to the rest of the config file. The fuzzy variable constructor, described below, will extract membership function definitions from the next few lines of the config file.

`fcFuzzyVar()` extracts private data members `szName`, `szUnits`, `NumOfMbfs`, `fMin`, and `fMax` from `InputStream`. It constructs one `fcMbf` instance for each membership function.

```
AddAtTail (new fcMbf (i,CfgFile));
```

NOTE: It is important that there be exactly `NumOfMbfs` membership function definitions following the fuzzy variable definition in the config file. Otherwise fcFuzzyVar() constructor will read an incorrect number of lines from the file.

Mbf Construction

fcMbf() reads one line from CfgFile, converts it to an input stream, and extracts the membership function's attributes from it. From points fP1, fP2, fP3, and fP4, which describe the membership function's shape, fcMbf() calculates fSlope1 (the slope of the left side) and fSlope2 (the slope of the right side), which will be used later by fcMbf::Fuzzify(). It also calculates the membership function's area, fArea, which will be used later by fcFuzzyVar::Defuzzify().

Figure 9.6 shows how InputVars and OutputVars will look after construction.

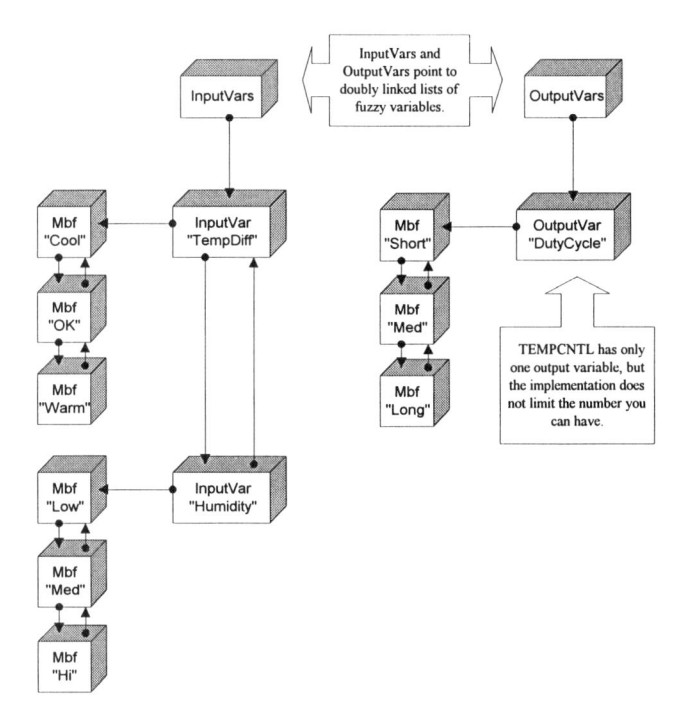

Figure 9.6 FuzzyEngine.InputVars and FuzzyEngine.OutputVars after construction.

Rulebase Construction

`FuzzyEngine`'s constructor creates `Rulebase` by calling its constructor:

    `fcRulebase("TEMPCNTL.RB", InputVars, OutputVars);`

`fcRulebase()` constructs one rule for each line in the rulebase file. Rulebase construction works similarly to fuzzy variable list construction: That is, `fcRulebase()` does not explicitly construct every rule. As it reads a line from the file it puts the line of text into an input stream, and then calls `fcRule(RuleStream, InputVars, OutputVars)` to construct the `Rule` object. The resulting `Rule` pointer is added to the tail of the rulebase.

Rule Construction

`fcRule()` constructs its `IfSide` and `ThenSide`:

    `IfSide = new fcIfSide (RuleStream, InputVars);`
    `ThenSide = new fcThenSide (RuleStream, OutputVars);`

`fcThenSide(RuleStream, OutputVars)` works identically to `fcIfSide(RuleStream, InputVars)`, except that when `fcRule()` calls `fcThenSide()`, it passes `OutputVars` as the second parameter instead of `InputVars`. Also, after `IfSide` construction, `RuleStream` no longer contains any antecedents, as they were all read from the stream during `IfSide` construction, as you'll see below.

IfSide Construction

Figure 9.7 shows what `RuleStream` looks like at the start of `fcIfSide`. `fcIfSide` immediately strips the rule's leading "IF" from `RuleStream`. Next it continues to construct antecedents and strip separators ("AND"s) until it encounters the rule's "THEN" separator. Each call to `fcIf()` results in one antecedent being stripped from `RuleStream`. At the completion of `fcIfSide()` construction, only the consequent fuzzy propositions remain, as illustrated in Figure 9.7.

Chapter 9: Object-Oriented Implementation of a Fuzzy Application

RuleStream at start of fcIfSide().

IF TempDiff IS Warm AND Humidity IS High THEN DutyCycle IS Long

TempDiff IS Warm AND Humidity IS High THEN DutyCycle IS Long

RuleStream after leading "IF" has been stripped.

AND Humidity IS High THEN DutyCycle IS Long

RuleStream after first call to fcIf.

DutyCycle IS Long

RuleStream after fcIfSide construction has completed. Only consequent fuzzy propositions remain.

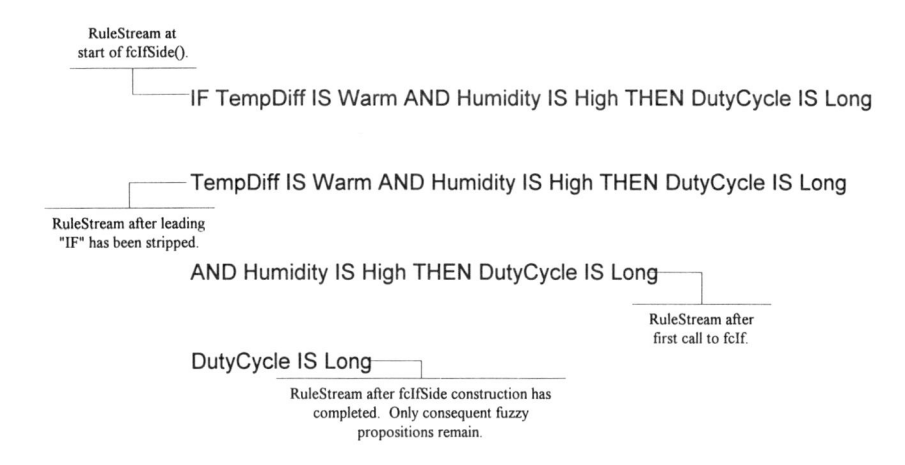

Figure 9.7: How `fcIfSide()` processes `RuleStream`.

ThenSide Construction

Before the call to `fcIfSide()`, `RuleStream` contained the text for the entire rule, and successive calls to `fcIf()` stripped off the antecedents, as well as the "THEN" field, leaving only the consequents. `fcThenSide` constructs consequents via calls to `fcThen()` and strips separators until `RuleStream` is empty.

While TEMPCNTL's rules each contain only one consequent, the *DutyCycle* output, `fcThenSide()` can support multiple consequents. It simply continues to construct consequents until it reads a zero-length string into `szSeparator`, signifying an empty `RuleStream`.

fcFuzzyProposition Construction

All fuzzy propositions, whether they be antecedents or consequents are constructed identically. Constructors `fcIf` and `fcThen` take two parameters: `RuleStream` and a list of input or output fuzzy variables. They each call parent class constructor `fcFuzzyProposition` to construct the antecedent or consequent object.

309

`fcFuzzyProposition` reads one complete antecedent from `RuleStream`.

```
RuleStream >> szFuzzyVarName   >> szIS >> szMbfLabel;
```

`szFuzzyVarName` is used to locate the pointer to the correct fuzzy variable in `FuzzyVarList`.

```
FuzzyVar =
    FuzzyVarList->FindFuzzyVar(szFuzzyVarName);
```

Once we have the `FuzzyVar` pointer we can use `szMbfLabel` to locate the correct membership function pointer.

```
Mbf = FuzzyVar->FindMbf(szMbfLabel);
```

The Relationship Between Rules and Membership Functions

Figure 9.8 shows `Rulebase` after construction. Pay particular attention to the rules' relationship to the fuzzy variables. There is one collection of membership functions for each `FuzzyVar`, and the rules' antecedents and consequents contain *pointers* to their `Mbfs`.

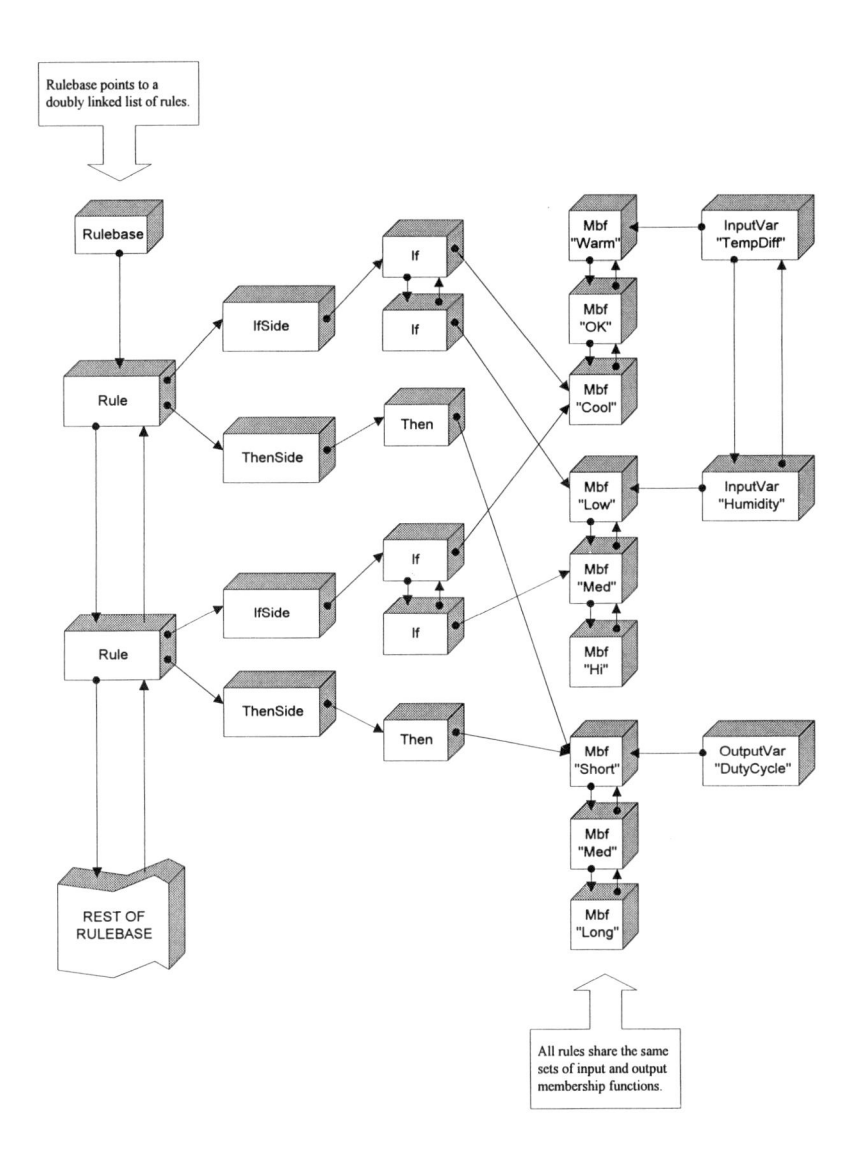

Chapter 9: Object-Oriented Implementation of a Fuzzy Application

Figure 9.8 `FuzzyEngine.Rulebase` *after construction. Two rules are expanded to demonstrate that rules share (point to) the same sets of input and output membership functions in memory.*

311

Application Control Flow

The Control Loop

`TEMPCNTL` enters the fuzzy application's control loop by calling:

```
FuzzyAppl.ControlLoop();
```

Recall from the design that the control loop consists of three steps: get a crisp input tuple, pass the tuple to the fuzzy engine for processing, and then apply (in this case display on the screen) the output from the fuzzy engine. Figure 8.13 from Chapter 8 translates to the following code:

```
while (TRUE)
   {
   if (!GetCrispInputs (TempSetpoint))
       return;
   istrstream OutputTuple =
       FuzzEngn->Run (InputTuple);
   DisplayOutputs (OutputTuple);
   }
```

`GetCrispInputs()` consists of some simple I/O requesting the current temperature and humidity readings from the user. The function uses `TempSetpoint` to calculate `TempDiff`, that is, the difference between the current temperature and the set point. `GetCrispInputs()` constructs input stream `InputTuple`, which is stored as an attribute of `FuzzyAppl`. If the user enters a 0 for the current temperature `GetCrispInputs()` will return `FALSE`, resulting in termination of the program.

The real work is initiated by the next function call in `fcFuzzyAppl::ControlLoop()`:

```
istrstream OutputTuple = FuzzEngn->Run (InputTuple);
```

This statement passes `InputTuple` to `FuzzEngn`, which processes it and returns the outputs in `OutputTuple`.

Running the Fuzzy Engine

We execute the fuzzy engine once each time through the control loop. The process described below comprises the intelligence of fuzzy logic. It is this process that is responsible for processing an input tuple (a current temperature and humidity reading) to produce an output tuple (comprised of only one crisp value, the duty cycle).

Before reading this section it is important that you completely understand Figure 9.8. It is most important to remember that the actual membership functions are constructed and managed by the fuzzy variables, and all the rules share them.

Fuzzy engine execution is initiated in

`fcFuzzyAppl::ControlLoop()` with the command:

`istrstream OutputTuple = FuzzEngn->Run (InputTuple);`

We elected to make `InputTuple` and `OutputTuple` data streams so that we can easily extract their contents with `operator>>`.

Fuzzify InputVars

Fuzzification computes the value of each input membership function's `fDegree` attribute. `fcFuzzyEngine::Run()` initiates fuzzification.

`InputVars.Fuzzify(InputVars, CrispInputs);`

The fuzzify message passes through the hierarchy until it reaches each input membership function object. There is one crisp input for each fuzzy variable, so for each fuzzy variable in the list, `InputVars.Fuzzify()` strips one crisp input from the `CrispInputs` tuple and passes it as a parameter to `FuzzyVar.Fuzzify()`. For example, the first value in `CrispInputs` is the `TempDiff` calculated by `FuzzyAppl.GetCrispInputs()`.

`FuzzyVar.Fuzzify()` tests the crisp input to determine whether it is within the universe of discourse, that is, whether its value lies between `FuzzyVar.fMin` and `FuzzyVar.fMax`. If it does not then it must be clipped to the universe of discourse as shown in Figure 9.9.

Fuzzy Logic for Real World Design

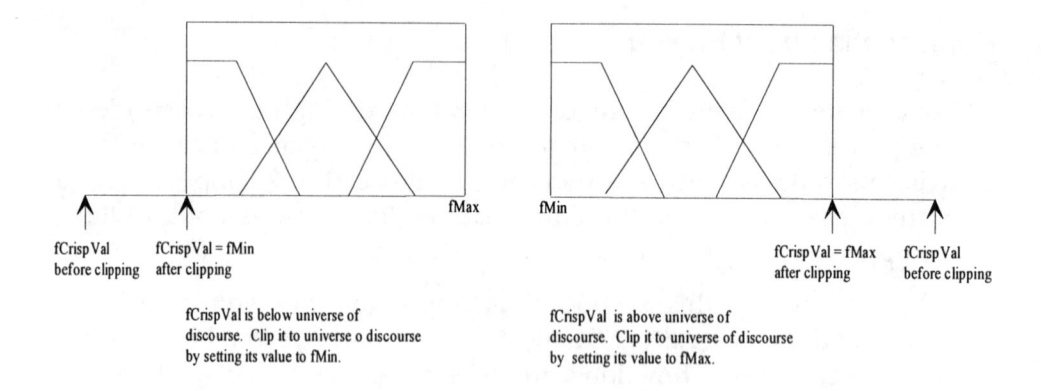

Figure 9.9: Clipping out-of-range crisp values to the universe of discourse.

`FuzzyVar.Fuzzify()` calls `Mbf.Fuzzify()` for each membership function in the collection. `Mbf.Fuzzify()` calculates `fCrispInput`'s degree of membership and stores it as private attribute `Mbf.fDegree`.

Calculating Degree of Truth

`Mbf.Fuzzify()` performs a series of tests, illustrated in Figure 9.10, to determine `fCrispVal`'s degree of truth.

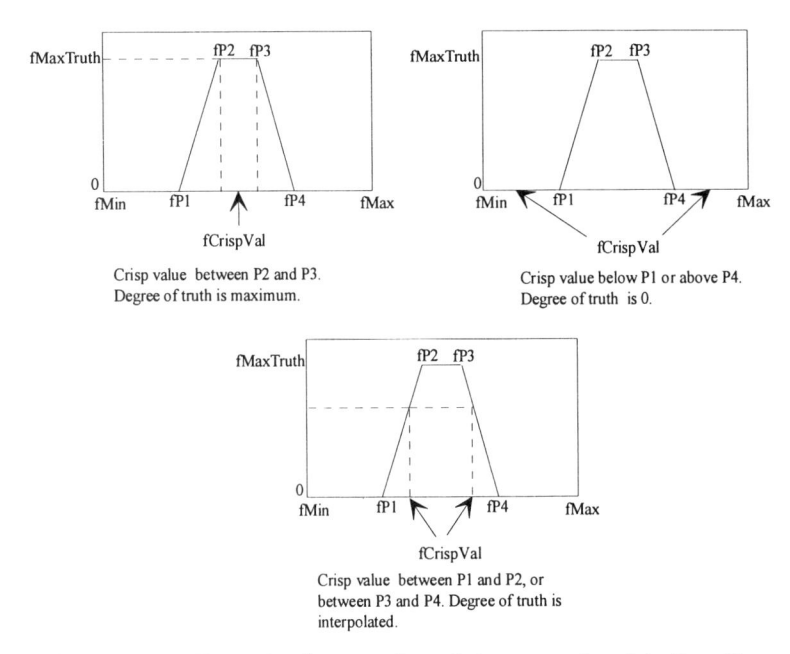

Chapter 9: Object-Oriented Implementation of a Fuzzy Application

Figure 9.10: Tests for degree of truth in a membership function.

First it checks to see whether `fCrispVal` is between `fP2` and `fP3`. If it is then `fDegree` is assigned the value of `fMaxTruth`. If the membership function is triangular (that is, if `fP2` equals `fP3`) this test will pass only if `fCrispVal` equals `fP2` equals `fP3`.

Next, `Mbf.Fuzzify()` checks to see whether `fCrispVal` is outside its bounds, that is, if it is less than `fP1` or greater than `fP4`. If so, then `fCrispVal` has no membership in `Mbf` and `fDegree` is assigned a value of 0.

Finally. if `fCrispVal` is within `Mbf`, but not between `fP2` and `fP3`, then it falls into the third category and `fDegree` must be interpolated. Figure 9.11 illustrates this concept.

315

Fuzzy Logic for Real World Design

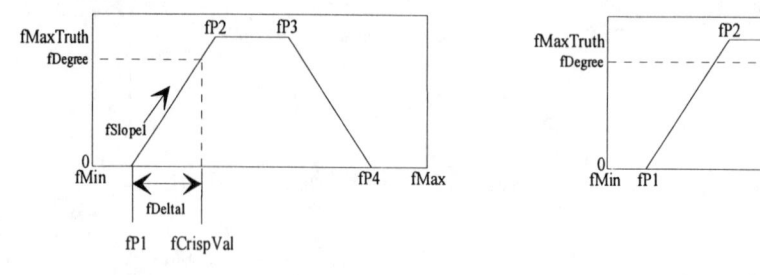

Figure 9.11: Calculating degree of membership using interpolation method. Recall that `fSlope1` *and* `fSlope2` *are private data members of* `fcMbf`.

At the end of fuzzification every membership function has calculated and stored as a private data member the degree of membership of the crisp input.

Evaluate Rules

Rulebase evaluation involves calculating the strength of every rule. A rule's strength is the result of evaluating its antecedents. Recall from Chapter 3 that there are many ways to evaluate rules, depending upon how you choose to interpret the AND operator. We implement "product" evaluation, which means that `IfSide` evaluates itself by *multiplying together* the `fDegree` values of each of its antecedents' `Mbf`s (Remember that an `If` object's `Mbf` is actually a pointer to an `Mbf` object in an input `FuzzyVar`. Refer to the diagram in Figure 9.8). In the previous step we calculated `fDegree` for every `Mbf`, so in this step `IfSide.Evaluate()` need only retrieve `fDegree`:

```
fDegree = iIfSide.Current()->GetMbf()->GetDegree();
```

and multiply it by the accumulated `fStrength`:

```
fStrength *= fDegree;
```

Chapter 9: Object-Oriented Implementation of a Fuzzy Application

The final result is returned to the rule, which stores it in data member fStrength.

Aggregate Rules

This method combines inference with aggregation. We have chosen "correlation product" inference, and "additive" aggregation. We gain computational efficiency here by aggregating only the rules that "fired" during evaluation. Those rules will have non-zero fStrengths. The gain can be significant in large systems where the fuzzy engine is run many times in succession. For example, **FuzzyLab**'s rulebase contains twenty-five rules. At most four rules will fire each control loop pass. Eliminating the other twenty-one rules from the aggregation process avoids useless calculations which could slow down the simulation.

Rules pass the job of inference/aggregation to their consequents' membership functions as follows:

```
Mbf->Aggregate(fStrength);
```

The membership function multiplies its area by fStrength. That's the inference. It then does the aggregation step by adding the resulting value into its total scaled area, fScaledArea. The whole process is actually very simple and is completed in just one line of code:

```
fScaledArea += fArea * fStrength;
```

The result, after aggregating the entire rulebase, is an aggregated scaled area for each output membership function.

Defuzzify OutputVars

Defuzzification results in one crisp value, fCrispVal, for each output fuzzy variable. We use "centroid" defuzzification. OutputVar.Defuzzify() uses two accumulator variables: fSumOfProducts and fSumOfAreas. The function loops through the variable's membership functions accumulating the two variables. Each membership function adds its fScaledArea to fSumOfAreas. It adds the result of its fScaledArea times its fCentroid to fSumOfProducts. Recall that as we constructed each Mbf object we also calculated its fArea and fCentroid data members. That means that all we have to do here is retrieve those values, saving ourselves from costly and redundant calculations.

As the loop processes each membership function, it also resets the membership function's fScaledArea to 0. The final fCrispVal is the result of dividing fSumOfProducts by fSumOfAreas.

How to Build Your Own Fuzzy Application Using Our Code

It is a very simple matter to build your own fuzzy application using the classes and objects that we have supplied. You do not have to make any changes whatsoever to the fcFuzzyEngine class hierarchy. The fuzzy logic intelligence is completely encapsulated in these classes. In fact, your changes need only affect FUZZAPPL.CPP, and, possibly, FUZZAPPL.H

Your only responsibilities will be to customize the input routine to gather inputs into a tuple and pass the tuple to the fuzzy engine for processing via one function call. The function will return an output tuple, from which you can extract and apply outputs to your system.

Create a Rulebase

Create a text file for your rulebase in the format shown in Figure 9.4.

Create Fuzzy Variables and Membership Functions

Create a text file for your fuzzy vars and membership functions in the format shown in Figure 9.2.

Chapter 9: Object-Oriented Implementation of a Fuzzy Application

Create application object with your rulebase and fuzzy variables.

You might start this step by copying and renaming `FUZZAPPL.CPP` and `FUZZAPPL.H`, since these are the two files that you will have to edit to customize your application*.

In `main()` change the line:

```
fcFuzzyAppl FuzzyAppl ("TEMPCNTL.CFG",
    "TEMPCNTL.RB");
```

to replace our filenames with the names of your own rulebase and fuzzy vars text files.

Customize I/O routines.

Customize `fcFuzzyAppl::GetInputs()` to retrieve inputs. Instead of collecting inputs from the keyboard, your application, for instance, might retrieve its inputs from a data file or database, or collect them from some mechanical device or from a simulation like **FuzzyLab** does.

Revise `fcFuzzyAppl::ApplyOutputs()` to report or apply outputs. For example, you may want to store outputs in a file or apply them to a mechanical system.

Add any other Customization

Change `fcFuzzyAppl`'s constructor to handle any initialization that your program may require. For example, ours requires the user to enter `TempSetpoint`. Also, add any necessary data or methods to

---

* Another approach would be to make `fcFuzzyAppl` a generic base class that encapsulates data and behavior that is common to all applications which use fuzzy logic. For example, all applications would want to embed the fuzzy engine, so the `fcFuzzyAppl` would have a `fcFuzzyEngine` member. You could then derive specific application objects from the base. In our case we could have declared `class fcTempCntlAppl : public fcFuzzyAppl`.

`fcFuzzyAppl`. We added private data member `TempSetpoint`. We did not require any additional methods.

Compile and Link

While your changes will affect only one or two files, you must also compile and link all of the fuzzy engine code, found in files `FUZZENGN.CPP`, `RULE.CPP`, `FUZZYVAR.CPP`, `MBF.CPP`, and `IFTHEN.CPP`.

You can find the C++ code listed in this chapter on your companion disk. The source on the disk is the same as the code presented here. We have also provided a `Borland C++ 4.0` project file `TEMPCNTL.IDE`, to which you should have to make at most minimal changes to accommodate your particular environment.

Enhancing our Code

The code presented in this chapter provides you with everything that you need to begin experimenting with your own fuzzy logic applications. However, space limitations prevented us from presenting some important features that we normally include in all of our own products. Below are some suggestions for how you can enhance our code.

Provide additional error checking

This would be especially useful during rule and fuzzy variable parsing to determine whether membership functions are trapezoidal, and rules and fuzzy variables are defined consistently. For example, you would want to handle the case in which a rule in the rules file contains a fuzzy proposition whose fuzzy variable or membership function components are not defined by the configuration file.

Add overloaded operators and copy semantics

We provided copy semantics and overloaded operators only where necessary. C++ classes, especially those which allocate memory dynamically, should have a copy constructor and an assignment operator. In our own products we also provide classes with output operator, Operator<<, so they can be printed to an output stream (like cout, for example).

Summary

This chapter has presented one possible implementation of the object-oriented design in Chapter 8. Specifically we showed how object-oriented classes translate to C++ data structures and code. While this implementation could just as easily have been done in Smalltalk, C, Object Pascal, Assembly Language or any other programming language, we have selected C++ because of its effeciency, suitability to object-oriented concepts, and its wide-ranging commercial support.

We encourage you to experiment with the code on the companion disk. You can use it in its current form or allow it to provide you with guidelines for creating your own implementation in whichever language you choose.

Chapter 10:
FuzzyLab Primer

FuzzyLab is an interactive program featuring a fuzzy control system and animated simulation written with the enthusiastic beginner in mind. The user is free to play with both the elements of fuzzy control and the simulated system under control to explore the fuzzy logic paradigm. **FuzzyLab** features high interactivity and rapid, multidimensional, graphical feedback so the user can see immediately the effects of their actions. **FuzzyLab** is written for the MS Windows operating system running on a PC compatible. The compressed files may be installed on your computer from the companion disk located inside the back cover of this book.

Installation

From MS Windows Program Manager select **File | Run A:\SETUP.EXE**. The setup program will lead you through a full or custom installation of the software on the companion disk.

Getting Started

FuzzyLab is designed to be your laboratory for exploring fuzzy logic. Your explorations will revolve around the fuzzy control of an "inverted-pendulum." We've provided an initial set of fuzzy rules and simulation parameters so that you can get started right away.

When you're ready you can make changes to the rules and parameters. The animated simulation provides you with feedback in a variety of ways so that you can learn how individual rules and parameters affect the controller's performance.

An inverted-pendulum consists of a "mass" and a "length". It is connected to a "base" around which it can rotate freely. The base contains a motor, which is used to drive the pendulum to the left or right.

An inverted pendulum is naturally unstable. Its tendency is to fall over to the left or right, rather than remain upright. It's like trying to balance a sharpened pencil on the end of your finger. You wind up moving your hand, and, if you're intent enough, your entire body, in a dozen different directions to compensate for the pencil's motion, and invariably you lose the battle anyway.

The controller's job is to balance the inverted-pendulum, that is, to keep it upright. If the pendulum is bopped to the left or right, the controller must make sure that it returns to and remains in its upright position. It does this by applying voltage to the motor, thus driving the Pendulum in either a positive (to the right) or negative (to the left) direction. Exactly how and when to apply voltage will be determined by the fuzzy rules that you write.

Before you begin playing with FuzzyLab we suggest you review the following sections:

What's on the Screen explains what is in each window on the screen..

Running the Simulation describes what happens in each window while the simulation is running.

Editing the Pendulum explains how to use the Pendulum Editor to edit the pendulum's mass and length, and how to adjust the pendulum controller's control loop.

Editing the Rulebase explains how to use the Rules Editor to add, delete, and replace rules in the rulebase.

Chapter 10: FuzzyLab Primer

Strategies for Exploring FuzzyLab details the steps you should take to get the most from FuzzyLab.

How the Pendulum Controller Works describes both how a real-life inverted pendulum works and how FuzzyLab simulates one graphically.

What's On the Screen

A description of each of the elements in **FuzzyLab**'s main window follows Figure 10.1.

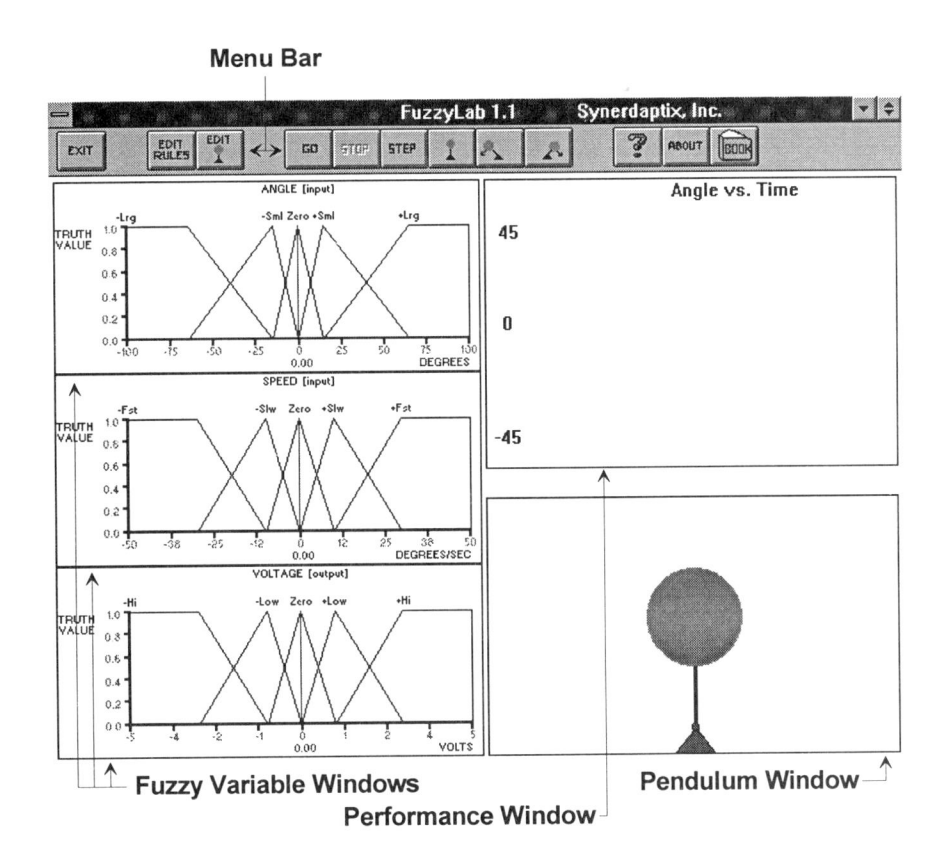

*Figure 10.1: The main window of **FuzzyLab** displaying the menu bar, pendulum window, performance window, and fuzzy variable windows.*

325

Menu Bar

At the top of the screen, below the title bar, is a row of menu buttons. You'll use the buttons to start, stop, and control the pendulum simulation, and to access the Pendulum Editor and Rules Editor. The function of each menu button is described below.

EDIT RULES

Bring up the Rules Editor to add, delete and replace rules in the rulebase.

EDIT

Bring up the Pendulum Editor to change the mass and length of the pendulum, and to change the length of the control loop.

GO

Start the simulation. Bops the pendulum once to the right to set it in motion. See Running the Simulation.

STEP

Run one "step" of the simulation. See How the Pendulum Controller Works.

STOP

Stop the simulation.

Resets Angle, Speed, and Voltage to 0. This can be done whether or not the simulation is running.

Bop the pendulum once to the left. This is equivalent to applying an instantaneous acceleration to the left. This is especially helpful in testing how well your fuzzy model corrects disturbances to the pendulum. This can be done whether or not the simulation is running

Bop the pendulum once to the right. This is equivalent to applying an instantaneous acceleration to the right. This is especially helpful in testing how well your fuzzy model corrects disturbances to the pendulum. This can be done whether or not the simulation is running.

Hypertext help is instantly available at any time the ? icon is displayed.

Fuzzy Variable Windows

On the left side of the screen are three windows which contain the fuzzy variables Angle, Speed, and Voltage. The fuzzy variables are the inputs (Angle and Speed) and output (Voltage) for the fuzzy model. Each fuzzy variable graph contains the following information:

Fuzzy Variable's Name: Appears at the top of the window. In brackets next to it will be the word "input" or "output" indicating which type of fuzzy variable it is.

Universe of Discourse: Represented by the x-axis. This is the range of crisp values over which the membership functions of a particular fuzzy variable are defined. Also, the units pertaining to a particular crisp input or output are located in the bottom right corner of the Fuzzy Variable Window.

Current Crisp Value: Appears in two ways: below the x-axis, as a number highlighted in red; and as a marker (a vertical red line)

on the graph. Both the marker and the number will be updated continually while the simulation is running.

Membership Functions: Each fuzzy variable is partitioned into five membership functions, which span the universe of discourse. A label is printed at the top of each membership function. The number of membership functions, five, was chosen as a typical starting point. Real-world fuzzy systems may use more or fewer membership functions, as the needs of the particular application dictate.

Truth Value: Represented by the y-axis. A truth value of 1 is the highest degree of truth in this fuzzy system, signifying complete membership in a fuzzy set.

Performance Window

At the top, right of the screen, below the menu bar, is a Performance Window, which graphs the pendulum angle versus time. It is updated continuously while the simulation is running. Once the simulation is started, a red line, indicating the past and present value of the pendulum angle, will continuously sweep across the Performance Window from left to right. When the red trace reaches the right edge of the window, the trace restarts at the left edge, ready for another sweep.

Pendulum Window

At the bottom right is the Pendulum Window, which contains the pendulum that will swing back and forth while the simulation is running. The pendulum consists of a mass (the green ball), a length (the black line), and a base (the gray triangle). On a typical computer, the pendulum window gets updated at a rate of about 20 frames per second, which produces a convincing animation.

Running the Simulation

When you hit GO on the menu bar the simulation begins and you will see activity in all of the windows on the screen.

Fuzzy Variable Windows

The markers and numeric displays will continue to update as the crisp values change. Note that while a new angle and speed are produced with each step of the simulation, voltage is an output of the fuzzy engine and thus will be updated only after the fuzzy model is run (see How the Pendulum Controller Works).

Don't worry if the Speed Window's marker sometimes disappears from the window. This happens when the pendulum is moving at a speed that exceeds the Speed's universe of discourse. The fuzzy engine will simply clip the out-of-range values to the nearest membership function (e.g., in the case of *Angle*, to *+Lrg* or *-Lrg*).

IMPORTANT! When no rules match a particular set of crisp inputs, the fuzzy engine will leave the output voltage unchanged. Watch for this condition as you begin to delete rules as it will often explain why the pendulum continues to stabilize even though you've deleted some crucial rules.

Performance Window

A red line will be drawn across the screen indicating the movement of the pendulum's angle over time. As the controller drives the pendulum back to its goal state (angle = 0 and speed = 0) the performance window shows exactly how the transition is made. For example, if the pendulum is oscillating, you will be able to see whether or not the amplitude of the oscillations is decreasing. A trace from a typical `rulebase` will show the angle approaching 0 with a decaying exponential characteristic. More finely tuned rulebases will exhibit a more rapid convergence to the goal state.

Pendulum Window

If the pendulum angle is 0 when the simulation is started, the pendulum will automatically be bopped once to the right to set it in motion. Otherwise the simulation will simply continue from the current angle. The pendulum will continue to swing back and forth until it is stabilized by your fuzzy model.

Editing the Pendulum

When you select ![EDIT] from the menu bar, the dialog box shown in Figure 10.2 appears.

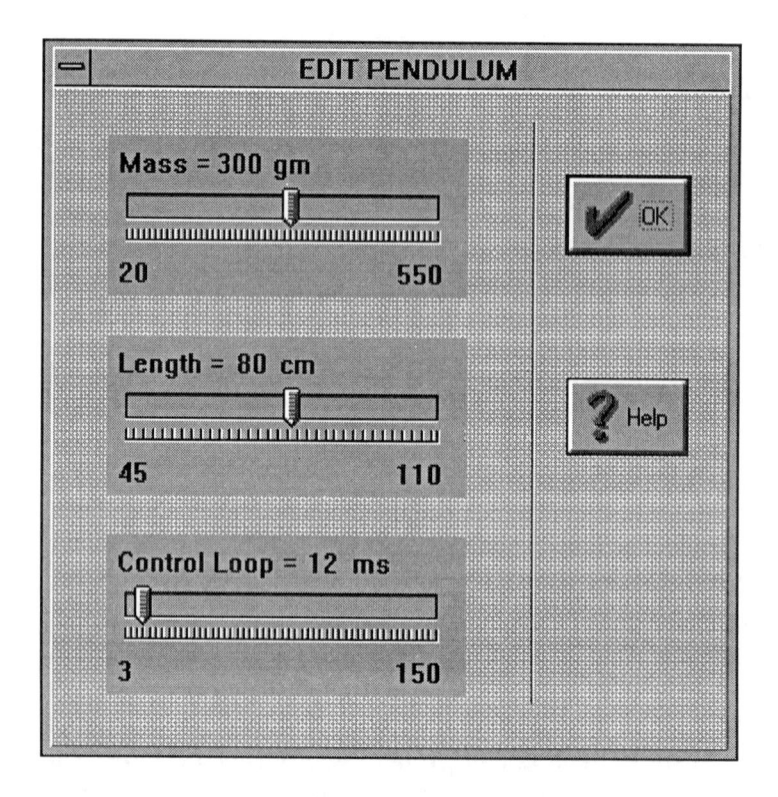

Figure 10.2. **FuzzyLab**'s *Pendulum Editor.*

You will use this dialog box to change the mass and length of the pendulum, and to adjust the length of the pendulum's control loop. The dialog box consists of three slider bars. On each slider is a button, which you click on and drag with the mouse to change the given parameter's value. In the case of the Mass and Length sliders, the pendulum is updated continually as you drag the slider button. This dialog box is placed above and to the left of center on the screen so that you can immediately see the results of your changes in the Pendulum Window.

Mass

This slider allows you to adjust the pendulum's mass. You may select values from 20 gm to 550 gm. The pendulum's mass (the green ball) will respond continually to your selections by getting larger or smaller.

Length

This slider allows you to adjust the pendulum's length. You may select values from 45 cm to 110 cm. The pendulum's length (the black line between the green ball and the base) will respond continually to your selection by getting longer or shorter.

Control Loop

This slider allows you to adjust the pendulum's control loop. You may select values from 3 ms to 150 ms. The section entitled How the Pendulum Controller Works will further your understanding of the implications of a changing control loop.

Fuzzy Logic for Real World Design

Editing the Rulebase

When you select **EDIT RULES** from the menu bar, the dialog box shown in Figure 10.3 appears.

```
EDIT RULES
IF Angle is +Lrg AND Speed is +Fst THEN Voltage is -Hi
IF Angle is +Lrg AND Speed is +Slw THEN Voltage is -Hi
IF Angle is +Lrg AND Speed is -Fst THEN Voltage is -Low
IF Angle is +Lrg AND Speed is -Slw THEN Voltage is -Hi
IF Angle is +Lrg AND Speed is Zero THEN Voltage is -Hi
IF Angle is +Sml AND Speed is +Fst THEN Voltage is -Hi
IF Angle is +Sml AND Speed is +Slw THEN Voltage is -Hi
IF Angle is +Sml AND Speed is -Fst THEN Voltage is Zero
IF Angle is +Sml AND Speed is -Slw THEN Voltage is Zero
IF Angle is +Sml AND Speed is Zero THEN Voltage is -Low
```

ANGLE	SPEED	VOLTAGE
● -Lrg	● -Fst	● -Hi
○ -Sml	○ -Slw	○ -Low
○ Zero	○ Zero	○ Zero
○ +Sml	○ +Slw	○ +Low
○ +Lrg	○ +Fst	○ +Hi

[ADD]   [DELETE]   [REPLACE]

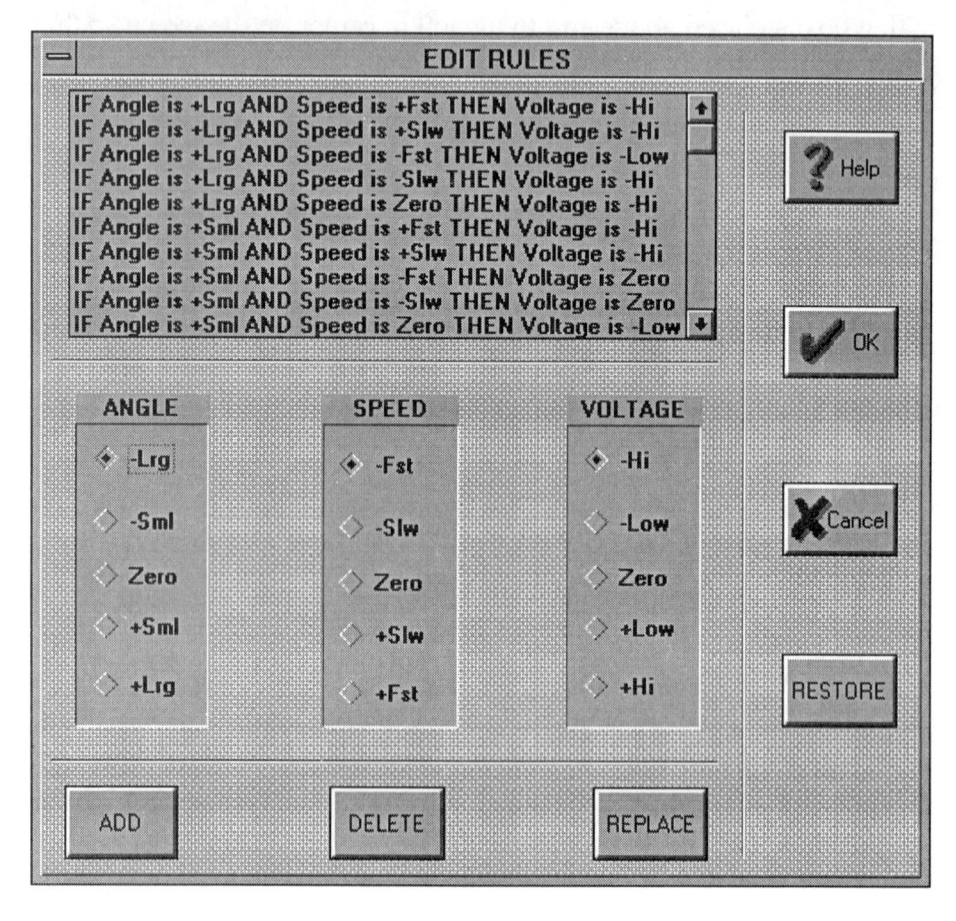

[Help]   [OK]   [Cancel]   [RESTORE]

*Figure 10.3. **FuzzyLab**'s Rules Editor.*

332

Chapter 10: FuzzyLab Primer

You will use this dialog box to browse your rulebase, and to add, delete, and replace rules, as well as restore the rulebase that originally came with FuzzyLab.

Before you exit FuzzyLab you'll be given a chance to save the changes that you have made to the rulebase.

At the top of the dialog box is a list box which contains the current rulebase. If there are too many rules to fit in the box you may use the scroller to browse the entire rulebase. You'll also use this list box to select rules for deletion.

At the bottom are three group boxes, one group box for each of the three fuzzy variables: Angle, Speed, and Voltage. Each group box contains a list of the membership functions for its fuzzy variable. You'll use these group boxes to compose the rules you want to add to the rulebase.

On the bottom are the buttons that you'll use to add, delete, and replace rules.

On the right are the buttons you'll use to accept or cancel and changes you've made, get help, or to get back the rulebase that we originally installed with FuzzyLab.

When you are done editing the rulebase you can select **OK** to accept your changes, or **CANCEL** to ignore any changes that you made while in the Rules Editor.

You can perform the following operations on the rulebase:

Delete a Rule

Select the rule that you want to delete by clicking on it in the list box. It will be highlighted. Click on the **DELETE** button. The rule will be deleted and the one below it highlighted.

IMPORTANT! When there is no rule to describe a given set of crisp inputs, the fuzzy engine will leave the output voltage unchanged. Watch for this condition as you begin to delete rules as it will often explain why the pendulum continues to stabilize even though you've deleted some crucial rules.

Add a Rule

Select one membership function from each fuzzy variable's group box. Then click on add. The rule will be added to the rulebase and to the list box. It will be highlighted in the list box. For example, if you want to add the rule:

IF Angle IS +Sml AND Speed IS +Lrg THEN Volts IS -Sml

Click on *+Sml* in *Angle*'s group box, click on *+Lrg* in *Speed*'s group box, and click on *-Sml* in *Volts*' group box. Click on ADD.

Replace a Rule

Select the rule that will be replaced by clicking on it in the list box. Don't hit **DELETE**. You might want to look at the selected rule while you're deciding which rule to add in its place. Next prepare the rule that will replace it by clicking on the new rule's membership functions in the group boxes. Click the **REPLACE** button, which will automatically delete the rule you highlighted in the list box, and add the new rule you have composed in the group boxes.

Restore Original Rulebase

When you first installed FuzzyLab we gave you rulebase of 25 rules which worked well at balancing the pendulum. If, at some point, you decide you want our rulebase back you can get it by clicking on the **RESTORE** button.

If you decide you didn't want to do that after all, just **CANCEL** out of the rules editor and you'll get your own rules back.

Strategies for Exploring FuzzyLab

FuzzyLab is designed to allow you to observe and "reverse-engineer" a fuzzy controller before you try to construct one yourself. We have provided you with a fuzzy system of 25 rules, and five membership functions for each of the three fuzzy variables: Angle, Speed, and Voltage. This initial setup works well at balancing the pendulum for a variety of lengths, masses, and control loops. The following tips should help to guide you in your exploration of FuzzyLab.

Experiment with the Menu Buttons

Start by exploring the effects of the Go, Step, Stop, Reset, Bop Left, and Bop Right buttons. Watch the effects in the Fuzzy Variable Windows as you bop the pendulum and as it returns to stability. Also notice how the Performance Window tracks the angular motion of the pendulum.

Experiment with the Pendulum Editor

Try selecting a variety of lengths and masses and observing their effect on the simulation. Although the motor is always the same strength, you'll notice that the fuzzy model is able to keep the pendulum balanced through a range of masses and lengths. It will, however, take a little longer to correct a longer, heavier pendulum than it will a shorter, lighter one.

Experiment with the control loop. This parameter determines the length of time between successive corrections to the pendulum. By lengthening the control loop you can vividly see how the pendulum passes from the region of stable control, to unstable control, and finally to total failure. Notice also how the stable regions of control are affected not only by the control loop, but also by the pendulum's mass and length.

Fuzzy Logic for Real World Design

Experiment with the Rules Editor

Investigate the effects of various rules by deleting them from the rulebase and observing subsequent pendulum behavior. In **FuzzyLab** you can experiment with impunity since the original rulebase may be restored any time you wish with a single click of a mouse button. For example, remove the rule:

IF Angle IS Zero AND Speed IS Zero THEN Volts IS Zero

The pendulum will still balance, however, the output voltage will no longer stay near zero.

This is a great place to experiment with **FuzzyLab**'s "step" feature. Use the Step button to run the simulation one display frame at a time. By observing the Fuzzy Variable Windows you'll quickly be able to determine which rules, if any, are active at any given time. Figure 10.4 helps illustrate this point.

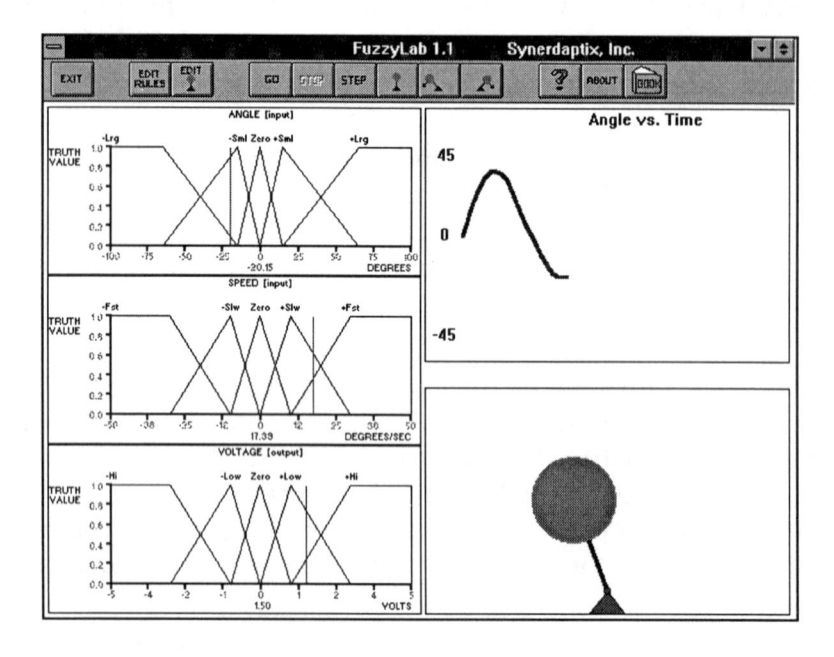

Figure 10.4: The vertical markers in each Fuzzy Variable Window indicate which membership functions are active for a specific set of crisp inputs.

The state of the pendulum captured by Figure 10.4 shows that the crisp angle = -20.15 degrees, and the speed = 17.39 degrees/sec. The vertical bar in the angle window shows that the crisp value belongs to the fuzzy set *-Sml* to a high degree and to the fuzzy set *-Lrg* to a small degree. Likewise the vertical bar in the speed window shows that the crisp input for speed belongs to the fuzzy set *+Slw* to a moderately high degree and to the fuzzy set *+Fst* to a lesser degree. Thus any rule that contains *Angle IS -Sml, Angle IS -Lrg, Speed IS +Slw,* or *Speed IS +Fst* may be active for this particular set of crisp inputs.

Perform another experiment to discover the fewest number of rules that will reliably balance the pendulum. Then, edit the pendulum and review your solution for stability. For tips on tuning a fuzzy controller see Chapter 5, *Fuzzy Engineering II*.

IMPORTANT! When there is no rule to describe a given set of crisp inputs, the fuzzy engine will leave the output voltage unchanged. Watch for this condition as you begin to delete rules, as it will often explain why the pendulum continues to stabilize even though you've deleted some crucial rules.

Delete and Reconstruct the Rulebase

When you feel that you have a clear understanding of fuzzy rules and their effects on the control of the pendulum you can use the Rules Editor to delete the rulebase, and then try to reconstruct it using what you have learned. First try writing rules to govern the pendulum when it is far away from upright (both positive and negative angles). Run the simulation to see what happens. (Most likely the pendulum angle oscillates from *+Lrg* to *-Lrg*). Next try writing a rule to govern the pendulum when it has reached the goal state (*angle* = 0 and *speed* = 0). Run the simulation to see what happens. Next try writing a few rules that help to dampen out the excessive speed that is causing the pendulum to oscillate. Remember that the response of the pendulum has a useful symmetry about the angle = 0 position, so the fuzzy rulebase should reflect that symmetry as well.

How the Pendulum Controller Works

The job of the pendulum controller is to drive the inverted pendulum to equilibrium (angle = 0 and speed = 0) by continually monitoring its angle and speed, and making corrections, in the form of voltage applied in a positive (to the right) or negative (to the left) direction. How often the inputs are read and the corrections applied depends on the length of the control loop that you select. The direction and strength of the corrections are determined by the rules that you write for your fuzzy model.

This simulation provides an accurate picture of what would happen if you built a real inverted pendulum and tried to control it using fuzzy logic. There are, of course, differences between a real-life inverted pendulum and the graphical simulation that you are using here. The differences are that we are simulating the control loop and, rather than taking readings from an actual mechanical device, we use equations of motion to compute changes to the pendulum's position and speed. In both the real world and our simulation, however, the voltages used to correct the pendulum will be calculated by a fuzzy engine.

Below are explanations of how a real-life control loop works and how we simulate one with this instructional tool. Understanding how we simulate the controller will help you to use the Step option to step through the simulation.

The Control Loop

In any controller, one control loop consists, at a minimum, of reading inputs from the target system, computing new outputs, and applying them back to the target. Time may also be taken during the loop to provide feedback to the user in the form of a computer display, etc.

Our pendulum control loop consists of reading the current values for angle and speed, running these values through the fuzzy model to produce a new voltage output, and then applying the voltage output to the pendulum's motor. As part of the simulation we allow you to adjust the length of the control loop, i.e., to select how much time it takes to read inputs, compute outputs, and apply the new voltage. See Editing the Pendulum.

The shorter the control loop, the more control you have over the pendulum, and, consequently, the more stable your pendulum will be. For example, a control loop of 3 ms means that you are updating the voltage to the motor 333 times a second, allowing it very little time to fall before you apply a correction to it. If, however, your control loop is 125 ms, you apply a new voltage to the motor only 8 times every second, leaving plenty of time for the pendulum to fall between corrections.

While the shorter control loop setting allows you finer control over your pendulum, the drawback is that, in a real world control system, a control loop of, for example, 3 ms, may not leave enough time to complete other tasks, such as managing other resources, displaying results to a user, etc.

Simulating the Control Loop

We animate the pendulum by setting a timer to tick at fixed intervals and redrawing all the windows on the screen for each tick. A timer tick is called a "step." During each step we use equations of motion to calculate the pendulum's new position. Because each iteration of the equations results in such a small movement of the pendulum we iterate them 17 times before we actually display the pendulum so that you can see some movement. Each iteration spans a time interval of 3 ms. That means that 51 ms of simulation occur each step.

How often the fuzzy model is run depends on the control loop that you select (see Editing the Pendulum). For example, if you have selected a control loop of 3 ms, a new output voltage will be generated by the fuzzy system for each iteration of the equations of motion. If you have selected a control loop of 150 ms, the fuzzy system will generate a new output voltage once for every fifty iterations of the equations of motion.

Appendix A:
Resources

We view this appendix as a developer's source book. We have collected in this appendix bibliographic, tool vendor, and on-line reference information that will help with all aspects of fuzzy model development and implementation.

The bibliographic entries contain references to both books and magazine articles. While the books present a more complete treatment of their subject, the magazine articles reflect the current state of applications of fuzzy logic in a colloquial form. While there have been many books published dealing with fuzzy logic in the last 15 years, most of the books appearing in our list have recent dates of publication. Hence most of the books listed are still in print and readily available from their respective publishers.

The second major section of this appendix covers fuzzy development tools and their respective vendors. These entries include addresses and phone numbers for both software resources and hardware resources. The charts at the end of the section summarize the capabilities of the tools available from various vendors. We use eleven categories to characterize the features of the various development tools. Once you have found a vendor whose tools address the needs of your application you can easily contact them using the telephone, fax, and mailing address information provided.

The on-line reference section lists various information resources available through both for-profit and not-for-profit computer networks.

The listings cover bulletin board services, electronic mailing lists, news groups, electronic forums, and Internet sources. Due to the nature of electronic information sites, they are more likely to change in content and location than any other resource in this appendix. Thus it is possible that some may already be out-of-date as you read this. However, using a search program that examines the information on the Internet will easily generate hundreds of matches to the key words *fuzzy logic*. You can then compile your own list of references.

Bibliography

Ajluni, C., "Neural Network/Fuzzy-Logic Technology Enables Intelligent and Fast Battery-Charger Circuit," *Electronic Design*, May 1, 1995, 38-40.

Coad, P., and Yourdon, E., *Object-Oriented Design*, New Jersey, Prentice-Hall, 1991.

Corbin, J., "A Fuzzy Logic-Based Financial Transaction System," *Embedded Systems Programming*, December, 1994, 24-29.

Coplien, J., *Advanced C++ Programming Styles and Idioms*, Massachusetts, 1992.

Cox, E., *Fuzzy Logic Applications in Business and Industry*, Charles River Media, 1995. ISBN 1-886801-01-0.

Cox, E., *The Fuzzy Systems Handbook*, Cambridge, MA, Academic Press, 1994.

Driankov, D., and Hellendoorn, H., and Reinfrank, M., *An Introduction to Fuzzy Control*, Berlin, Springer-Verlag, 1993.

Dubois, D., and Prade, H., *Fuzzy Sets and Systems: Theory and Applications. Mathematics in Science and Engineering, Vol 144*, San Diego, CA, Academic Press, 1980.

Fares, M., and Kaminska, B., "Exploring Test Space with Fuzzy Decision Making," *IEEE Design & Test of Computers*, Fall, 1994, 17-27.

Folger, and Klir, *Fuzzy Sets, Uncertainty, and Information*, New Jersey, Prentice-Hall, 1988.

Kolts, B. S., "Fuzzy Logic Improves Small-Lot Data Analysis," *EE-Evaluation Engineering*, May, 1994, 30-36.

Korn, G.A., "Neural Networks and Fuzzy-Logic Control on Personal Computers and Workstations", MIT Press, Cambridge, MA, 1995.

Kosko, B., and Isaka, S., "Fuzzy Logic," *Scientific American*, July, 1993, 76-81.

Kosko, B., *Fuzzy Thinking*, New York, Hyperion, 1993.

Kosko, B., *Neural Networks and Fuzzy Systems*, New Jersey, Prentice-Hall, 1992.

Legg, G., "Transmission's Fuzzy Logic Keeps You On Track," *EDN*, December, 1993, 60-63.

Lytle, D., "A Nova in Your Pocket?" *Photonics Spectra*, September, 1993, 38-39.

McNeill, F. M., and Thro, E., *Fuzzy Logic: A Practical Approach*, Cambridge, MA, Academic Press, 1994.

McNeill, D., and Freiberger, P., *Fuzzy Logic*, New York, Simon & Schuster, 1993.

Meyers, S., *Effective C++: 50 Specific Ways to Improve Your Programs and Designs*, Massachusetts, Addison esley, 1992.

Nijmeijer, H., van der Schaft, A. J., *Nonlinear Dynamical Control Systems*, New York, Springer-Verlag, 1990.

Pedrycz, W., *Fuzzy Control and Fuzzy Systems*, New York, John Wiley & Sons, 1993.

Rumbaugh, J., et. al., *Object-Oriented Modeling and Design*, New Jersey, Prentice-Hall, 1991.

Smets, P., Mamdani, E. H., Dubois, D., and Prade, H., *Non-Standard Logics for Automated Reasoning*, London, Academic Press Limited, 1988.

Sugeno (Ed.), *Industrial Applications of Fuzzy Control*, North Holland, 1985.

Terano, T., Asai, K., and Sugeno, M., *Applied Fuzzy Systems*, Cambridge, MA, Academic Press, 1994.

Von Altrock, C., *Fuzzy Logic & Neuro-Fuzzy Applications Explained*, New Jersey, Prentice Hall, 1995. ISBN 0-133684-65-2.

Wang, L.-X., *Adaptive Fuzzy Systems and Control*, New Jersey, Prentice-Hall, 1994.

Wang, L.-X., and Mendel, J. M., "Generating Fuzzy Rules from Numerical Data, with Applications," USC-SIPI Report No. 169. Signal and Image Processing Institute, University of Southern CA, Los Angeles, CA, 1991.

Welstead, S. T., *Neural Network and Fuzzy Logic Applications in C/C++*, New York, John Wiley & Sons, 1994.

Williams, Tom, "Fuzzy Logic Simplifies Complex Control Problems," *Computer Design*, March 1, 1991, 90-102.

Yager, R., and Filev, D. P., *Essentials of Fuzzy Modeling and Control*, New York, John Wiley & Sons, 1994.

Yager, R. E., and Zadeh, L. (Eds.), *An Introduction to Fuzzy Logic Applications in Intelligent Systems*, Norwell, MA, Kluwer Academic Publishers, 1992.

Yager, R. R., Ovchinnikov, S., Tong, R. M., and Nguyen, H.T., (Eds.), *Fuzzy Sets and Applications: Selected Papers by L. A. Zadeh*, New York, John Wiley & Sons, 1987.

Zadeh, L. A., and Kacprzyk, J. (Eds.), *Fuzzy Logic for the Management of Uncertainty*, New York., John Wiley & Sons, 1992.

Zimmerman, H. J., *Fuzzy Set Theory--and Its Applications*, Norwell, MA, Kluwer Academic Publishers, 1985.

Zimmerman, H. J., *Fuzzy Sets, Decision Making, and Expert Systems*, Norwell, MA, Kluwer Academic Publishers, 1987.

Fuzzy Logic Development Tools

The intent of this section is to convey the capabilities of the fuzzy development tools that are available at the time this book is published. We expect the tools listed in this section to remain a representative sampling for several years. In addition, due to the still expanding market for fuzzy tools, we expect to see feature enhancements and additional functionality from all tool vendors listed.

Product information is organized in terms of the functionality offered by the tools of a particular vendor. The functionalities listed for a particular vendor may cover several separate products or just a single software package. Thus the information presented here is an indication of the core functionality offered by each vendor rather than a review of individual products.

The following charts identify eleven functional areas. Those areas and their descriptions are:

Analysis and Debugging

While virtually every vendor provides a means to design the rules and fuzzy sets comprising a fuzzy model, not all provide the analysis or debugging tools useful in examining the exact behavior of that model. Without analysis and debugging tools it may be difficult and time consuming to get the best performance from a complex fuzzy model.

The most common approach to analysis is to provide a graphing tool that plots the output of the fuzzy model for all possible combinations of inputs. This approach works particularly well for one-and two-input fuzzy models. The most common approach to debugging allows the designer to work backwards step-by-step from the crisp output of the model to reveal which rules contributed and the degrees of their contributions.

Some vendors have developed fuzzy logic "toolboxes" as additions to their existing mathematical modeling or embedded control design tools. In such cases the analysis and debugging tools have already been built into the larger design tool and can be quite extensive in capability.

Automated Fuzzy Model Generation

A few vendors have recently introduced tools that generate fuzzy sets and rules from representative data. This process builds a fuzzy model from data. The intent is to build a valid fuzzy model that mimics the input-output relationships embodied in the data. Most tools base their model building algorithms on variants of basic neural network technology.

While these techniques are not guaranteed to work on every conceivable set of data, they do enhance the fuzzy set definition and rule writing process if representative data is available from the system of interest.

Code Generation

A tool with code generation capability will translate your fuzzy model into another computer language. The resulting computer code can then be easily integrated into a larger application. The two most common languages to translate into are C and assembly. Assembly language code generation is targeted to specific 8 and 16-bit microcontrollers. Since assembly language code is specific to a microcontroller, you must make sure that the vendor's tools support your particular microcontroller. C language code generation is much more portable since C compilers exist and are readily available for virtually all processors that you might consider for any application. Other languages supported include ADA, BASIC, C++, and FORTRAN.

DLL/DDE Support

DLL (dynamic link library) and DDE (dynamic data exchange) are features which interact specifically with the MS Windows family of PC operating systems. These features are useful when building and running fuzzy models in a PC environment because they enhance a tools ability to interact with other Windows applications. The DLL features allow, for example, the same single fuzzy engine to perform the calculations for many different fuzzy models. The DDE features are especially suited to information systems fuzzy models that, for example, may take input from a spreadsheet program and send output back to that spreadsheet to be graphed.

Education

Both software and human facilitated education are covered in this category. Of the educational software that exists, most is targeted from the beginner to intermediate level of knowledge. While the scope of the concepts addressed by the software is wide, it may not be deep. Other educational alternatives include organized classes and focused seminars. Classes and focused seminars offer excellent ways to achieve the depth of understanding that you require.

Embedded Systems

This category identifies a vendor as supplying tools that generate code to run your fuzzy models directly on a microcontroller in an embedded system. Some tools in this category provide enhanced debugging and analysis of fuzzy models in the embedded environment. As with code generation, you must make sure that the tools work with your specific microcontroller.

Design Services

Some vendors offer a range of consulting and design services. These services range from expert counseling to collaborative brainstorming to full development of turnkey solutions. Contracting outside expertise can be a cost-effective way to jump start a development team on a new project or bring a different perspective to bear on a particularly challenging project.

Hardware Accelerator

Hardware acceleration deals with the special integrated circuits used to speed some or all of the fuzzy inference mechanisms. The vendors in this category offer either complete chips or designs which can be included in your custom chip design. Note: digital signal processors may be regarded as hardware accelerators in the sense that they speed up fuzzy computations, however, they do not fit this category since they are in fact general purpose algorithm processors.

Information Systems

This category identifies features that are particularly useful in the development of information system models. In contrast to most embedded control models, information system models may need to handle fuzzy arithmetic, complex fuzzy set shapes, hedges, database access, and flexible inference methods. These additional features: 1) allow a more complete representation of human expertise in the fuzzy domain, 2) address the need for greater flexibility in obtaining crisp inputs from a variety of sources including human operators and traditional databases, and 3) provide enhanced, interactive graphical and textual output targeted to a human operator.

Integrated Toolset

Some vendors offer a variety of fuzzy logic components as part of a comprehensive and integrated suite of products. To make this column a vendor must offer tools that seamlessly integrate fuzzy model development with analysis, debugging, simulation, and code generation capability. In addition to convenience, an integrated toolset can offer increased productivity through the model development, testing, tuning and implementation phases.

Simulation Capability

Tuning a fuzzy model is an iterative process in which the model is adjusted and then tested. The testing can be carried out on the actual final apparatus or, more conveniently, on a simulation of the final apparatus. Typically, in a given amount of time, much more tuning can be performed with a simulation than with the final apparatus. Thus a tool that integrates the ability to simulate the final apparatus with fuzzy model development offers a significant productivity enhancement. Tools in this category offer a level of simulation support that varies but is nonetheless quite useful.

Spreadsheet Interface

The paradigm of our world is one where the need for continual learning is imposed upon us both intentionally and as a natural by-product of the information age. In the software industry this paradigm is driven by new operating systems and application software that constantly changes in the pursuit of improved feature/function content. The rate at which we must learn just to keep up is high. Some vendors have recognized this phenomena and decided to give the software users a break from the break-neck pace. They have introduced fuzzy tools that supply a user interface with which many people are already familiar. A spreadsheet interface to a tool usually means that the beginner can obtain useful results faster.

Fuzzy Logic for Real World Design

Vendor	Analysis/ Debuggin	Automated Fuzzy Model Generation	Code Generation	DLL/DDE Support	Education	Embedded Systems
Adaptive Logic						Yes
Aptronix	Yes		Yes			Yes
ByteCraft			Yes			Yes
Byte Dynamics	Yes		Yes			
Fuzzy Systems Engineering	Yes		Yes	Yes		Yes
FuziWare				Yes		
FuzzySoft						
Hitachi America						
HIWARE AG	Yes		Yes			Yes
Huntington Advanced Technology						
HyperLogic	Yes	Yes	Yes	Yes		Yes
IIS					Yes	
Inform Software	Yes	Yes	Yes	Yes		Yes
Integrated Systems	Yes		Yes	Yes		
Intelligent Machines						
Kemp-Carraway	Yes					
The MathWorks	Yes	Yes	Yes			
Metus Systems	Yes				Yes	
Microchip Technology	Yes		Yes		Yes	Yes
MIT- Management Intelligenter Technologien	Yes	Yes				
Modico			Yes			
Motorola	Yes	Yes	Yes		Yes	Yes
National Semiconductor	Yes	Yes	Yes			Yes
NeuralWare	Yes		Yes			Yes
Nicesoft				Yes		
Omron Electronics	Yes			Yes		
Ortech	Yes		Yes			Yes
SGS-Thomson	Yes					Yes
Synerdaptix					Yes	
Siemens Comp.						Yes
VLSI Technology						Yes

Appendix A: Resources

Vendor	Design Services	Hardware Accelerator	Information Systems	Integrated Toolset	Simulation Capability	Spreadsheet Interface
Adaptive Logic		Yes				
Aptronix	Yes				Yes	
ByteCraft						
Byte Dynamics					Yes	
Fuzzy Systems Engineering			Yes	Yes		
FuziWare			Yes			Yes
FuzzySoft						
Hitachi America						
HIWARE AG					Yes	
Huntington Advanced Technology	Yes					
HyperLogic					Yes	
IIS						
Inform Software	Yes			Yes	Yes	
Integrated Systems				Yes		
Intelligent Machines				Yes		
Kemp-Carraway			Yes			
The MathWorks				Yes	Yes	
Metus Systems	Yes		Yes			
Microchip Technology						
MIT- Management Intelligenter Technologien						
Modico						
Motorola						
National Semiconductor						
NeuralWare			Yes			
Nicesoft			Yes			
Omron Electronics		Yes	Yes	Yes	Yes	
Ortech		Yes	Yes	Yes	Yes	
SGS-Thomson		Yes		Yes	Yes	
Synerdaptix	Yes					
Siemens Components		Yes				
VLSI Technology		Yes				

351

Vendor Reference

Adaptive Logic
800 Charcot Ave., Suite 112
San Jose, CA 95131
Tel: (408) 383-7200
Fax: (408) 383-7201

Aptronix, Inc.
P.O. Box 640248
San Jose, CA 95164-0248
Tel: (408) 428-1888
Fax: (408) 428-1884
FuzzyNet BBS: (408) 428-1883

ByteCraft Limited
421 King Street North
Waterloo, Ontario
N2J 2E4
Tel: (519) 888-6911
Fax: (519) 746-6751
BBS: (519) 888-7626

Byte Dynamics, Inc.
14608 E. Olympic Ave.
Spokane, WA 99216
Tel: (800) 233-2983
Fax: (509) 926-6130

Fuzzy Systems Engineering,
12223 Wilsey Way
Poway, CA 92064
Tel: (619) 748-7384

FuziWare, Inc.
P. O. Box 11287
Knoxville, TN 37939-1287
Tel: (800) 472-6183
Fax: (615) 588-9487

FuzzySoft AG
GTS Trautzl GmbH
Gottlieb-Diamler-Str. 9
W-2358 Kaltenkirchen, GERMANY
Tel: 49 4191-8711
Fax: 49 4191-88665

Hitachi America Ltd
Brisbane, CA
Tel: (415) 589-8300

HIWARE AG
Gundeldingerstrasse 432
CH-4053 Basel
Switzerland
Tel: ++41-61-331-7151
Fax: ++41-61-331-1054

Huntington Advanced Technology
883 Santa Cruz Ave., Suite 31
Menlo Park, CA 94025-4608
Tel: (415) 325-7554
Internet: brubaker@cup.portal.com

HyperLogic
1855 East Valley Parkway, Suite 210
PO Box 300010
Escondido, CA 92027
Tel: (619) 746-2765
Fax: (619) 746-4089

Inform GmbH
Constantin Von Altrack
Pascalstr. 23
D-5100 Aachen
Germany

Inform Software Corporation
1840 Oak Avenue
Evanston, IL 60201
Tel: (800) 929-2815
Fax: (708) 866-1839

Intelligent Inference Systems Corp.
P.O. Box 2908
Sunnyvale, CA 94087
Tel: (408) 730-8345
Fax: (408) 730-8550
Email: iiscorp@netcom.com

Integrated Systems, Inc.
3260 Jay Street
Santa Clara, CA 95054-3309
Tel: (408) 980-1500
Fax: (408) 980-0400

Intelligent Machines, Inc.
1153 Bordeaux Drive
Sunnyvale, CA 94089
Tel: (408) 745-0881
Fax: (408) 745-6408

Kemp-Carraway Heart Institute
1600 Carraway Boulevard
Birmingham, AL 35234
Tel: (800) 909-0809
Fax: (205) 250-8958
Internet: williamsiler@delphi.com

The MathWorks, Inc.
PO Box 6100
Holliston, MA 01746-9752
Tel: (508) 653-1415
Fax: (508) 653-6284

The Metus Systems Group
1 Griggs Ln
Chappaqua, NY 10514
Tel: (914) 238-0647

Microchip Technology, Inc.
2355 West Chandler Blvd.
Chandler, AZ 85224

MIT- Management Intelligenter Technologien GmbH
Promenade 9
52076 Aachen
Germany
Tel: +49 / 2408 / 9 45 80
Fax: +49 / 2408 / 9 45 82

Modico, Inc.
Box 8485
Knoxville, TN 37996
Tel: (615) 531-7008

Motorola Technologies Group
6501 William Cannon Drive West
Austin, TX 78735-8598
Tel: (512) 505-8823
Fax: (512) 505-8100
BBS: (512) 891-3733
Internet: fuzzy.info@ae.sps.mot.com

National Semiconductor
2900 Semiconductor Drive
Santa Clara, CA 95052-8090
Tel: (800) 272-9959

NeuralWare, Inc.
Penn Center West IV
Pittsburgh, PA 15276
Tel: (412) 787-8222
Fax: (412) 787-8220

Nicesoft Corp.
9215 Ashton Ridge
Austin, TX 78750
Tel: (512) 331-9027
Fax: (512) 219-5837

Omron Electronics, Inc.
One East Commerce Dr.
Schaumburg, IL 60173
Tel: (708) 843-7900
Fax: (708) 843-7787

Ortech Engineering, Inc.
17000 El Camino Real #208
Houston, TX 77058
Tel: (713) 480-8904
Fax: (713) 480-8906

SGS-Thomson Microelectronics
Via C. Olivetti 2
20041 Agrate Brianza (MI) ITALY
Tel: 39-39-6355031
Fax: 39-39-6056315

Synerdaptix, Inc.
P. O. Box 775
Suwanee, GA 30174
CompuServe: 75104,3236

Siemens Components, Inc.
Santa Clara, CA
Tel: (408) 980-4518

VLSI Technology, Inc.
San Jose, CA
Tel: (408) 434-7673

On-Line Resources

The amount of information that is now available in electronic form is huge and growing rapidly. One need not be a visionary to imagine a time in the not-too-distant future when entire contents of libraries will only be a few key presses away. Today, however, the existing electronic information sources form only a patchwork compendium. Different information sites require different access methods. Some simply require a dial-up modem. Others require a reliable connection to the Internet. At present, the cutting edge of information distribution is the World Wide Web (www). The www is a hypertext information delivery vehicle. Your interface to the www is a browser program that knows how to negotiate the Internet to retrieve the multimedia documents that you find interesting. In reading the following fuzzy logic references keep in mind that this list is only a fraction of the total available electronically.

Appendix A: Resources

Bulletin Board Servers

A bulletin board server (BBS) is accessed with a dial-up modem. In addition to the phone number, four parameters indicate the baud rate, parity, number of data bits, and number of stop bits required to establish communication.

Aptronix FuzzyNET BBS:
Tel: (408) 261-1883, 1200-9600 N/8/1

Electronic Design News (EDN) BBS:
Tel: (617) 558-4241, 1200-9600 N/8/1

Motorola Freeware BBS:
Tel: (512) 891-3733, 1200-9600 E/7/1

The Turning Point BBS:
Tel: (512) 219-7828/7848, DS/HST 1200-19,200 N/8/1

Electronic Mail Servers

An electronic mail (Email) server allows anyone with access to a unique Internet Email address to subscribe to an electronic mailing list. New documents are then automatically Emailed to every address on the mailing list.

Aptronix FuzzyNET Email Server:
To receive instructions on how to access the server, send the following message to fuzzynet@aptronix.com:

begin
help
end

Tim Butler's Fuzzy Logic Email Server:
To receive instructions on how to access the server, send the following message to rnalib@its.bldrdoc.gov:

@@ help

355

File Transfer Repositories

A file transfer repository is also known as an ftp site. These sites allow an anonymous user to login to a remote computer system for the purpose of retrieving files stored on the remote computer system. These systems commonly archive documents and executable software that have been placed in the public domain by their respective authors. File transfers from these sites typically take place to and from valid Internet addresses.

Carnegie-Mellon AI FTP Repository:
address: ftp.cs.cmu.edu
directory: /user/ai/areas/fuzzy

Ostfold Regional College Fuzzy Logic Anonymous FTP Repository:
address: ftp.dhhalden.no
directory: /pub/Fuzzy/

Tim Butler's Fuzzy Logic Anonymous FTP Repository:
address: ntia.its.bldrdoc.gov
directory: /pub/fuzzy

"General Purpose Fuzzy Reasoning Library":
address: utsun.s.u-tokyo.ac.jp
directory: /fj/fj.sources/v25/2577.Z

C++ class library for Fuzzy Logic:
address: ftp.dfv.rwth-aachen.de
directory: /pub/CNCL

FuzzyCLIPS 6.02a Expert System Builder:
address: ai.iit.nrc.ca
directory: /pub/fzclips/

FuNeGen-1.0 Fuzzy Neural System:
address: obelix.microelectronic.e-technik.th-darmstadt.de
directory: /pub/neurofuzzy/

NEFCON-I (NEural Fuzzy CONtroller) Simulation Environment:
address: ibr.cs.tu-bs.de
directory: /pub/local/nefcon/

C++ Fuzzy Arithmetic Library:
address: mathct.dipmat.unict.it
directory: fuzzy

World Wide Web Documents

World Wide Web (www) documents present the latest and greatest addition to the electronic information gallery. Most www documents provide links to other documents on the Internet. The key information in the following references is the URL listed for each reference. The URL tells your www browser exactly where to look on the Internet for the document of interest.

Berkeley Initiative in Soft Computing:

URL: http://http.cs.berkeley.edu/projects/Bisc/bisc.welcome.html

Center for Fuzzy Logic and Intelligent Systems Research at Texas A&M:

URL: http://www.cs.tamu.edu/research/CFL/

Complex Systems Page at Australian National University:

URL: http://life.anu.edu.au/complex_systems/complex.html

Frequently Asked Questions in Fuzzy Logic:

URL: http://www.cs.cmu.edu/Web/Groups/AI/html/faqs/ai/fuzzy/part1/faq.html

Intelligent Fuzzy Systems Laboratory at University of Toronto:

URL: http://analogy.ie.utoronto.ca/fuzzy.html

Knowledge System Lab of the National Research Council of Canada:

URL: http://ai.iit.nrc.ca/fuzzy/fuzzy.html

Laboratory for Computational Neuroscience at University of Pittsburgh:
URL: http://www.neuronet.pitt.edu/lcn

North American Fuzzy Information Processing Society:
URL: http://seraphim.csee.usf.edu/nafips.html

Soft Computing Group at Milan University:
URL: http://www.dsi.unimi.it/.../index.html

Appendix B:
Cross Compiling the Fuzzy Kernel

Target: Motorola 68HC11

In this section we will look at compiling the fuzzy kernel for the Motorola 68HC11 8-bit microcontroller. For a complete discussion of the design and ANSI C code implementation of the kernel refer to Chapter 6.

Four routines comprise the core of the fuzzy kernel. Those routines are: *Init_Fuzzy_Kernel*, *Fuzzify_Inputs*, *Eval_Rulebase*, and *Defuzzify_Rulebase*. A shareware development tool called the SMALLC compiler [79] produced the following listings. Although the SMALLC compiler supports only a subset of the ANSI C standard, we were able to successfully generate 68HC11 assembly code with little additional effort.

---

[79] The SMALLC compiler that produces 68HC11 assembly code can be obtained from the Motorola Freeware electronic bulletin board service. Phone (512) 891-3733.

Fuzzy Logic for Real World Design

The fuzzy kernel in this code listing uses MIN intersection, MAX aggregation, and MAX defuzzification. The assembly code size for the four main routines is 1898 bytes. The CONFIG_BLOCK and RULE_BLOCK specifically encode a two input, one output, and twenty-five rule fuzzy model requiring 260 bytes of ROM combined.

```
0001 0000                           ORG 0
0002                               _Num_of_I
0003 0000 00 00                     FILL 0,2*1
0004                               _Num_of_O
0005 0002 00 00                     FILL 0,2*1
0006                               _Mbf_Degr
0007 0004 00 00 00 00 00 00         FILL 0,2*255
     00 00 00 00 00 00
     00 00 00 00 00 00
     00 00 00 00 00 00
     00 00 00 00 00 00
     00 00 00 00 00 00
     00 00 00 00 00 00
     00 00 00 00 00 00
     00 00 00 00 00 00
     00 00 00 00 00 00
     00 00 00 00
0008                               _Crisp_In
0009 0202 00 00 00 00 00 00         FILL 0,2*8
     00 00 00 00 00 00
     00 00 00 00
0010                               _Crisp_Ou
0011 0212 00 00 00 00 00 00         FILL 0,2*4
     00 00
0012                               _cfg_inde
0013 021a 00 00                     FILL 0,2*1
0014                               _m_index
0015 021c 00 00                     FILL 0,2*1
0016                               _j
0017 021e 00                        FILL 0,1
0018                               _Num_of_R
0019 021f 00                        FILL 0,1
0020 e000                           ORG -8192
0021                               _CONFIG_B
0022 e000 00 02 00 01 00 05         FDB 2,1,5,0,0,11468,27852,-1,3,11468
     00 00 00 00 2c cc
     6c cc ff ff 00 03
     2c cc
0023 e014 6c cc 6c cc 7f ff         FDB 27852,27852,32767,3,13,27852,32767,32767,-27853,13
     00 03 00 0d 6c cc
     7f ff 7f ff 93 33
     00 0d
0024 e028 00 0d 7f ff 93 33         FDB 13,32767,-27853,-27853,-11469,13,3,-27853,-11469,-1
     93 33 d3 33 00 0d
     00 03 93 33 d3 33
     ff ff
0025 e03c ff ff 00 03 ff ff         FDB -1,3,-1,5,0,0,13107,26214,-1,5
     00 05 00 00 00 00
     33 33 66 66 ff ff
     00 05
```

360

Appendix B: Cross Compiling the Fuzzy Kernel

```
0026 e050 33 33 66 66 66 66        FDB     13107,26214,26214,32767,5,10,26214,32767,32767,-26215
     7f ff 00 05 00 0a
     66 66 7f ff 7f ff
     99 99
0027 e064 00 0a 00 09 7f ff        FDB     10,9,32767,-26215,-26215,-13108,9,5,-26215,-13108
     99 99 99 99 cc cc
     00 09 00 05 99 99
     cc cc
0028 e078 ff ff ff ff 00 05        FDB     -1,-1,5,-1,5,10194,24029,-32768,-24030,-10195
     ff ff 00 05 27 d2
     5d dd 80 00 a2 22
     d8 2d
0029                                _RULE_BLO
0030 e08c 19 02 05 ff 0d ff        FCB     25,2,5,-1,13,-1,2,6,-1,13
     02 06 ff 0d
0031 e096 ff 02 07 ff 0c ff        FCB     -1,2,7,-1,12,-1,2,8,-1,11
     02 08 ff 0b
0032 e0a0 ff 02 09 ff 0b ff        FCB     -1,2,9,-1,11,-1,0,5,-1,14
     00 05 ff 0e
0033 e0aa ff 00 06 ff 0e ff        FCB     -1,0,6,-1,14,-1,0,7,-1,14
     00 07 ff 0e
0034 e0b4 ff 00 08 ff 0e ff        FCB     -1,0,8,-1,14,-1,0,9,-1,13
     00 09 ff 0d
0035 e0be ff 01 05 ff 0e ff        FCB     -1,1,5,-1,14,-1,1,6,-1,14
     01 06 ff 0e
0036 e0c8 ff 01 07 ff 0d ff        FCB     -1,1,7,-1,13,-1,1,8,-1,12
     01 08 ff 0c
0037 e0d2 ff 01 09 ff 0c ff        FCB     -1,1,9,-1,12,-1,3,5,-1,12
     03 05 ff 0c
0038 e0dc ff 03 06 ff 0c ff        FCB     -1,3,6,-1,12,-1,3,7,-1,11
     03 07 ff 0b
0039 e0e6 ff 03 08 ff 0a ff        FCB     -1,3,8,-1,10,-1,3,9,-1,10
     03 09 ff 0a
0040 e0f0 ff 04 05 ff 0b ff        FCB     -1,4,5,-1,11,-1,4,6,-1,10
     04 06 ff 0a
0041 e0fa ff 04 07 ff 0a ff        FCB     -1,4,7,-1,10,-1,4,8,-1,10
     04 08 ff 0a
0042 e104 ff 04 09 ff 0a ff        FCB     -1,4,9,-1,10,-1


0043                                _Init_Fuz
0044 e10a 18 3c                    PSHY
0045 e10c 18 30                    TSY
0046 e10e 18 3c                    PSHY
0047 e110 18 30                    TSY
0048 e112 fc e0 00                 LDD     _CONFIG_B
0049 e115 dd 00                    STD     _Num_of_I
0050 e117 cc e0 00                 LDD     #_CONFIG_B
0051 e11a c3 00 02                 ADDD    #2
0052 e11d 37                       PSHB
0053 e11e 36                       PSHA
0054 e11f 38                       PULX
0055 e120 ec 00                    LDD     0,X
0056 e122 dd 02                    STD     _Num_of_O
0057 e124 4f                       CLRA
0058 e125 f6 e0 8c                 LDAB    _RULE_BLO
0059 e128 f7 02 1f                 STAB    _Num_of_R
0060 e12b cc 00 00                 LDD     #0
0061 e12e f7 02 1e                 STAB    _j
```

361

Fuzzy Logic for Real World Design

```
0062                              _4
0063 e131 4f                      CLRA
0064 e132 f6 02 1e                LDAB _j
0065 e135 37                      PSHB
0066 e136 36                      PSHA
0067 e137 38                      PULX
0068 e138 cc 00 ff                LDD #255
0069 e13b bd e9 f8                JSR __ULT
0070 e13e 1a 83 00 00             CPD #0
0071 e142 26 03                   BNE *+5
0072 e144 7e e1 75                JMP _3
0073 e147 7e e1 5a                JMP _5
0074                              _2
0075 e14a 4f                      CLRA
0076 e14b f6 02 1e                LDAB _j
0077 e14e c3 00 01                ADDD #1
0078 e151 f7 02 1e                STAB _j
0079 e154 83 00 01                SUBD #1
0080 e157 7e e1 31                JMP _4
0081                              _5
0082 e15a cc 00 04                LDD #_Mbf_Degr
0083 e15d 37                      PSHB
0084 e15e 36                      PSHA
0085 e15f 4f                      CLRA
0086 e160 f6 02 1e                LDAB _j
0087 e163 38                      PULX
0088 e164 05                      LSLD
0089 e165 3c                      PSHX
0090 e166 30                      TSX
0091 e167 e3 00                   ADDD 0,X
0092 e169 38                      PULX
0093 e16a 37                      PSHB
0094 e16b 36                      PSHA
0095 e16c 38                      PULX
0096 e16d cc 00 00                LDD #0
0097 e170 ed 00                   STD 0,X
0098 e172 7e e1 4a                JMP _2
0099                              _3
0100 e175 cc 00 00                LDD #0
0101 e178 f7 02 1e                STAB _j
0102                              _8
0103 e17b 4f                      CLRA
0104 e17c f6 02 1e                LDAB _j
0105 e17f 37                      PSHB
0106 e180 36                      PSHA
0107 e181 38                      PULX
0108 e182 cc 00 08                LDD #8
0109 e185 bd e9 f8                JSR __ULT
0110 e188 1a 83 00 00             CPD #0
0111 e18c 26 03                   BNE *+5
0112 e18e 7e e1 bf                JMP _7
0113 e191 7e e1 a4                JMP _9
0114                              _6
0115 e194 4f                      CLRA
0116 e195 f6 02 1e                LDAB _j
0117 e198 c3 00 01                ADDD #1
0118 e19b f7 02 1e                STAB _j
0119 e19e 83 00 01                SUBD #1
0120 e1a1 7e e1 7b                JMP _8
0121                              _9
0122 e1a4 cc 02 02                LDD #_Crisp_In
0123 e1a7 37                      PSHB
```

Appendix B: Cross Compiling the Fuzzy Kernel

```
0124 e1a8 36                   PSHA
0125 e1a9 4f                   CLRA
0126 e1aa f6 02 1e             LDAB _j
0127 e1ad 38                   PULX
0128 e1ae 05                   LSLD
0129 e1af 3c                   PSHX
0130 e1b0 30                   TSX
0131 e1b1 e3 00                ADDD 0,X
0132 e1b3 38                   PULX
0133 e1b4 37                   PSHB
0134 e1b5 36                   PSHA
0135 e1b6 38                   PULX
0136 e1b7 cc 00 00             LDD #0
0137 e1ba ed 00                STD 0,X
0138 e1bc 7e e1 94             JMP _6
0139                      _7
0140 e1bf cc 00 00             LDD #0
0141 e1c2 f7 02 1e             STAB _j
0142                      _12
0143 e1c5 4f                   CLRA
0144 e1c6 f6 02 1e             LDAB _j
0145 e1c9 37                   PSHB
0146 e1ca 36                   PSHA
0147 e1cb 38                   PULX
0148 e1cc cc 00 04             LDD #4
0149 e1cf bd e9 f8             JSR __ULT
0150 e1d2 1a 83 00 00          CPD #0
0151 e1d6 26 03                BNE *+5
0152 e1d8 7e e2 09             JMP _11
0153 e1db 7e e1 ee             JMP _13
0154                      _10
0155 e1de 4f                   CLRA
0156 e1df f6 02 1e             LDAB _j
0157 e1e2 c3 00 01             ADDD #1
0158 e1e5 f7 02 1e             STAB _j
0159 e1e8 83 00 01             SUBD #1
0160 e1eb 7e e1 c5             JMP _12
0161                      _13
0162 e1ee cc 02 12             LDD #_Crisp_Ou
0163 e1f1 37                   PSHB
0164 e1f2 36                   PSHA
0165 e1f3 4f                   CLRA
0166 e1f4 f6 02 1e             LDAB _j
0167 e1f7 38                   PULX
0168 e1f8 05                   LSLD
0169 e1f9 3c                   PSHX
0170 e1fa 30                   TSX
0171 e1fb e3 00                ADDD 0,X
0172 e1fd 38                   PULX
0173 e1fe 37                   PSHB
0174 e1ff 36                   PSHA
0175 e200 38                   PULX
0176 e201 cc 00 00             LDD #0
0177 e204 ed 00                STD 0,X
0178 e206 7e e1 de             JMP _10
0179                      _11
0180 e209 18 38                PULY
0181 e20b 18 35                TYS
0182 e20d 18 38                PULY
0183 e20f 39                   RTS
```

363

Fuzzy Logic for Real World Design

```
0184                          _Fuzzify_
0185 e210 18 3c                    PSHY
0186 e212 18 30                    TSY
0187 e214 30                       TSX
0188 e215 8f                       XGDX
0189 e216 c3 ff f9                 ADDD #-7
0190 e219 8f                       XGDX
0191 e21a 35                       TXS
0192 e21b 18 3c                    PSHY
0193 e21d 18 30                    TSY
0194 e21f cc 00 02                 LDD  #2
0195 e222 fd 02 1a                 STD  _cfg_inde
0196 e225 cc 00 00                 LDD  #0
0197 e228 fd 02 1c                 STD  _m_index
0198 e22b cc 00 00                 LDD  #0
0199 e22e f7 02 1e                 STAB _j
0200                          _17
0201 e231 4f                       CLRA
0202 e232 f6 02 1e                 LDAB _j
0203 e235 37                       PSHB
0204 e236 36                       PSHA
0205 e237 dc 00                    LDD  _Num_of_I
0206 e239 38                       PULX
0207 e23a bd e9 f8                 JSR  __ULT
0208 e23d 1a 83 00 00              CPD  #0
0209 e241 26 03                    BNE  *+5
0210 e243 7e e5 33                 JMP  _16
0211 e246 7e e2 59                 JMP  _18
0212                          _15
0213 e249 4f                       CLRA
0214 e24a f6 02 1e                 LDAB _j
0215 e24d c3 00 01                 ADDD #1
0216 e250 f7 02 1e                 STAB _j
0217 e253 83 00 01                 SUBD #1
0218 e256 7e e2 31                 JMP  _17
0219                          _18
0220 e259 cc e0 00                 LDD  #_CONFIG_B
0221 e25c 37                       PSHB
0222 e25d 36                       PSHA
0223 e25e fc 02 1a                 LDD  _cfg_inde
0224 e261 38                       PULX
0225 e262 05                       LSLD
0226 e263 3c                       PSHX
0227 e264 30                       TSX
0228 e265 e3 00                    ADDD 0,X
0229 e267 38                       PULX
0230 e268 37                       PSHB
0231 e269 36                       PSHA
0232 e26a 38                       PULX
0233 e26b ec 00                    LDD  0,X
0234 e26d 8f                       XGDX
0235 e26e 18 3c                    PSHY
0236 e270 32                       PULA
0237 e271 33                       PULB
0238 e272 c3 00 08                 ADDD #8
0239 e275 8f                       XGDX
0240 e276 e7 00                    STAB 0,X
0241 e278 cc 00 00                 LDD  #0
0242 e27b 18 ed 06                 STD  6,Y
0243                          _21
0244 e27e 18 ec 06                 LDD  6,Y
0245 e281 37                       PSHB
```

Appendix B: Cross Compiling the Fuzzy Kernel

```
0246 e282 36              PSHA
0247 e283 4f              CLRA
0248 e284 18 e6 09        LDAB 8+1,Y
0249 e287 38              PULX
0250 e288 bd e9 f8        JSR __ULT
0251 e28b 1a 83 00 00     CPD #0
0252 e28f 26 03           BNE *+5
0253 e291 7e e5 24        JMP _20
0254 e294 7e e2 a6        JMP _22
0255                  _19
0256 e297 18 ec 06        LDD 6,Y
0257 e29a c3 00 01        ADDD #1
0258 e29d 18 ed 06        STD  6,Y
0259 e2a0 83 00 01        SUBD #1
0260 e2a3 7e e2 7e        JMP _21
0261                  _22
0262 e2a6 cc 02 02        LDD #_Crisp_In
0263 e2a9 37              PSHB
0264 e2aa 36              PSHA
0265 e2ab 4f              CLRA
0266 e2ac f6 02 1e        LDAB _j
0267 e2af 38              PULX
0268 e2b0 05              LSLD
0269 e2b1 3c              PSHX
0270 e2b2 30              TSX
0271 e2b3 e3 00           ADDD 0,X
0272 e2b5 38              PULX
0273 e2b6 37              PSHB
0274 e2b7 36              PSHA
0275 e2b8 38              PULX
0276 e2b9 ec 00           LDD 0,X
0277 e2bb 37              PSHB
0278 e2bc 36              PSHA
0279 e2bd cc e0 00        LDD #_CONFIG_B
0280 e2c0 37              PSHB
0281 e2c1 36              PSHA
0282 e2c2 fc 02 1a        LDD _cfg_inde
0283 e2c5 c3 00 01        ADDD #1
0284 e2c8 38              PULX
0285 e2c9 05              LSLD
0286 e2ca 3c              PSHX
0287 e2cb 30              TSX
0288 e2cc e3 00           ADDD 0,X
0289 e2ce 38              PULX
0290 e2cf 37              PSHB
0291 e2d0 36              PSHA
0292 e2d1 38              PULX
0293 e2d2 ec 00           LDD 0,X
0294 e2d4 38              PULX
0295 e2d5 bd e9 dd        JSR __ULE
0296 e2d8 1a 83 00 00     CPD #0
0297 e2dc 27 03           BEQ *+5
0298 e2de 7e e3 22        JMP _24
0299 e2e1 cc 02 02        LDD #_Crisp_In
0300 e2e4 37              PSHB
0301 e2e5 36              PSHA
0302 e2e6 4f              CLRA
0303 e2e7 f6 02 1e        LDAB _j
0304 e2ea 38              PULX
0305 e2eb 05              LSLD
0306 e2ec 3c              PSHX
0307 e2ed 30              TSX
```

365

Fuzzy Logic for Real World Design

```
0308 e2ee e3 00            ADDD 0,X
0309 e2f0 38               PULX
0310 e2f1 37               PSHB
0311 e2f2 36               PSHA
0312 e2f3 38               PULX
0313 e2f4 ec 00            LDD 0,X
0314 e2f6 37               PSHB
0315 e2f7 36               PSHA
0316 e2f8 cc e0 00         LDD #_CONFIG_B
0317 e2fb 37               PSHB
0318 e2fc 36               PSHA
0319 e2fd fc 02 1a         LDD _cfg_inde
0320 e300 c3 00 04         ADDD #4
0321 e303 38               PULX
0322 e304 05               LSLD
0323 e305 3c               PSHX
0324 e306 30               TSX
0325 e307 e3 00            ADDD 0,X
0326 e309 38               PULX
0327 e30a 37               PSHB
0328 e30b 36               PSHA
0329 e30c 38               PULX
0330 e30d ec 00            LDD 0,X
0331 e30f 38               PULX
0332 e310 bd e9 b9         JSR __UGE
0333 e313 1a 83 00 00      CPD #0
0334 e317 27 03            BEQ *+5
0335 e319 7e e3 22         JMP _24
0336 e31c cc 00 00         LDD #0
0337 e31f 7e e3 25         JMP _25
0338                                            _24
0339 e322 cc 00 01         LDD #1
0340                                            _25
0341 e325 1a 83 00 00      CPD #0
0342 e329 26 03            BNE *+5
0343 e32b 7e e3 51         JMP _23
0344 e32e cc 00 04         LDD #_Mbf_Degr
0345 e331 37               PSHB
0346 e332 36               PSHA
0347 e333 fc 02 1c         LDD _m_index
0348 e336 c3 00 01         ADDD #1
0349 e339 fd 02 1c         STD  _m_index
0350 e33c 83 00 01         SUBD #1
0351 e33f 38               PULX
0352 e340 05               LSLD
0353 e341 3c               PSHX
0354 e342 30               TSX
0355 e343 e3 00            ADDD 0,X
0356 e345 38               PULX
0357 e346 37               PSHB
0358 e347 36               PSHA
0359 e348 38               PULX
0360 e349 cc 00 00         LDD #0
0361 e34c ed 00            STD 0,X
0362 e34e 7e e5 18         JMP _26
0363                                            _23
0364 e351 cc 02 02         LDD #_Crisp_In
0365 e354 37               PSHB
0366 e355 36               PSHA
0367 e356 4f               CLRA
0368 e357 f6 02 1e         LDAB _j
0369 e35a 38               PULX
```

Appendix B: Cross Compiling the Fuzzy Kernel

```
0370 e35b 05              LSLD
0371 e35c 3c              PSHX
0372 e35d 30              TSX
0373 e35e e3 00           ADDD 0,X
0374 e360 38              PULX
0375 e361 37              PSHB
0376 e362 36              PSHA
0377 e363 38              PULX
0378 e364 ec 00           LDD 0,X
0379 e366 37              PSHB
0380 e367 36              PSHA
0381 e368 cc e0 00        LDD #_CONFIG_B
0382 e36b 37              PSHB
0383 e36c 36              PSHA
0384 e36d fc 02 1a        LDD _cfg_inde
0385 e370 c3 00 03        ADDD #3
0386 e373 38              PULX
0387 e374 05              LSLD
0388 e375 3c              PSHX
0389 e376 30              TSX
0390 e377 e3 00           ADDD 0,X
0391 e379 38              PULX
0392 e37a 37              PSHB
0393 e37b 36              PSHA
0394 e37c 38              PULX
0395 e37d ec 00           LDD 0,X
0396 e37f 38              PULX
0397 e380 bd e9 dd        JSR __ULE
0398 e383 1a 83 00 00     CPD #0
0399 e387 26 03           BNE *+5
0400 e389 7e e3 cd        JMP _28
0401 e38c cc 02 02        LDD #_Crisp_In
0402 e38f 37              PSHB
0403 e390 36              PSHA
0404 e391 4f              CLRA
0405 e392 f6 02 1e        LDAB _j
0406 e395 38              PULX
0407 e396 05              LSLD
0408 e397 3c              PSHX
0409 e398 30              TSX
0410 e399 e3 00           ADDD 0,X
0411 e39b 38              PULX
0412 e39c 37              PSHB
0413 e39d 36              PSHA
0414 e39e 38              PULX
0415 e39f ec 00           LDD 0,X
0416 e3a1 37              PSHB
0417 e3a2 36              PSHA
0418 e3a3 cc e0 00        LDD #_CONFIG_B
0419 e3a6 37              PSHB
0420 e3a7 36              PSHA
0421 e3a8 fc 02 1a        LDD _cfg_inde
0422 e3ab c3 00 02        ADDD #2
0423 e3ae 38              PULX
0424 e3af 05              LSLD
0425 e3b0 3c              PSHX
0426 e3b1 30              TSX
0427 e3b2 e3 00           ADDD 0,X
0428 e3b4 38              PULX
0429 e3b5 37              PSHB
0430 e3b6 36              PSHA
0431 e3b7 38              PULX
```

```
0432 e3b8 ec 00              LDD 0,X
0433 e3ba 38                 PULX
0434 e3bb bd e9 b9           JSR __UGE
0435 e3be 1a 83 00 00        CPD #0
0436 e3c2 26 03              BNE *+5
0437 e3c4 7e e3 cd           JMP _28
0438 e3c7 cc 00 01           LDD #1
0439 e3ca 7e e3 d0           JMP _29
0440                   _28
0441 e3cd cc 00 00           LDD #0
0442                   _29
0443 e3d0 1a 83 00 00        CPD #0
0444 e3d4 26 03              BNE *+5
0445 e3d6 7e e3 fc           JMP _27
0446 e3d9 cc 00 04           LDD #_Mbf_Degr
0447 e3dc 37                 PSHB
0448 e3dd 36                 PSHA
0449 e3de fc 02 1c           LDD _m_index
0450 e3e1 c3 00 01           ADDD #1
0451 e3e4 fd 02 1c           STD _m_index
0452 e3e7 83 00 01           SUBD #1
0453 e3ea 38                 PULX
0454 e3eb 05                 LSLD
0455 e3ec 3c                 PSHX
0456 e3ed 30                 TSX
0457 e3ee e3 00              ADDD 0,X
0458 e3f0 38                 PULX
0459 e3f1 37                 PSHB
0460 e3f2 36                 PSHA
0461 e3f3 38                 PULX
0462 e3f4 cc ff ff           LDD #-1
0463 e3f7 ed 00              STD 0,X
0464 e3f9 7e e5 18           JMP _30
0465                   _27
0466 e3fc cc 02 02           LDD #_Crisp_In
0467 e3ff 37                 PSHB
0468 e400 36                 PSHA
0469 e401 4f                 CLRA
0470 e402 f6 02 1e           LDAB _j
0471 e405 38                 PULX
0472 e406 05                 LSLD
0473 e407 3c                 PSHX
0474 e408 30                 TSX
0475 e409 e3 00              ADDD 0,X
0476 e40b 38                 PULX
0477 e40c 37                 PSHB
0478 e40d 36                 PSHA
0479 e40e 38                 PULX
0480 e40f ec 00              LDD 0,X
0481 e411 37                 PSHB
0482 e412 36                 PSHA
0483 e413 cc e0 00           LDD #_CONFIG_B
0484 e416 37                 PSHB
0485 e417 36                 PSHA
0486 e418 fc 02 1a           LDD _cfg_inde
0487 e41b c3 00 02           ADDD #2
0488 e41e 38                 PULX
0489 e41f 05                 LSLD
0490 e420 3c                 PSHX
0491 e421 30                 TSX
0492 e422 e3 00              ADDD 0,X
0493 e424 38                 PULX
```

Appendix B: Cross Compiling the Fuzzy Kernel

```
0494  e425  37               PSHB
0495  e426  36               PSHA
0496  e427  38               PULX
0497  e428  ec 00            LDD 0,X
0498  e42a  38               PULX
0499  e42b  bd e9 f8         JSR __ULT
0500  e42e  1a 83 00 00      CPD #0
0501  e432  26 03            BNE *+5
0502  e434  7e e4 a9         JMP _31
0503  e437  cc 00 04         LDD #_Mbf_Degr
0504  e43a  37               PSHB
0505  e43b  36               PSHA
0506  e43c  fc 02 1c         LDD _m_index
0507  e43f  c3 00 01         ADDD #1
0508  e442  fd 02 1c         STD _m_index
0509  e445  83 00 01         SUBD #1
0510  e448  38               PULX
0511  e449  05               LSLD
0512  e44a  3c               PSHX
0513  e44b  30               TSX
0514  e44c  e3 00            ADDD 0,X
0515  e44e  38               PULX
0516  e44f  37               PSHB
0517  e450  36               PSHA
0518  e451  cc 02 02         LDD #_Crisp_In
0519  e454  37               PSHB
0520  e455  36               PSHA
0521  e456  4f               CLRA
0522  e457  f6 02 1e         LDAB _j
0523  e45a  38               PULX
0524  e45b  05               LSLD
0525  e45c  3c               PSHX
0526  e45d  30               TSX
0527  e45e  e3 00            ADDD 0,X
0528  e460  38               PULX
0529  e461  37               PSHB
0530  e462  36               PSHA
0531  e463  38               PULX
0532  e464  ec 00            LDD 0,X
0533  e466  37               PSHB
0534  e467  36               PSHA
0535  e468  cc e0 00         LDD #_CONFIG_B
0536  e46b  37               PSHB
0537  e46c  36               PSHA
0538  e46d  fc 02 1a         LDD _cfg_inde
0539  e470  c3 00 01         ADDD #1
0540  e473  38               PULX
0541  e474  05               LSLD
0542  e475  3c               PSHX
0543  e476  30               TSX
0544  e477  e3 00            ADDD 0,X
0545  e479  38               PULX
0546  e47a  37               PSHB
0547  e47b  36               PSHA
0548  e47c  38               PULX
0549  e47d  ec 00            LDD 0,X
0550  e47f  38               PULX
0551  e480  8f               XGDX
0552  e481  3c               PSHX
0553  e482  30               TSX
0554  e483  a3 00             SUBD 0,X
0555  e485  38               PULX
```

Fuzzy Logic for Real World Design

```
0556 e486 37                PSHB
0557 e487 36                PSHA
0558 e488 cc e0 00          LDD #_CONFIG_B
0559 e48b 37                PSHB
0560 e48c 36                PSHA
0561 e48d fc 02 1a          LDD _cfg_inde
0562 e490 c3 00 05          ADDD #5
0563 e493 38                PULX
0564 e494 05                LSLD
0565 e495 3c                PSHX
0566 e496 30                TSX
0567 e497 e3 00             ADDD 0,X
0568 e499 38                PULX
0569 e49a 37                PSHB
0570 e49b 36                PSHA
0571 e49c 38                PULX
0572 e49d ec 00             LDD 0,X
0573 e49f 38                PULX
0574 e4a0 bd e9 7d          JSR __MUL12
0575 e4a3 38                PULX
0576 e4a4 ed 00             STD 0,X
0577 e4a6 7e e5 18          JMP _32
0578                  _31
0579 e4a9 cc 00 04          LDD #_Mbf_Degr
0580 e4ac 37                PSHB
0581 e4ad 36                PSHA
0582 e4ae fc 02 1c          LDD _m_index
0583 e4b1 c3 00 01          ADDD #1
0584 e4b4 fd 02 1c          STD _m_index
0585 e4b7 83 00 01          SUBD #1
0586 e4ba 38                PULX
0587 e4bb 05                LSLD
0588 e4bc 3c                PSHX
0589 e4bd 30                TSX
0590 e4be e3 00             ADDD 0,X
0591 e4c0 38                PULX
0592 e4c1 37                PSHB
0593 e4c2 36                PSHA
0594 e4c3 cc e0 00          LDD #_CONFIG_B
0595 e4c6 37                PSHB
0596 e4c7 36                PSHA
0597 e4c8 fc 02 1a          LDD _cfg_inde
0598 e4cb c3 00 04          ADDD #4
0599 e4ce 38                PULX
0600 e4cf 05                LSLD
0601 e4d0 3c                PSHX
0602 e4d1 30                TSX
0603 e4d2 e3 00             ADDD 0,X
0604 e4d4 38                PULX
0605 e4d5 37                PSHB
0606 e4d6 36                PSHA
0607 e4d7 38                PULX
0608 e4d8 ec 00             LDD 0,X
0609 e4da 37                PSHB
0610 e4db 36                PSHA
0611 e4dc cc 02 02          LDD #_Crisp_In
0612 e4df 37                PSHB
0613 e4e0 36                PSHA
0614 e4e1 4f                CLRA
0615 e4e2 f6 02 1e          LDAB _j
0616 e4e5 38                PULX
0617 e4e6 05                LSLD
```

```
0618 e4e7 3c              PSHX
0619 e4e8 30              TSX
0620 e4e9 e3 00           ADDD 0,X
0621 e4eb 38              PULX
0622 e4ec 37              PSHB
0623 e4ed 36              PSHA
0624 e4ee 38              PULX
0625 e4ef ec 00           LDD 0,X
0626 e4f1 38              PULX
0627 e4f2 8f              XGDX
0628 e4f3 3c              PSHX
0629 e4f4 30              TSX
0630 e4f5 a3 00            SUBD 0,X
0631 e4f7 38              PULX
0632 e4f8 37              PSHB
0633 e4f9 36              PSHA
0634 e4fa cc e0 00        LDD #_CONFIG_B
0635 e4fd 37              PSHB
0636 e4fe 36              PSHA
0637 e4ff fc 02 1a        LDD _cfg_inde
0638 e502 c3 00 06        ADDD #6
0639 e505 38              PULX
0640 e506 05              LSLD
0641 e507 3c              PSHX
0642 e508 30              TSX
0643 e509 e3 00           ADDD 0,X
0644 e50b 38              PULX
0645 e50c 37              PSHB
0646 e50d 36              PSHA
0647 e50e 38              PULX
0648 e50f ec 00           LDD 0,X
0649 e511 38              PULX
0650 e512 bd e9 7d        JSR    __MUL12
0651 e515 38              PULX
0652 e516 ed 00           STD 0,X
0653                _32
0654                _30
0655                _26
0656 e518 fc 02 1a        LDD _cfg_inde
0657 e51b c3 00 06        ADDD #6
0658 e51e fd 02 1a        STD _cfg_inde
0659 e521 7e e2 97        JMP _19
0660                _20
0661 e524 fc 02 1a        LDD _cfg_inde
0662 e527 c3 00 01        ADDD #1
0663 e52a fd 02 1a        STD _cfg_inde
0664 e52d 83 00 01        SUBD #1
0665 e530 7e e2 49        JMP _15
0666                _16
0667 e533 fc 02 1a        LDD _cfg_inde
0668 e536 18 ed 04        STD  4,Y
0669 e539 fc 02 1c        LDD _m_index
0670 e53c 18 ed 02        STD  2,Y
0671 e53f cc 00 00        LDD #0
0672 e542 f7 02 1e        STAB _j
0673                _35
0674 e545 4f              CLRA
0675 e546 f6 02 1e        LDAB _j
0676 e549 37              PSHB
0677 e54a 36              PSHA
0678 e54b dc 02           LDD _Num_of_O
0679 e54d 38              PULX
```

```
0680 e54e bd e9 f8           JSR __ULT
0681 e551 1a 83 00 00        CPD #0
0682 e555 26 03              BNE *+5
0683 e557 7e e5 ec           JMP _34
0684 e55a 7e e5 6d           JMP _36
0685                  _33
0686 e55d 4f                 CLRA
0687 e55e f6 02 1e           LDAB _j
0688 e561 c3 00 01           ADDD #1
0689 e564 f7 02 1e           STAB _j
0690 e567 83 00 01           SUBD #1
0691 e56a 7e e5 45           JMP _35
0692                  _36
0693 e56d cc e0 00           LDD #_CONFIG_B
0694 e570 37                 PSHB
0695 e571 36                 PSHA
0696 e572 fc 02 1a           LDD _cfg_inde
0697 e575 38                 PULX
0698 e576 05                 LSLD
0699 e577 3c                 PSHX
0700 e578 30                 TSX
0701 e579 e3 00              ADDD 0,X
0702 e57b 38                 PULX
0703 e57c 37                 PSHB
0704 e57d 36                 PSHA
0705 e57e 38                 PULX
0706 e57f ec 00              LDD 0,X
0707 e581 8f                 XGDX
0708 e582 18 3c              PSHY
0709 e584 32                 PULA
0710 e585 33                 PULB
0711 e586 c3 00 08           ADDD #8
0712 e589 8f                 XGDX
0713 e58a e7 00              STAB 0,X
0714 e58c cc 00 00           LDD #0
0715 e58f 18 ed 06           STD 6,Y
0716                  _39
0717 e592 18 ec 06           LDD 6,Y
0718 e595 37                 PSHB
0719 e596 36                 PSHA
0720 e597 4f                 CLRA
0721 e598 18 e6 09           LDAB 8+1,Y
0722 e59b 38                 PULX
0723 e59c bd e9 f8           JSR __ULT
0724 e59f 1a 83 00 00        CPD #0
0725 e5a3 26 03              BNE *+5
0726 e5a5 7e e5 e9           JMP _38
0727 e5a8 7e e5 ba           JMP _40
0728                  _37
0729 e5ab 18 ec 06           LDD 6,Y
0730 e5ae c3 00 01           ADDD #1
0731 e5b1 18 ed 06           STD 6,Y
0732 e5b4 83 00 01           SUBD #1
0733 e5b7 7e e5 92           JMP _39
0734                  _40
0735 e5ba cc 00 04           LDD #_Mbf_Degr
0736 e5bd 37                 PSHB
0737 e5be 36                 PSHA
0738 e5bf fc 02 1c           LDD _m_index
0739 e5c2 c3 00 01           ADDD #1
0740 e5c5 fd 02 1c           STD _m_index
0741 e5c8 83 00 01           SUBD #1
```

Appendix B: Cross Compiling the Fuzzy Kernel

```
0742  e5cb 38                    PULX
0743  e5cc 05                    LSLD
0744  e5cd 3c                    PSHX
0745  e5ce 30                    TSX
0746  e5cf e3 00                 ADDD 0,X
0747  e5d1 38                    PULX
0748  e5d2 37                    PSHB
0749  e5d3 36                    PSHA
0750  e5d4 38                    PULX
0751  e5d5 cc 00 00              LDD #0
0752  e5d8 ed 00                 STD 0,X
0753  e5da fc 02 1a              LDD _cfg_inde
0754  e5dd c3 00 01              ADDD #1
0755  e5e0 fd 02 1a              STD _cfg_inde
0756  e5e3 83 00 01              SUBD #1
0757  e5e6 7e e5 ab              JMP _37
0758                       _38
0759  e5e9 7e e5 5d              JMP _33
0760                       _34
0761  e5ec 18 ec 04              LDD 4,Y
0762  e5ef fd 02 1a              STD _cfg_inde
0763  e5f2 18 ec 02              LDD 2,Y
0764  e5f5 fd 02 1c              STD _m_index
0765  e5f8 18 38                 PULY
0766  e5fa 18 35                 TYS
0767  e5fc 18 38                 PULY
0768  e5fe 39                    RTS


0769                       _Eval_Rul
0770  e5ff 18 3c                 PSHY
0771  e601 18 30                 TSY
0772  e603 30                    TSX
0773  e604 8f                    XGDX
0774  e605 c3 ff fb              ADDD #-5
0775  e608 8f                    XGDX
0776  e609 35                    TXS
0777  e60a 18 3c                 PSHY
0778  e60c 18 30                 TSY
0779  e60e cc 00 01              LDD #1
0780  e611 18 ed 02              STD 2,Y
0781  e614 cc 00 00              LDD #0
0782  e617 f7 02 1e              STAB _j
0783                       _44
0784  e61a 4f                    CLRA
0785  e61b f6 02 1e              LDAB _j
0786  e61e 37                    PSHB
0787  e61f 36                    PSHA
0788  e620 4f                    CLRA
0789  e621 f6 02 1f              LDAB _Num_of_R
0790  e624 38                    PULX
0791  e625 bd e9 f8              JSR __ULT
0792  e628 1a 83 00 00           CPD #0
0793  e62c 26 03                 BNE *+5
0794  e62e 7e e7 4b              JMP _43
0795  e631 7e e6 44              JMP _45
0796                       _42
0797  e634 4f                    CLRA
0798  e635 f6 02 1e              LDAB _j
0799  e638 c3 00 01              ADDD #1
0800  e63b f7 02 1e              STAB _j
0801  e63e 83 00 01              SUBD #1
```

373

```
0802 e641 7e e6 1a            JMP  _44
0803                    _45
0804 e644 cc ff ff            LDD  #-1
0805 e647 18 ed 05            STD  5,Y
0806                    _46
0807 e64a cc e0 8c            LDD  #_RULE_BLO
0808 e64d 37                  PSHB
0809 e64e 36                  PSHA
0810 e64f 18 ec 02            LDD  2,Y
0811 e652 c3 00 01            ADDD #1
0812 e655 18 ed 02            STD  2,Y
0813 e658 83 00 01            SUBD #1
0814 e65b 38                  PULX
0815 e65c 3c                  PSHX
0816 e65d 30                  TSX
0817 e65e e3 00               ADDD 0,X
0818 e660 38                  PULX
0819 e661 37                  PSHB
0820 e662 36                  PSHA
0821 e663 38                  PULX
0822 e664 4f                  CLRA
0823 e665 e6 00               LDAB 0,X
0824 e667 8f                  XGDX
0825 e668 18 3c               PSHY
0826 e66a 32                  PULA
0827 e66b 33                  PULB
0828 e66c c3 00 04            ADDD #4
0829 e66f 8f                  XGDX
0830 e670 e7 00               STAB 0,X
0831 e672 37                  PSHB
0832 e673 36                  PSHA
0833 e674 38                  PULX
0834 e675 cc 00 ff            LDD  #255
0835 e678 bd ea 0a            JSR  __NE
0836 e67b 1a 83 00 00         CPD  #0
0837 e67f 26 03               BNE  *+5
0838 e681 7e e6 cc            JMP  _47
0839 e684 18 ec 05            LDD  5,Y
0840 e687 37                  PSHB
0841 e688 36                  PSHA
0842 e689 cc 00 04            LDD  #_Mbf_Degr
0843 e68c 37                  PSHB
0844 e68d 36                  PSHA
0845 e68e 4f                  CLRA
0846 e68f 18 e6 05            LDAB 4+1,Y
0847 e692 38                  PULX
0848 e693 05                  LSLD
0849 e694 3c                  PSHX
0850 e695 30                  TSX
0851 e696 e3 00               ADDD 0,X
0852 e698 38                  PULX
0853 e699 37                  PSHB
0854 e69a 36                  PSHA
0855 e69b 38                  PULX
0856 e69c ec 00               LDD  0,X
0857 e69e 38                  PULX
0858 e69f bd e9 cb            JSR  __UGT
0859 e6a2 1a 83 00 00         CPD  #0
0860 e6a6 26 03               BNE  *+5
0861 e6a8 7e e6 c3            JMP  _48
0862 e6ab cc 00 04            LDD  #_Mbf_Degr
0863 e6ae 37                  PSHB
```

Appendix B: Cross Compiling the Fuzzy Kernel

```
0864 e6af 36              PSHA
0865 e6b0 4f              CLRA
0866 e6b1 18 e6 05        LDAB 4+1,Y
0867 e6b4 38              PULX
0868 e6b5 05              LSLD
0869 e6b6 3c              PSHX
0870 e6b7 30              TSX
0871 e6b8 e3 00           ADDD 0,X
0872 e6ba 38              PULX
0873 e6bb 37              PSHB
0874 e6bc 36              PSHA
0875 e6bd 38              PULX
0876 e6be ec 00           LDD 0,X
0877 e6c0 7e e6 c6        JMP _49
0878                  _48
0879 e6c3 18 ec 05        LDD 5,Y
0880                  _49
0881 e6c6 18 ed 05        STD  5,Y
0882 e6c9 7e e6 4a        JMP _46
0883                  _47
0884                  _50
0885 e6cc cc e0 8c        LDD #_RULE_BLO
0886 e6cf 37              PSHB
0887 e6d0 36              PSHA
0888 e6d1 18 ec 02        LDD 2,Y
0889 e6d4 c3 00 01        ADDD #1
0890 e6d7 18 ed 02        STD  2,Y
0891 e6da 83 00 01        SUBD #1
0892 e6dd 38              PULX
0893 e6de 3c              PSHX
0894 e6df 30              TSX
0895 e6e0 e3 00           ADDD 0,X
0896 e6e2 38              PULX
0897 e6e3 37              PSHB
0898 e6e4 36              PSHA
0899 e6e5 38              PULX
0900 e6e6 4f              CLRA
0901 e6e7 e6 00           LDAB 0,X
0902 e6e9 8f              XGDX
0903 e6ea 18 3c           PSHY
0904 e6ec 32              PULA
0905 e6ed 33              PULB
0906 e6ee c3 00 04        ADDD #4
0907 e6f1 8f              XGDX
0908 e6f2 e7 00           STAB 0,X
0909 e6f4 37              PSHB
0910 e6f5 36              PSHA
0911 e6f6 38              PULX
0912 e6f7 cc 00 ff        LDD #255
0913 e6fa bd ea 0a        JSR __NE
0914 e6fd 1a 83 00 00     CPD #0
0915 e701 26 03           BNE *+5
0916 e703 7e e7 48        JMP _51
0917 e706 cc 00 04        LDD #_Mbf_Degr
0918 e709 37              PSHB
0919 e70a 36              PSHA
0920 e70b 4f              CLRA
0921 e70c 18 e6 05        LDAB 4+1,Y
0922 e70f 38              PULX
0923 e710 05              LSLD
0924 e711 3c              PSHX
0925 e712 30              TSX
```

375

```
0926 e713 e3 00              ADDD 0,X
0927 e715 38                 PULX
0928 e716 37                 PSHB
0929 e717 36                 PSHA
0930 e718 38                 PULX
0931 e719 ec 00              LDD 0,X
0932 e71b 37                 PSHB
0933 e71c 36                 PSHA
0934 e71d 18 ec 05           LDD 5,Y
0935 e720 38                 PULX
0936 e721 bd e9 f8           JSR __ULT
0937 e724 1a 83 00 00        CPD #0
0938 e728 26 03              BNE *+5
0939 e72a 7e e7 45           JMP _52
0940 e72d cc 00 04           LDD #_Mbf_Degr
0941 e730 37                 PSHB
0942 e731 36                 PSHA
0943 e732 4f                 CLRA
0944 e733 18 e6 05           LDAB 4+1,Y
0945 e736 38                 PULX
0946 e737 05                 LSLD
0947 e738 3c                 PSHX
0948 e739 30                 TSX
0949 e73a e3 00              ADDD 0,X
0950 e73c 38                 PULX
0951 e73d 37                 PSHB
0952 e73e 36                 PSHA
0953 e73f 18 ec 05           LDD 5,Y
0954 e742 38                 PULX
0955 e743 ed 00              STD 0,X
0956                   _52
0957 e745 7e e6 cc           JMP _50
0958                   _51
0959 e748 7e e6 34           JMP _42
0960                   _43
0961 e74b 18 38              PULY
0962 e74d 18 35              TYS
0963 e74f 18 38              PULY
0964 e751 39                 RTS


0965                   _Defuzzif
0966 e752 18 3c              PSHY
0967 e754 18 30              TSY
0968 e756 30                 TSX
0969 e757 8f                 XGDX
0970 e758 c3 ff f6           ADDD #-10
0971 e75b 8f                 XGDX
0972 e75c 35                 TXS
0973 e75d 18 3c              PSHY
0974 e75f 18 30              TSY
0975 e761 cc 00 00           LDD #0
0976 e764 f7 02 1e           STAB _j
0977                   _56
0978 e767 4f                 CLRA
0979 e768 f6 02 1e           LDAB _j
0980 e76b 37                 PSHB
0981 e76c 36                 PSHA
0982 e76d dc 02              LDD _Num_of_O
0983 e76f 38                 PULX
0984 e770 bd e9 f8           JSR __ULT
0985 e773 1a 83 00 00        CPD #0
```

Appendix B: Cross Compiling the Fuzzy Kernel

```
0986 e777 26 03              BNE  *+5
0987 e779 7e e8 67            JMP  _55
0988 e77c 7e e7 8f            JMP  _57
0989                    _54
0990 e77f 4f                  CLRA
0991 e780 f6 02 1e            LDAB _j
0992 e783 c3 00 01            ADDD #1
0993 e786 f7 02 1e            STAB _j
0994 e789 83 00 01            SUBD #1
0995 e78c 7e e7 67            JMP  _56
0996                    _57
0997 e78f cc 00 00            LDD  #0
0998 e792 18 ed 06            STD  6,Y
0999 e795 cc 00 00            LDD  #0
1000 e798 18 ed 04            STD  4,Y
1001 e79b cc e0 00            LDD  #_CONFIG_B
1002 e79e 37                  PSHB
1003 e79f 36                  PSHA
1004 e7a0 fc 02 1a            LDD  _cfg_inde
1005 e7a3 c3 00 01            ADDD #1
1006 e7a6 fd 02 1a            STD  _cfg_inde
1007 e7a9 83 00 01            SUBD #1
1008 e7ac 38                  PULX
1009 e7ad 05                  LSLD
1010 e7ae 3c                  PSHX
1011 e7af 30                  TSX
1012 e7b0 e3 00               ADDD 0,X
1013 e7b2 38                  PULX
1014 e7b3 37                  PSHB
1015 e7b4 36                  PSHA
1016 e7b5 38                  PULX
1017 e7b6 ec 00               LDD  0,X
1018 e7b8 18 ed 08            STD  8,Y
1019 e7bb cc 00 00            LDD  #0
1020 e7be 18 ed 0a            STD  10,Y
1021                    _60
1022 e7c1 18 ec 0a            LDD  10,Y
1023 e7c4 37                  PSHB
1024 e7c5 36                  PSHA
1025 e7c6 18 ec 08            LDD  8,Y
1026 e7c9 38                  PULX
1027 e7ca bd e9 f8            JSR  __ULT
1028 e7cd 1a 83 00 00         CPD  #0
1029 e7d1 26 03               BNE  *+5
1030 e7d3 7e e8 64            JMP  _59
1031 e7d6 7e e7 e8            JMP  _61
1032                    _58
1033 e7d9 18 ec 0a            LDD  10,Y
1034 e7dc c3 00 01            ADDD #1
1035 e7df 18 ed 0a            STD  10,Y
1036 e7e2 83 00 01            SUBD #1
1037 e7e5 7e e7 c1            JMP  _60
1038                    _61
1039 e7e8 cc 00 04            LDD  #_Mbf_Degr
1040 e7eb 37                  PSHB
1041 e7ec 36                  PSHA
1042 e7ed fc 02 1c            LDD  _m_index
1043 e7f0 c3 00 01            ADDD #1
1044 e7f3 fd 02 1c            STD  _m_index
1045 e7f6 83 00 01            SUBD #1
1046 e7f9 38                  PULX
1047 e7fa 05                  LSLD
```

377

Fuzzy Logic for Real World Design

```
1048 e7fb 3c                   PSHX
1049 e7fc 30                   TSX
1050 e7fd e3 00                ADDD 0,X
1051 e7ff 38                   PULX
1052 e800 37                   PSHB
1053 e801 36                   PSHA
1054 e802 38                   PULX
1055 e803 ec 00                LDD  0,X
1056 e805 18 ed 06             STD  6,Y
1057 e808 18 ec 06             LDD  6,Y
1058 e80b 37                   PSHB
1059 e80c 36                   PSHA
1060 e80d 18 ec 04             LDD  4,Y
1061 e810 38                   PULX
1062 e811 bd e9 cb             JSR  __UGT
1063 e814 1a 83 00 00          CPD  #0
1064 e818 26 03                BNE  *+5
1065 e81a 7e e8 55             JMP  _62
1066 e81d 18 ec 06             LDD  6,Y
1067 e820 18 ed 04             STD  4,Y
1068 e823 cc 02 12             LDD  #_Crisp_Ou
1069 e826 37                   PSHB
1070 e827 36                   PSHA
1071 e828 4f                   CLRA
1072 e829 f6 02 1e             LDAB _j
1073 e82c 38                   PULX
1074 e82d 05                   LSLD
1075 e82e 3c                   PSHX
1076 e82f 30                   TSX
1077 e830 e3 00                ADDD 0,X
1078 e832 38                   PULX
1079 e833 37                   PSHB
1080 e834 36                   PSHA
1081 e835 cc e0 00             LDD  #_CONFIG_B
1082 e838 37                   PSHB
1083 e839 36                   PSHA
1084 e83a fc 02 1a             LDD  _cfg_inde
1085 e83d c3 00 01             ADDD #1
1086 e840 fd 02 1a             STD  _cfg_inde
1087 e843 83 00 01             SUBD #1
1088 e846 38                   PULX
1089 e847 05                   LSLD
1090 e848 3c                   PSHX
1091 e849 30                   TSX
1092 e84a e3 00                ADDD 0,X
1093 e84c 38                   PULX
1094 e84d 37                   PSHB
1095 e84e 36                   PSHA
1096 e84f 38                   PULX
1097 e850 ec 00                LDD  0,X
1098 e852 38                   PULX
1099 e853 ed 00                STD  0,X
1100                      _62
1101 e855 fc 02 1a             LDD  _cfg_inde
1102 e858 c3 00 01             ADDD #1
1103 e85b fd 02 1a             STD  _cfg_inde
1104 e85e 83 00 01             SUBD #1
1105 e861 7e e7 d9             JMP  _58
1106                      _59
1107 e864 7e e7 7f             JMP  _54
1108                      _55
1109 e867 18 38                PULY
```

Appendix B: Cross Compiling the Fuzzy Kernel

```
1110  e869 18 35                TYS
1111  e86b 18 38                PULY
1112  e86d 39                   RTS


1113  0006              ARG1    EQU   6
1114  0008              ARG2    EQU   8
1115                    _CCARGC
1116  e86e 18 3c                PSHY
1117  e870 18 30                TSY
1118  e872 18 3c                PSHY
1119  e874 18 30                TSY
1120  e876 3c                   PSHX
1121  e877 32                   PULA
1122  e878 33                   PULB
1123  e879 18 38                PULY
1124  e87b 18 35                TYS
1125  e87d 18 38                PULY
1126  e87f 39                   RTS
1127                    _e_int
1128  e880 18 3c                PSHY
1129  e882 18 30                TSY
1130  e884 18 3c                PSHY
1131  e886 18 30                TSY
1132  e888 0e                   CLI
1133  e889 18 38                PULY
1134  e88b 18 35                TYS
1135  e88d 18 38                PULY
1136  e88f 39                   RTS
1137                    _d_int
1138  e890 18 3c                PSHY
1139  e892 18 30                TSY
1140  e894 18 3c                PSHY
1141  e896 18 30                TSY
1142  e898 0f                   SEI
1143  e899 18 38                PULY
1144  e89b 18 35                TYS
1145  e89d 18 38                PULY
1146  e89f 39                   RTS
1147                    _bit_set
1148  e8a0 18 3c                PSHY
1149  e8a2 18 30                TSY
1150  e8a4 18 3c                PSHY
1151  e8a6 18 30                TSY
1152  e8a8 cd ee 08             LDX   ARG2,Y
1153  e8ab 18 e6 07             LDAB  ARG1+1,Y
1154  e8ae ea 00                ORAB  0,X
1155  e8b0 e7 00                STAB  0,X
1156  e8b2 18 38                PULY
1157  e8b4 18 35                TYS
1158  e8b6 18 38                PULY
1159  e8b8 39                   RTS
1160                    _bit_clr
1161  e8b9 18 3c                PSHY
1162  e8bb 18 30                TSY
1163  e8bd 18 3c                PSHY
1164  e8bf 18 30                TSY
1165  e8c1 cd ee 08             LDX   ARG2,Y
1166  e8c4 18 e6 07             LDAB  ARG1+1,Y
1167  e8c7 53                   COMB
1168  e8c8 e4 00                ANDB  0,X
1169  e8ca e7 00                STAB  0,X
```

379

```
1170 e8cc 18 38                  PULY
1171 e8ce 18 35                  TYS
1172 e8d0 18 38                  PULY
1173 e8d2 39                     RTS
1174                    _pokeb
1175 e8d3 18 3c                  PSHY
1176 e8d5 18 30                  TSY
1177 e8d7 18 3c                  PSHY
1178 e8d9 18 30                  TSY
1179 e8db cd ee 08               LDX     ARG2,Y
1180 e8de 18 e6 07               LDAB    ARG1+1,Y
1181 e8e1 e7 00                  STAB    0,X
1182 e8e3 18 38                  PULY
1183 e8e5 18 35                  TYS
1184 e8e7 18 38                  PULY
1185 e8e9 39                     RTS
1186                    _poke
1187 e8ea 18 3c                  PSHY
1188 e8ec 18 30                  TSY
1189 e8ee 18 3c                  PSHY
1190 e8f0 18 30                  TSY
1191 e8f2 cd ee 08               LDX ARG2,Y
1192 e8f5 18 ec 06               LDD ARG1,Y
1193 e8f8 ed 00                  STD 0,X
1194 e8fa 18 38                  PULY
1195 e8fc 18 35                  TYS
1196 e8fe 18 38                  PULY
1197 e900 39                     RTS
1198                    _eepokeb
1199 e901 18 3c                  PSHY
1200 e903 18 30                  TSY
1201 e905 18 3c                  PSHY
1202 e907 18 30                  TSY
1203 e909 cd ee 08               LDX     ARG2,Y
1204 e90c 18 e6 07               LDAB    ARG1+1,Y
1205 e90f e6 00                  LDAB    0,X
1206 e911 c1 ff                  CMPB    #$FF
1207 e913 27 14                  BEQ     PROG
1208 e915 c6 16                  LDAB    #$16
1209 e917 f7 b0 3b               STAB    $B03B
1210 e91a c6 ff                  LDAB    #$FF
1211 e91c e7 00                  STAB    0,X
1212 e91e c6 17                  LDAB    #$17
1213 e920 f7 b0 3b               STAB    $B03B
1214 e923 bd e9 43               JSR     DLY12
1215 e926 7f b0 3b               CLR     $B03B
1216 e929 c6 02         PROG     LDAB    #$02
1217 e92b f7 b0 3b               STAB    $B03B
1218 e92e cd ee 08               LDX     ARG2,Y
1219 e931 18 e6 07               LDAB    ARG1+1,Y
1220 e934 e7 00                  STAB    0,X
1221 e936 c6 03                  LDAB    #$03
1222 e938 f7 b0 3b               STAB    $B03B
1223 e93b bd e9 43               JSR     DLY12
1224 e93e 7f b0 3b               CLR     $B03B
1225 e941 20 09                  BRA     EXIT
1226 e943 3c         DLY12       PSHX
1227 e944 ce 08 00               LDX     #$0800
1228 e947 09         DLOOP       DEX
1229 e948 26 fd                  BNE     DLOOP
1230 e94a 38                     PULX
1231 e94b 39                     RTS
```

Appendix B: Cross Compiling the Fuzzy Kernel

```
1232 e94c 01                    EXIT    NOP
1233 e94d 18 38                         PULY
1234 e94f 18 35                         TYS
1235 e951 18 38                         PULY
1236 e953 39                            RTS
1237                            _peekb
1238 e954 18 3c                         PSHY
1239 e956 18 30                         TSY
1240 e958 18 3c                         PSHY
1241 e95a 18 30                         TSY
1242 e95c cd ee 06                      LDX     ARG1,Y
1243 e95f 4f                            CLRA
1244 e960 e6 00                         LDAB    0,X
1245 e962 18 38                         PULY
1246 e964 18 35                         TYS
1247 e966 18 38                         PULY
1248 e968 39                            RTS
1249                            _peek
1250 e969 18 3c                         PSHY
1251 e96b 18 30                         TSY
1252 e96d 18 3c                         PSHY
1253 e96f 18 30                         TSY
1254 e971 cd ee 06                      LDX ARG1,Y
1255 e974 ec 00                         LDD 0,X
1256 e976 18 38                         PULY
1257 e978 18 35                         TYS
1258 e97a 18 38                         PULY
1259 e97c 39                            RTS
1260
****************************************************
1261                    ***     START OF LOW LEVEL LIBRARY ROUTINES   ****
1262                    ***     PARTS BORROWED FROM MOTOROLA, OTHERS  ****
1263                    ***     RE-WRITTEN BY MATT TAYLOR             ****
1264
****************************************************
1265                            ********
1266                            * MUL12
1267                            __MUL12
1268 e97d 3c                            PSHX
1269 e97e 3c                            PSHX
1270 e97f 37                            PSHB
1271 e980 36                            PSHA
1272 e981 30                            TSX
1273 e982 a6 03                         LDAA    3,X
1274 e984 3d                            MUL
1275 e985 ed 04                         STD     4,X
1276 e987 ec 01                         LDD     1,X
1277 e989 3d                            MUL
1278 e98a eb 04                         ADDB    4,X
1279 e98c e7 04                         STAB    4,X
1280 e98e e6 03                         LDAB    3,X
1281 e990 3d                            MUL
1282 e991 eb 04                         ADDB    4,X
1283 e993 e7 04                         STAB    4,X
1284 e995 38                            PULX
1285 e996 38                            PULX
1286 e997 32                            PULA
1287 e998 33                            PULB
1288 e999 39                            RTS
1289                            ********
1290                            * ASR12
1291                            __ASR12
```

381

Fuzzy Logic for Real World Design

```
1292 e99a 8f                   XGDX
1293                      __ASRLOOP
1294 e99b 8c 00 00             CPX   #0
1295 e99e 27 05                BEQ   __ASRDONE
1296 e9a0 47                   ASRA
1297 e9a1 56                   RORB
1298 e9a2 09                   DEX
1299 e9a3 20 f6                BRA   __ASRLOOP
1300                      __ASRDONE
1301 e9a5 39                   RTS
1302                      ********
1303                      * ASL12
1304                      __ASL12
1305 e9a6 8f                   XGDX
1306                      __ASLLOOP
1307 e9a7 8c 00 00             CPX   #0
1308 e9aa 27 04                BEQ   __ASLDONE
1309 e9ac 05                   ASLD
1310 e9ad 09                   DEX
1311 e9ae 20 f7                BRA   __ASLLOOP
1312                      __ASLDONE
1313 e9b0 39                   RTS
1314                      ********
1315                      * __GE
1316                      __GE
1317 e9b1 3c                   PSHX
1318 e9b2 1a a3 00             CPD   0,X
1319 e9b5 2d 5c                BLT   __TRUE
1320 e9b7 20 5f                BRA   __FALSE
1321                      ********
1322                      * __UGE
1323                      __UGE
1324 e9b9 3c                   PSHX
1325 e9ba 30                   TSX
1326 e9bb 1a a3 00             CPD   0,X
1327 e9be 25 53                BLO   __TRUE
1328 e9c0 20 56                BRA   __FALSE
1329                      ********
1330                      * __GT
1331                      __GT
1332 e9c2 3c                   PSHX
1333 e9c3 30                   TSX
1334 e9c4 1a a3 00             CPD   0,X
1335 e9c7 2f 4a                BLE   __TRUE
1336 e9c9 20 4d                BRA   __FALSE
1337                      ********
1338                      * __UGT
1339                      __UGT
1340 e9cb 3c                   PSHX
1341 e9cc 30                   TSX
1342 e9cd 1a a3 00             CPD   0,X
1343 e9d0 23 41                BLS   __TRUE
1344 e9d2 20 44                BRA   __FALSE
1345                      ********
1346                      * __LE
1347                      __LE
1348 e9d4 3c                   PSHX
1349 e9d5 30                   TSX
1350 e9d6 1a a3 00             CPD   0,X
1351 e9d9 2e 38                BGT   __TRUE
1352 e9db 20 3b                BRA   __FALSE
1353                      ********
```

382

```
1354                        *    __ULE
1355                        __ULE
1356  e9dd  3c                   PSHX
1357  e9de  30                   TSX
1358  e9df  1a a3 00             CPD    0,X
1359  e9e2  22 2f                BHI    __TRUE
1360  e9e4  20 32                BRA    __FALSE
1361                        ********
1362                        *    __LNEG
1363                        __LNEG
1364  e9e6  3c                   PSHX
1365  e9e7  1a 83 00 00          CPD    #0
1366  e9eb  26 2b                BNE    __FALSE
1367  e9ed  20 24                BRA    __TRUE
1368                        ********
1369                        *    __LT
1370                        __LT
1371  e9ef  3c                   PSHX
1372  e9f0  30                   TSX
1373  e9f1  1a a3 00             CPD    0,X
1374  e9f4  2c 1d                BGE    __TRUE
1375  e9f6  20 20                BRA    __FALSE
1376                        ********
1377                        *    __ULT
1378                        __ULT
1379  e9f8  3c                   PSHX
1380  e9f9  30                   TSX
1381  e9fa  1a a3 00             CPD    0,X
1382  e9fd  24 14                BHS    __TRUE
1383  e9ff  20 17                BRA    __FALSE
1384                        ********
1385                        *    __EQ
1386                        __EQ
1387  ea01  3c                   PSHX
1388  ea02  30                   TSX
1389  ea03  1a a3 00             CPD    0,X
1390  ea06  27 0b                BEQ    __TRUE
1391  ea08  20 0e                BRA    __FALSE
1392                        ********
1393                        *    __NE
1394                        __NE
1395  ea0a  3c                   PSHX
1396  ea0b  30                   TSX
1397  ea0c  1a a3 00             CPD    0,X
1398  ea0f  26 02                BNE    __TRUE
1399  ea11  20 05                BRA    __FALSE
1400                        ********
1401                        *__TRUE
1402                        __TRUE
1403  ea13  38                   PULX
1404  ea14  cc 00 01             LDD    #1
1405  ea17  39                   RTS
1406                        ********
1407                        *__FALSE
1408                        __FALSE
1409  ea18  38                   PULX
1410  ea19  cc 00 00             LDD    #0
1411  ea1c  39                   RTS
1412                        ********
1413                        *    __SWITCH
1414                        __SWITCH
1415  ea1d  38                   PULX
```

383

```
1416                          __XLOOP
1417 ea1e 3c                    PSHX
1418 ea1f ee 00                 LDX   0,X
1419 ea21 8c 00 00              CPX   #0
1420 ea24 38                    PULX
1421 ea25 27 0b                 BEQ   __XDEFAULT
1422 ea27 1a a3 02              CPD   2,X
1423 ea2a 27 0a                 BEQ   __XMATCH
1424 ea2c 08                    INX
1425 ea2d 08                    INX
1426 ea2e 08                    INX
1427 ea2f 08                    INX
1428 ea30 20 ec                 BRA   __XLOOP
1429                          __XDEFAULT
1430 ea32 08                    INX
1431 ea33 08                    INX
1432 ea34 6e 00                 JMP   0,X
1433                          __XMATCH
1434 ea36 ee 00                 LDX   0,X
1435 ea38 6e 00                 JMP   0,X
```

Target: Motorola 68HC16

In this section we will look at compiling the fuzzy kernel for the Motorola 68HC16 a 16-bit microcontroller. For a complete discussion of the design and ANSI C code implementation of the kernel refer to Chapter 6.

Four routines comprise the core of the fuzzy kernel. Those routines are: *Init_Fuzzy_Kernel*, *Fuzzify_Inputs*, *Eval_Rulebase*, and *Defuzzify_Rulebase*. A development tool offered by Archimedes Software[80] produced the following listings. The assembly code size for these four routines is 1014 bytes.

The assembly code listing below reflect the fuzzy kernel compiled for MIN intersection, MAX aggregation, and weighted average defuzzification.

```
Procedure _Init_Fuzzy_Kernel At address: 0 code size: 122
dec PC:    0 00000000 3410           PSHM    Z
dec PC:    2 00000002 276F           TSZ
dec PC:    4 00000004 37FF00000000   MOVW    CONFIG_BLOCK,Num_of_Inputs
dec PC:   10 0000000A 37FF00020000   MOVW    CONFIG_BLOCK:2,Num_of_Outputs
dec PC:   16 00000010 37FE00000000   MOVB    RULE_BLOCK,Num_of_Rules
dec PC:   22 00000016 17350000       CLR     j
```

---

[80]Archimedes Software, 2159 Union Street, San Francisco, CA 94123. Phone (415) 567-4010.

Appendix B: Cross Compiling the Fuzzy Kernel

```
dec PC:      26 0000001A B00E           BRA     @14 ; abs = $00002E
dec PC:      28 0000001C 17F50000       LDAB    j
dec PC:      32 00000020 3705           CLRA
dec PC:      34 00000022 27F4           ASLD
dec PC:      36 00000024 37CC           XGDX
dec PC:      38 00000026 27050000       CLRW    Mbf_Degree,X
dec PC:      42 0000002A 17330000       INC     j
dec PC:      46 0000002E 17750000       LDAA    j
dec PC:      50 00000032 780F           CMPA    #15
dec PC:      52 00000034 B5E2           BCS     @-30 ; abs = $00001C
dec PC:      54 00000036 17350000       CLR     j
dec PC:      58 0000003A B00E           BRA     @14 ; abs = $00004E
dec PC:      60 0000003C 17F50000       LDAB    j
dec PC:      64 00000040 3705           CLRA
dec PC:      66 00000042 27F4           ASLD
dec PC:      68 00000044 37CC           XGDX
dec PC:      70 00000046 27050000       CLRW    Crisp_Input,X
dec PC:      74 0000004A 17330000       INC     j
dec PC:      78 0000004E 17750000       LDAA    j
dec PC:      82 00000052 7802           CMPA    #2
dec PC:      84 00000054 B5E2           BCS     @-30 ; abs = $00003C
dec PC:      86 00000056 17350000       CLR     j
dec PC:      90 0000005A B00E           BRA     @14 ; abs = $00006E
dec PC:      92 0000005C 17F50000       LDAB    j
dec PC:      96 00000060 3705           CLRA
dec PC:      98 00000062 27F4           ASLD
dec PC:     100 00000064 37CC           XGDX
dec PC:     102 00000066 27050000       CLRW    Crisp_Output,X
dec PC:     106 0000006A 17330000       INC     j
dec PC:     110 0000006E 17750000       LDAA    j
dec PC:     114 00000072 7801           CMPA    #1
dec PC:     116 00000074 B5E2           BCS     @-30 ; abs = $00005C
dec PC:     118 00000076 3504           PULM    Z
dec PC:     120 00000078 27F7           RTS


Procedure _Fuzzify_Inputs At address: 0 code size: 514
dec PC:       0 00000000 3FF8           AIS     #-8
dec PC:       2 00000002 3410           PSHM    Z
dec PC:       4 00000004 276F           TSZ
dec PC:       6 00000006 37B50002       LDD     #2
dec PC:      10 0000000A 37FA0000       STD     cfg_index
dec PC:      14 0000000E 27350000       CLRW    m_index
dec PC:      18 00000012 17350000       CLR     j
dec PC:      22 00000016 37800170       LBRA    @0368 ; abs = $00018C
dec PC:      26 0000001A 37F50000       LDD     cfg_index
dec PC:      30 0000001E 27F4           ASLD
dec PC:      32 00000020 37CC           XGDX
dec PC:      34 00000022 17450001       LDAA    CONFIG_BLOCK:1,X
dec PC:      38 00000026 6A09           STAA    num_of_mbfs
dec PC:      40 00000028 27250007       CLRW    k
dec PC:      44 0000002C 37800148       LBRA    @0328 ; abs = $00017A
dec PC:      48 00000030 37F50000       LDD     cfg_index
dec PC:      52 00000034 FC01           ADDD    #1
dec PC:      54 00000036 27F4           ASLD
dec PC:      56 00000038 37CC           XGDX
dec PC:      58 0000003A 17CC0000       LDX     CONFIG_BLOCK,X
dec PC:      62 0000003E 17F50000       LDAB    j
dec PC:      66 00000042 3705           CLRA
dec PC:      68 00000044 27F4           ASLD
dec PC:      70 00000046 37DC           XGDY
dec PC:      72 00000048 175C0000       CPX     Crisp_Input,Y
```

385

Fuzzy Logic for Real World Design

```
dec PC:      76 0000004C B41A         BCC    @26 ; abs = $00006C
dec PC:      78 0000004E 37F50000     LDD    cfg_index
dec PC:      82 00000052 FC04         ADDD   #4
dec PC:      84 00000054 27F4         ASLD
dec PC:      86 00000056 37CC         XGDX
dec PC:      88 00000058 17CC0000     LDX    CONFIG_BLOCK,X
dec PC:      92 0000005C 17F50000     LDAB   j
dec PC:      96 00000060 3705         CLRA
dec PC:      98 00000062 27F4         ASLD
dec PC:     100 00000064 37DC         XGDY
dec PC:     102 00000066 175C0000     CPX    Crisp_Input,Y
dec PC:     106 0000006A B210         BHI    @16 ; abs = $000080
dec PC:     108 0000006C 37F50000     LDD    m_index
dec PC:     112 00000070 27330000     INCW   m_index
dec PC:     116 00000074 27F4         ASLD
dec PC:     118 00000076 37CC         XGDX
dec PC:     120 00000078 27050000     CLRW   Mbf_Degree,X
dec PC:     124 0000007C 378000EA     LBRA   @0234 ; abs = $00016C
dec PC:     128 00000080 37F50000     LDD    cfg_index
dec PC:     132 00000084 FC03         ADDD   #3
dec PC:     134 00000086 27F4         ASLD
dec PC:     136 00000088 37CC         XGDX
dec PC:     138 0000008A 17CC0000     LDX    CONFIG_BLOCK,X
dec PC:     142 0000008E 17F50000     LDAB   j
dec PC:     146 00000092 3705         CLRA
dec PC:     148 00000094 27F4         ASLD
dec PC:     150 00000096 37DC         XGDY
dec PC:     152 00000098 175C0000     CPX    Crisp_Input,Y
dec PC:     156 0000009C B532         BCS    @50 ; abs = $0000D4
dec PC:     158 0000009E 37F50000     LDD    cfg_index
dec PC:     162 000000A2 FC02         ADDD   #2
dec PC:     164 000000A4 27F4         ASLD
dec PC:     166 000000A6 37CC         XGDX
dec PC:     168 000000A8 17CC0000     LDX    CONFIG_BLOCK,X
dec PC:     172 000000AC 17F50000     LDAB   j
dec PC:     176 000000B0 3705         CLRA
dec PC:     178 000000B2 27F4         ASLD
dec PC:     180 000000B4 37DC         XGDY
dec PC:     182 000000B6 175C0000     CPX    Crisp_Input,Y
dec PC:     186 000000BA B214         BHI    @20 ; abs = $0000D4
dec PC:     188 000000BC 37F50000     LDD    m_index
dec PC:     192 000000C0 27330000     INCW   m_index
dec PC:     196 000000C4 27F4         ASLD
dec PC:     198 000000C6 37CC         XGDX
dec PC:     200 000000C8 37B5FFFF     LDD    #-1
dec PC:     204 000000CC 37CA0000     STD    Mbf_Degree,X
dec PC:     208 000000D0 37800096     LBRA   @0150 ; abs = $00016C
dec PC:     212 000000D4 37F50000     LDD    cfg_index
dec PC:     216 000000D8 FC02         ADDD   #2
dec PC:     218 000000DA 27F4         ASLD
dec PC:     220 000000DC 37CC         XGDX
dec PC:     222 000000DE 17CC0000     LDX    CONFIG_BLOCK,X
dec PC:     226 000000E2 17F50000     LDAB   j
dec PC:     230 000000E6 3705         CLRA
dec PC:     232 000000E8 27F4         ASLD
dec PC:     234 000000EA 37DC         XGDY
dec PC:     236 000000EC 175C0000     CPX    Crisp_Input,Y
dec PC:     240 000000F0 B33A         BLS    @58 ; abs = $000130
dec PC:     242 000000F2 37F50000     LDD    cfg_index
dec PC:     246 000000F6 FC01         ADDD   #1
dec PC:     248 000000F8 27F4         ASLD
dec PC:     250 000000FA 37CC         XGDX
```

Appendix B: Cross Compiling the Fuzzy Kernel

```
dec PC:   252 000000FC 17F50000     LDAB   j
dec PC:   256 00000100 3705         CLRA
dec PC:   258 00000102 27F4         ASLD
dec PC:   260 00000104 37DC         XGDY
dec PC:   262 00000106 37D50000     LDD    Crisp_Input,Y
dec PC:   266 0000010A 37C00000     SUBD   CONFIG_BLOCK,X
dec PC:   270 0000010E 37750000     LDE    cfg_index
dec PC:   274 00000112 7C05         ADDE   #5
dec PC:   276 00000114 2774         ASLE
dec PC:   278 00000116 374C         XGEX
dec PC:   280 00000118 37450000     LDE    CONFIG_BLOCK,X
dec PC:   284 0000011C 3725         EMUL
dec PC:   286 0000011E 37750000     LDE    m_index
dec PC:   290 00000122 27330000     INCW   m_index
dec PC:   294 00000126 2774         ASLE
dec PC:   296 00000128 374C         XGEX
dec PC:   298 0000012A 37CA0000     STD    Mbf_Degree,X
dec PC:   302 0000012E B038         BRA    @56 ; abs = $00016C
dec PC:   304 00000130 37F50000     LDD    cfg_index
dec PC:   308 00000134 FC04         ADDD   #4
dec PC:   310 00000136 27F4         ASLD
dec PC:   312 00000138 37CC         XGDX
dec PC:   314 0000013A 17F50000     LDAB   j
dec PC:   318 0000013E 3705         CLRA
dec PC:   320 00000140 27F4         ASLD
dec PC:   322 00000142 37DC         XGDY
dec PC:   324 00000144 37C50000     LDD    CONFIG_BLOCK,X
dec PC:   328 00000148 37D00000     SUBD   Crisp_Input,Y
dec PC:   332 0000014C 37750000     LDE    cfg_index
dec PC:   336 00000150 7C06         ADDE   #6
dec PC:   338 00000152 2774         ASLE
dec PC:   340 00000154 374C         XGEX
dec PC:   342 00000156 37450000     LDE    CONFIG_BLOCK,X
dec PC:   346 0000015A 3725         EMUL
dec PC:   348 0000015C 37750000     LDE    m_index
dec PC:   352 00000160 27330000     INCW   m_index
dec PC:   356 00000164 2774         ASLE
dec PC:   358 00000166 374C         XGEX
dec PC:   360 00000168 37CA0000     STD    Mbf_Degree,X
dec PC:   364 0000016C 37F50000     LDD    cfg_index
dec PC:   368 00000170 FC06         ADDD   #6
dec PC:   370 00000172 37FA0000     STD    cfg_index
dec PC:   374 00000176 27230007     INCW   k
dec PC:   378 0000017A E509         LDAB   num_of_mbfs
dec PC:   380 0000017C 3705         CLRA
dec PC:   382 0000017E A807         CPD    k
dec PC:   384 00000180 3782FEAA     LBHI   @-0342 ; abs = $000030
dec PC:   388 00000184 27330000     INCW   cfg_index
dec PC:   392 00000188 17330000     INC    j
dec PC:   396 0000018C 17F50000     LDAB   j
dec PC:   400 00000190 3705         CLRA
dec PC:   402 00000192 37F80000     CPD    Num_of_Inputs
dec PC:   406 00000196 3785FE7E     LBCS   @-0386 ; abs = $00001A
dec PC:   410 0000019A 37F50000     LDD    cfg_index
dec PC:   414 0000019E AA05         STD    tmp1
dec PC:   416 000001A0 37F50000     LDD    m_index
dec PC:   420 000001A4 AA03         STD    tmp2
dec PC:   422 000001A6 17350000     CLR    j
dec PC:   426 000001AA B034         BRA    @52 ; abs = $0001E4
dec PC:   428 000001AC 37F50000     LDD    cfg_index
dec PC:   432 000001B0 27F4         ASLD
dec PC:   434 000001B2 37CC         XGDX
```

387

Fuzzy Logic for Real World Design

```
dec PC:   436 000001B4 17450001    LDAA    CONFIG_BLOCK:1,X
dec PC:   440 000001B8 6A09        STAA    num_of_mbfs
dec PC:   442 000001BA 27250007    CLRW    k
dec PC:   446 000001BE B014        BRA     @20 ; abs = $0001D8
dec PC:   448 000001C0 37F50000    LDD     m_index
dec PC:   452 000001C4 27330000    INCW    m_index
dec PC:   456 000001C8 27F4        ASLD
dec PC:   458 000001CA 37CC        XGDX
dec PC:   460 000001CC 27050000    CLRW    Mbf_Degree,X
dec PC:   464 000001D0 27330000    INCW    cfg_index
dec PC:   468 000001D4 27230007    INCW    k
dec PC:   472 000001D8 E509        LDAB    num_of_mbfs
dec PC:   474 000001DA 3705        CLRA
dec PC:   476 000001DC A807        CPD     k
dec PC:   478 000001DE B2DC        BHI     @-36 ; abs = $0001C0
dec PC:   480 000001E0 17330000    INC     j
dec PC:   484 000001E4 17F50000    LDAB    j
dec PC:   488 000001E8 3705        CLRA
dec PC:   490 000001EA 37F80000    CPD     Num_of_Outputs
dec PC:   494 000001EE B5B8        BCS     @-72 ; abs = $0001AC
dec PC:   496 000001F0 A505        LDD     tmp1
dec PC:   498 000001F2 37FA0000    STD     cfg_index
dec PC:   502 000001F6 A503        LDD     tmp2
dec PC:   504 000001F8 37FA0000    STD     m_index
dec PC:   508 000001FC EE00        LDZ     0,Z
dec PC:   510 000001FE 3F0A        AIS     #10
dec PC:   512 00000200 27F7        RTS


Procedure _Eval_Rulebase At address: 0 code size: 182
dec PC:     0 00000000 3FF8        AIS     #-8
dec PC:     2 00000002 3410        PSHM    Z
dec PC:     4 00000004 276F        TSZ
dec PC:     6 00000006 37B50001    LDD     #1
dec PC:    10 0000000A AA03        STD     rule_index
dec PC:    12 0000000C 17350000    CLR     j
dec PC:    16 00000010 3780008E    LBRA    @0142 ; abs = $0000A4
dec PC:    20 00000014 37B5FFFF    LDD     #-1
dec PC:    24 00000018 AA08        STD     rule_strength:2
dec PC:    26 0000001A 27250006    CLRW    rule_strength
dec PC:    30 0000001E B030        BRA     @48 ; abs = $000054
dec PC:    32 00000020 E505        LDAB    index
dec PC:    34 00000022 3705        CLRA
dec PC:    36 00000024 27F4        ASLD
dec PC:    38 00000026 37CC        XGDX
dec PC:    40 00000028 37C50000    LDD     Mbf_Degree,X
dec PC:    44 0000002C 37BC0000    LDX     #0
dec PC:    48 00000030 6C06        CPX     rule_strength
dec PC:    50 00000032 B6FE        BNE     @-02 ; abs = $000036
dec PC:    52 00000034 A808        CPD     rule_strength:2
dec PC:    54 00000036 B504        BCS     @04 ; abs = $000040
dec PC:    56 00000038 A508        LDD     rule_strength:2
dec PC:    58 0000003A 37650006    LDE     rule_strength
dec PC:    62 0000003E B00A        BRA     @10 ; abs = $00004E
dec PC:    64 00000040 E505        LDAB    index
dec PC:    66 00000042 3705        CLRA
dec PC:    68 00000044 27F4        ASLD
dec PC:    70 00000046 37CC        XGDX
dec PC:    72 00000048 37C50000    LDD     Mbf_Degree,X
dec PC:    76 0000004C 2775        CLRE
dec PC:    78 0000004E AA08        STD     rule_strength:2
dec PC:    80 00000050 376A0006    STE     rule_strength
```

388

Appendix B: Cross Compiling the Fuzzy Kernel

```
dec PC:      84 00000054 A503           LDD     rule_index
dec PC:      86 00000056 27230003       INCW    rule_index
dec PC:      90 0000005A 37CC           XGDX
dec PC:      92 0000005C 17450000       LDAA    RULE_BLOCK,X
dec PC:      96 00000060 6A05           STAA    index
dec PC:      98 00000062 78FF           CMPA    #255
dec PC:     100 00000064 B6B6           BNE     @-74 ; abs = $000020
dec PC:     102 00000066 B022           BRA     @34 ; abs = $00008E
dec PC:     104 00000068 E505           LDAB    index
dec PC:     106 0000006A 3705           CLRA
dec PC:     108 0000006C 27F4           ASLD
dec PC:     110 0000006E 37CC           XGDX
dec PC:     112 00000070 37C50000       LDD     Mbf_Degree,X
dec PC:     116 00000074 37BC0000       LDX     #0
dec PC:     120 00000078 6C06           CPX     rule_strength
dec PC:     122 0000007A B6FE           BNE     @-02 ; abs = $00007E
dec PC:     124 0000007C A808           CPD     rule_strength:2
dec PC:     126 0000007E B40A           BCC     @10 ; abs = $00008E
dec PC:     128 00000080 E505           LDAB    index
dec PC:     130 00000082 3705           CLRA
dec PC:     132 00000084 27F4           ASLD
dec PC:     134 00000086 37CC           XGDX
dec PC:     136 00000088 A508           LDD     rule_strength:2
dec PC:     138 0000008A 37CA0000       STD     Mbf_Degree,X
dec PC:     142 0000008E A503           LDD     rule_index
dec PC:     144 00000090 27230003       INCW    rule_index
dec PC:     148 00000094 37CC           XGDX
dec PC:     150 00000096 17450000       LDAA    RULE_BLOCK,X
dec PC:     154 0000009A 6A05           STAA    index
dec PC:     156 0000009C 78FF           CMPA    #255
dec PC:     158 0000009E B6C4           BNE     @-60 ; abs = $000068
dec PC:     160 000000A0 17330000       INC     j
dec PC:     164 000000A4 17750000       LDAA    j
dec PC:     168 000000A8 17780000       CMPA    Num_of_Rules
dec PC:     172 000000AC 3785FF62       LBCS    @-0158 ; abs = $000014
dec PC:     176 000000B0 EE00           LDZ     0,Z
dec PC:     178 000000B2 3F0A           AIS     #10
dec PC:     180 000000B4 27F7           RTS

Procedure _Defuzzify_Rulebase At address: 0 code size: 196
dec PC:       0 00000000 3FF0           AIS     #-16
dec PC:       2 00000002 3410           PSHM    Z
dec PC:       4 00000004 276F           TSZ
dec PC:       6 00000006 17350000       CLR     j
dec PC:      10 0000000A 378000A0       LBRA    @0160 ; abs = $0000B0
dec PC:      14 0000000E 2725000C       CLRW    temp1:2
dec PC:      18 00000012 2725000A       CLRW    temp1
dec PC:      22 00000016 27250008       CLRW    temp2:2
dec PC:      26 0000001A 27250006       CLRW    temp2
dec PC:      30 0000001E 37F50000       LDD     cfg_index
dec PC:      34 00000022 27330000       INCW    cfg_index
dec PC:      38 00000026 27F4           ASLD
dec PC:      40 00000028 37CC           XGDX
dec PC:      42 0000002A 37C50000       LDD     CONFIG_BLOCK,X
dec PC:      46 0000002E AA0E           STD     num_mbfs
dec PC:      48 00000030 27250010       CLRW    k
dec PC:      52 00000034 B04E           BRA     @78 ; abs = $000088
dec PC:      54 00000036 37F50000       LDD     m_index
dec PC:      58 0000003A 27330000       INCW    m_index
dec PC:      62 0000003E 27F4           ASLD
dec PC:      64 00000040 37CC           XGDX
```

389

Fuzzy Logic for Real World Design

```
dec PC:   66  00000042  37C50000    LDD    Mbf_Degree,X
dec PC:   70  00000046  AA04        STD    degree:2
dec PC:   72  00000048  27250002    CLRW   degree
dec PC:   76  0000004C  3765000C    LDE    temp1:2
dec PC:   80  00000050  A50A        LDD    temp1
dec PC:   82  00000052  37610004    ADDE   degree:2
dec PC:   86  00000056  A302        ADCD   degree
dec PC:   88  00000058  376A000C    STE    temp1:2
dec PC:   92  0000005C  AA0A        STD    temp1
dec PC:   94  0000005E  37F50000    LDD    cfg_index
dec PC:   98  00000062  27330000    INCW   cfg_index
dec PC:  102  00000066  27F4        ASLD
dec PC:  104  00000068  37CC        XGDX
dec PC:  106  0000006A  37C50000    LDD    CONFIG_BLOCK,X
dec PC:  110  0000006E  2775        CLRE
dec PC:  112  00000070  275E        TZY
dec PC:  114  00000072  3D02        AIY    @degree
dec PC:  116  00000074  FA000000    JSR    _LMULU
dec PC:  120  00000078  A108        ADDD   temp2:2
dec PC:  122  0000007A  37630006    ADCE   temp2
dec PC:  126  0000007E  AA08        STD    temp2:2
dec PC:  128  00000080  376A0006    STE    temp2
dec PC:  132  00000084  27230010    INCW   k
dec PC:  136  00000088  A510        LDD    k
dec PC:  138  0000008A  A80E        CPD    num_mbfs
dec PC:  140  0000008C  B5A4        BCS    @-92 ; abs = $000036
dec PC:  142  0000008E  A508        LDD    temp2:2
dec PC:  144  00000090  37650006    LDE    temp2
dec PC:  148  00000094  275E        TZY
dec PC:  150  00000096  3D0A        AIY    @temp1
dec PC:  152  00000098  FA000000    JSR    _LDIVU
dec PC:  156  0000009C  277B        TDE
dec PC:  158  0000009E  17F50000    LDAB   j
dec PC:  162  000000A2  3705        CLRA
dec PC:  164  000000A4  27F4        ASLD
dec PC:  166  000000A6  37CC        XGDX
dec PC:  168  000000A8  374A0000    STE    Crisp_Output,X
dec PC:  172  000000AC  17330000    INC    j
dec PC:  176  000000B0  17F50000    LDAB   j
dec PC:  180  000000B4  3705        CLRA
dec PC:  182  000000B6  37F80000    CPD    Num_of_Outputs
dec PC:  186  000000BA  3785FF4E    LBCS   @-0178 ; abs = $00000E
dec PC:  190  000000BE  EE00        LDZ    0,Z
dec PC:  192  000000C0  3F12        AIS    #18
dec PC:  194  000000C2  27F7        RTS
```

Target: Intel 80C196

In this section we will look at compiling the fuzzy kernel for the Intel 80C196 16-bit microcontroller. For a complete discussion of the design and ANSI C code implementation of the kernel refer to Chapter 6.

Four routines comprise the core of the fuzzy kernel. Those routines are: *Init_Fuzzy_Kernel, Fuzzify_Inputs, Eval_Rulebase,* and

Appendix B: Cross Compiling the Fuzzy Kernel

Defuzzify_Rulebase. A development tool offered by IAR Systems[81] produced the following listings. Variable declarations can be seen before the assembly code. The configuration and rule blocks are declared at the end of the listings. The assembly code size for the four main routines is 1386 bytes. The configuration and rule blocks consume another 265 bytes.

The assembly code listings below reflect the fuzzy kernel options of MIN intersection, MAX aggregation, and weighted average defuzzification.

```
\   0000                       NAME     fz80196(16)
\   0000                       RSEG     CODE(1)
\   0000                       RSEG     DATA(1)
\   0000                       RSEG     CONST(1)
\   0000                       RSEG     ZVECT(1)
\   0000                       PUBLIC   CONFIG_BLOCK
\   0000                       PUBLIC   Crisp_Input
\   0000                       PUBLIC   Crisp_Output
\   0000                       PUBLIC   Defuzzify_Rulebase
\   0000                       PUBLIC   Eval_Rulebase
\   0000                       PUBLIC   Fuzzify_Inputs
\   0000                       PUBLIC   Init_Fuzzy_Kernel
\   0000                       PUBLIC   Mbf_Degree
\   0000                       PUBLIC   Num_of_Inputs
\   0000                       PUBLIC   Num_of_Outputs
\   0000                       PUBLIC   Num_of_Rules
\   0000                       PUBLIC   RULE_BLOCK
\   0000                       PUBLIC   cfg_index
\   0000                       PUBLIC   j
\   0000                       PUBLIC   m_index
\   0000                       EXTERN   ?UL_MUL_L02
\   0000                       EXTERN   ?UL_DIV_SWAP_L02
\   0000                       EXTERN   ?CL8096_3_01_L00
\   0000                       RSEG     CODE
\   0000             $REG8096.INC


\   0000                       Init_Fuzzy_Kernel:
\   0000   A3010000             LD     R0,CONFIG_BLOCK[0]
\   0004   1C
\   0005   C3010000             ST     R0,Num_of_Inputs[0]
\   0009   1C
\   000A   A3010200             LD     R0,CONFIG_BLOCK+2[0]
\   000E   1C
\   000F   C3010200             ST     R0,Num_of_Outputs[0]
\   0013   1C
\   0014   B3018C00             LDB    R0,RULE_BLOCK[0]
\   0018   1C
\   0019   C7011F02             STB    R0,Num_of_Rules[0]
\   001D   1C
```

---

[81]IAR Systems Software, Inc. One Maritime Plaza, Suite 1770, San Francisco, CA 94111. Phone (415) 765-5500.

391

Fuzzy Logic for Real World Design

```
\   001E   C7011E02            STB     0,j[0]
\   0022   00
\   0023              ?0003:
\   0023   B3011E02            LDB     R0,j[0]
\   0027   1C
\   0028   99FF1C              CMPB    R0,#255
\   002B   DB20                JC      ?0002
\   002D              ?0004:
\   002D   A104001C            LD      R0,#Mbf_Degree
\   0031   AF011E02            LDBZE   R4,j[0]
\   0035   20
\   0036   642020              ADD     R4,R4
\   0039   64201C              ADD     R0,R4
\   003C   C21C00              ST      0,[R0]
\   003F   B3011E02            LDB     R0,j[0]
\   0043   1C
\   0044   171C                INCB    R0
\   0046   C7011E02            STB     R0,j[0]
\   004A   1C
\   004B   27D6                SJMP    ?0003
\   004D              ?0002:
\   004D   C7011E02            STB     0,j[0]
\   0051   00
\   0052              ?0007:
\   0052   B3011E02            LDB     R0,j[0]
\   0056   1C
\   0057   99081C              CMPB    R0,#8
\   005A   DB20                JC      ?0006
\   005C              ?0008:
\   005C   A102021C            LD      R0,#Crisp_Input
\   0060   AF011E02            LDBZE   R4,j[0]
\   0064   20
\   0065   642020              ADD     R4,R4
\   0068   64201C              ADD     R0,R4
\   006B   C21C00              ST      0,[R0]
\   006E   B3011E02            LDB     R0,j[0]
\   0072   1C
\   0073   171C                INCB    R0
\   0075   C7011E02            STB     R0,j[0]
\   0079   1C
\   007A   27D6                SJMP    ?0007
\   007C              ?0006:
\   007C   C7011E02            STB     0,j[0]
\   0080   00
\   0081              ?0011:
\   0081   B3011E02            LDB     R0,j[0]
\   0085   1C
\   0086   99041C              CMPB    R0,#4
\   0089   DB20                JC      ?0010
\   008B              ?0012:
\   008B   A112021C            LD      R0,#Crisp_Output
\   008F   AF011E02            LDBZE   R4,j[0]
\   0093   20
\   0094   642020              ADD     R4,R4
\   0097   64201C              ADD     R0,R4
\   009A   C21C00              ST      0,[R0]
\   009D   B3011E02            LDB     R0,j[0]
\   00A1   1C
\   00A2   171C                INCB    R0
\   00A4   C7011E02            STB     R0,j[0]
\   00A8   1C
\   00A9   27D6                SJMP    ?0011
```

Appendix B: Cross Compiling the Fuzzy Kernel

```
\   00AB                ?0010:
\   00AB   F0                       RET


\   00AC                Fuzzify_Inputs:
\   00AC   69080018              SUB       SP,#8
\   00B0   BD021C                LDBSE     R0,#2
\   00B3   C3011A02              ST        R0,cfg_index[0]
\   00B7   1C
\   00B8   C3011C02              ST        0,m_index[0]
\   00BC   00
\   00BD   C7011E02              STB       0,j[0]
\   00C1   00
\   00C2                ?0015:
\   00C2   AF011E02              LDBZE     R0,j[0]
\   00C6   1C
\   00C7   8B010000              CMP       R0,Num_of_Inputs[0]
\   00CB   1C
\   00CC   D303                  JNC       $+5
\   00CE   E7F501                LJMP      ?0014
\   00D1                ?0016:
\   00D1   A3011A02              LD        R0,cfg_index[0]
\   00D5   1C
\   00D6   641C1C                ADD       R0,R0
\   00D9   6500001C              ADD       R0,#CONFIG_BLOCK
\   00DD   B21C1C                LDB       R0,[R0]
\   00E0   C6181C                STB       R0,[SP]
\   00E3   C3180200              ST        0,2[SP]
\   00E7                ?0019:
\   00E7   A318021C              LD        R0,2[SP]
\   00EB   AE1820                LDBZE     R4,[SP]
\   00EE   88201C                CMP       R0,R4
\   00F1   D303                  JNC       $+5
\   00F3   E7B601                LJMP      ?0018
\   00F6                ?0020:
\   00F6   A102021C              LD        R0,#Crisp_Input
\   00FA   AF011E02              LDBZE     R4,j[0]
\   00FE   20
\   00FF   642020                ADD       R4,R4
\   0102   64201C                ADD       R0,R4
\   0105   A21C1C                LD        R0,[R0]
\   0108   A3011A02              LD        R4,cfg_index[0]
\   010C   20
\   010D   642020                ADD       R4,R4
\   0110   65020020              ADD       R4,#CONFIG_BLOCK+2
\   0114   8A201C                CMP       R0,[R4]
\   0117   D123                  JNH       ?0022
\   0119   A102021C              LD        R0,#Crisp_Input
\   011D   AF011E02              LDBZE     R4,j[0]
\   0121   20
\   0122   642020                ADD       R4,R4
\   0125   64201C                ADD       R0,R4
\   0128   A21C1C                LD        R0,[R0]
\   012B   A3011A02              LD        R4,cfg_index[0]
\   012F   20
\   0130   642020                ADD       R4,R4
\   0133   65080020              ADD       R4,#CONFIG_BLOCK+8
\   0137   8A201C                CMP       R0,[R4]
\   013A   D31D                  JNC       ?0023
\   013C                ?0024:
\   013C                ?0025:
\   013C                ?0022:
```

Fuzzy Logic for Real World Design

```
\   013C  A104001C              LD      R0,#Mbf_Degree
\   0140  A3011C02              LD      R4,m_index[0]
\   0144  20
\   0145  0720                  INC     R4
\   0147  C3011C02              ST      R4,m_index[0]
\   014B  20
\   014C  0520                  DEC     R4
\   014E  642020                ADD     R4,R4
\   0151  64201C                ADD     R0,R4
\   0154  C21C00                ST      0,[R0]
\   0157  213A                  SJMP    ?0026
\   0159                ?0023:
\   0159  A102021C              LD      R0,#Crisp_Input
\   015D  AF011E02              LDBZE   R4,j[0]
\   0161  20
\   0162  642020                ADD     R4,R4
\   0165  64201C                ADD     R0,R4
\   0168  A21C1C                LD      R0,[R0]
\   016B  A3011A02              LD      R4,cfg_index[0]
\   016F  20
\   0170  642020                ADD     R4,R4
\   0173  65060020              ADD     R4,#CONFIG_BLOCK+6
\   0177  8A201C                CMP     R0,[R4]
\   017A  D944                  JH      ?0028
\   017C  A102021C              LD      R0,#Crisp_Input
\   0180  AF011E02              LDBZE   R4,j[0]
\   0184  20
\   0185  642020                ADD     R4,R4
\   0188  64201C                ADD     R0,R4
\   018B  A21C1C                LD      R0,[R0]
\   018E  A3011A02              LD      R4,cfg_index[0]
\   0192  20
\   0193  642020                ADD     R4,R4
\   0196  65040020              ADD     R4,#CONFIG_BLOCK+4
\   019A  8A201C                CMP     R0,[R4]
\   019D  D321                  JNC     ?0028
\   019F                ?0030:
\   019F                ?0029:
\   019F                ?0027:
\   019F  A104001C              LD      R0,#Mbf_Degree
\   01A3  A3011C02              LD      R4,m_index[0]
\   01A7  20
\   01A8  0720                  INC     R4
\   01AA  C3011C02              ST      R4,m_index[0]
\   01AE  20
\   01AF  0520                  DEC     R4
\   01B1  642020                ADD     R4,R4
\   01B4  64201C                ADD     R0,R4
\   01B7  A1FFFF20              LD      R4,#-1
\   01BB  C21C20                ST      R4,[R0]
\   01BE  20D3                  SJMP    ?0031
\   01C0                ?0028:
\   01C0  A102021C              LD      R0,#Crisp_Input
\   01C4  AF011E02              LDBZE   R4,j[0]
\   01C8  20
\   01C9  642020                ADD     R4,R4
\   01CC  64201C                ADD     R0,R4
\   01CF  A21C1C                LD      R0,[R0]
\   01D2  A3011A02              LD      R4,cfg_index[0]
\   01D6  20
\   01D7  642020                ADD     R4,R4
\   01DA  65040020              ADD     R4,#CONFIG_BLOCK+4
```

Appendix B: Cross Compiling the Fuzzy Kernel

```
\       01DE    8A201C                      CMP     R0,[R4]
\       01E1    DB57                        JC      ?0033
\       01E3                    ?0032:
\       01E3    A102021C                    LD      R0,#Crisp_Input
\       01E7    AF011E02                    LDBZE   R4,j[0]
\       01EB    20
\       01EC    642020                      ADD     R4,R4
\       01EF    64201C                      ADD     R0,R4
\       01F2    A21C1C                      LD      R0,[R0]
\       01F5    A3011A02                    LD      R4,cfg_index[0]
\       01F9    20
\       01FA    642020                      ADD     R4,R4
\       01FD    65020020                    ADD     R4,#CONFIG_BLOCK+2
\       0201    A22020                      LD      R4,[R4]
\       0204    68201C                      SUB     R0,R4
\       0207    A3011A02                    LD      R4,cfg_index[0]
\       020B    20
\       020C    642020                      ADD     R4,R4
\       020F    650A0020                    ADD     R4,#CONFIG_BLOCK+10
\       0213    A22020                      LD      R4,[R4]
\       0216    6C201C                      MULU    R0,R4
\       0219    A1040020                    LD      R4,#Mbf_Degree
\       021D    C81C                        PUSH    R0
\       021F    A3011C02                    LD      R0,m_index[0]
\       0223    1C
\       0224    071C                        INC     R0
\       0226    C3011C02                    ST      R0,m_index[0]
\       022A    1C
\       022B    051C                        DEC     R0
\       022D    641C1C                      ADD     R0,R0
\       0230    64201C                      ADD     R0,R4
\       0233    CC20                        POP     R4
\       0235    C21C20                      ST      R4,[R0]
\       0238    2059                        SJMP    ?0034
\       023A                    ?0033:
\       023A    A3011A02                    LD      R0,cfg_index[0]
\       023E    1C
\       023F    641C1C                      ADD     R0,R0
\       0242    6508001C                    ADD     R0,#CONFIG_BLOCK+8
\       0246    A21C1C                      LD      R0,[R0]
\       0249    A1020220                    LD      R4,#Crisp_Input
\       024D    C81C                        PUSH    R0
\       024F    AF011E02                    LDBZE   R0,j[0]
\       0253    1C
\       0254    641C1C                      ADD     R0,R0
\       0257    64201C                      ADD     R0,R4
\       025A    CC20                        POP     R4
\       025C    A21C1C                      LD      R0,[R0]
\       025F    681C20                      SUB     R4,R0
\       0262    A3011A02                    LD      R0,cfg_index[0]
\       0266    1C
\       0267    641C1C                      ADD     R0,R0
\       026A    650C001C                    ADD     R0,#CONFIG_BLOCK+12
\       026E    A21C1C                      LD      R0,[R0]
\       0271    6C201C                      MULU    R0,R4
\       0274    A1040020                    LD      R4,#Mbf_Degree
\       0278    C81C                        PUSH    R0
\       027A    A3011C02                    LD      R0,m_index[0]
\       027E    1C
\       027F    071C                        INC     R0
\       0281    C3011C02                    ST      R0,m_index[0]
\       0285    1C
```

395

Fuzzy Logic for Real World Design

```
\    0286   051C                DEC     R0
\    0288   641C1C              ADD     R0,R0
\    028B   64201C              ADD     R0,R4
\    028E   CC20                POP     R4
\    0290   C21C20              ST      R4,[R0]
\    0293           ?0034:
\    0293           ?0031:
\    0293           ?0026:
\    0293   BD061C              LDBSE   R0,#6
\    0296   67011A02            ADD     R0,cfg_index[0]
\    029A   1C
\    029B   C3011A02            ST      R0,cfg_index[0]
\    029F   1C
\    02A0   A318021C            LD      R0,2[SP]
\    02A4   071C                INC     R0
\    02A6   C318021C            ST      R0,2[SP]
\    02AA   263B                SJMP    ?0019
\    02AC           ?0018:
\    02AC   A3011A02            LD      R0,cfg_index[0]
\    02B0   1C
\    02B1   071C                INC     R0
\    02B3   C3011A02            ST      R0,cfg_index[0]
\    02B7   1C
\    02B8   B3011E02            LDB     R0,j[0]
\    02BC   1C
\    02BD   171C                INCB    R0
\    02BF   C7011E02            STB     R0,j[0]
\    02C3   1C
\    02C4   25FC                SJMP    ?0015
\    02C6           ?0014:
\    02C6   A3011A02            LD      R0,cfg_index[0]
\    02CA   1C
\    02CB   C318041C            ST      R0,4[SP]
\    02CF   A3011C02            LD      R0,m_index[0]
\    02D3   1C
\    02D4   C318061C            ST      R0,6[SP]
\    02D8   C7011E02            STB     0,j[0]
\    02DC   00
\    02DD           ?0036:
\    02DD   AF011E02            LDBZE   R0,j[0]
\    02E1   1C
\    02E2   8B010200            CMP     R0,Num_of_Outputs[0]
\    02E6   1C
\    02E7   DB63                JC      ?0035
\    02E9           ?0037:
\    02E9   A3011A02            LD      R0,cfg_index[0]
\    02ED   1C
\    02EE   641C1C              ADD     R0,R0
\    02F1   6500001C            ADD     R0,#CONFIG_BLOCK
\    02F5   B21C1C              LDB     R0,[R0]
\    02F8   C6181C              STB     R0,[SP]
\    02FB   C3180200            ST      0,2[SP]
\    02FF           ?0040:
\    02FF   A318021C            LD      R0,2[SP]
\    0303   AE1820              LDBZE   R4,[SP]
\    0306   88201C              CMP     R0,R4
\    0309   DB33                JC      ?0039
\    030B           ?0041:
\    030B   A104001C            LD      R0,#Mbf_Degree
\    030F   A3011C02            LD      R4,m_index[0]
\    0313   20
\    0314   0720                INC     R4
```

396

Appendix B: Cross Compiling the Fuzzy Kernel

```
\   0316   C3011C02              ST      R4,m_index[0]
\   031A   20
\   031B   0520                  DEC     R4
\   031D   642020                ADD     R4,R4
\   0320   64201C                ADD     R0,R4
\   0323   C21C00                ST      0,[R0]
\   0326   A3011A02              LD      R0,cfg_index[0]
\   032A   1C
\   032B   071C                  INC     R0
\   032D   C3011A02              ST      R0,cfg_index[0]
\   0331   1C
\   0332   A318021C              LD      R0,2[SP]
\   0336   071C                  INC     R0
\   0338   C318021C              ST      R0,2[SP]
\   033C   27C1                  SJMP    ?0040
\   033E                 ?0039:
\   033E   B3011E02              LDB     R0,j[0]
\   0342   1C
\   0343   171C                  INCB    R0
\   0345   C7011E02              STB     R0,j[0]
\   0349   1C
\   034A   2791                  SJMP    ?0036
\   034C                 ?0035:
\   034C   A318041C              LD      R0,4[SP]
\   0350   C3011A02              ST      R0,cfg_index[0]
\   0354   1C
\   0355   A318061C              LD      R0,6[SP]
\   0359   C3011C02              ST      R0,m_index[0]
\   035D   1C
\   035E   65080018              ADD     SP,#8
\   0362   F0                    RET


\   0363                 Eval_Rulebase:
\   0363   69080018              SUB     SP,#8
\   0367   BD011C                LDBSE   R0,#1
\   036A   C318061C              ST      R0,6[SP]
\   036E   C7011E02              STB     0,j[0]
\   0372   00
\   0373                 ?0044:
\   0373   B3011E02              LDB     R0,j[0]
\   0377   1C
\   0378   9B011F02              CMPB    R0,Num_of_Rules[0]
\   037C   1C
\   037D   D303                  JNC     $+5
\   037F   E7D000                LJMP    ?0043
\   0382                 ?0045:
\   0382   A1FFFF1C              LD      R0,#-1
\   0386   011E                  CLR     R2
\   0388   C2181C                ST      R0,[SP]
\   038B   C318021E              ST      R2,2[SP]
\   038F                 ?0048:
\   038F   A18C001C              LD      R0,#RULE_BLOCK
\   0393   A3180620              LD      R4,6[SP]
\   0397   0720                  INC     R4
\   0399   C3180620              ST      R4,6[SP]
\   039D   0520                  DEC     R4
\   039F   64201C                ADD     R0,R4
\   03A2   B21C1C                LDB     R0,[R0]
\   03A5   C718041C              STB     R0,4[SP]
\   03A9   99FF1C                CMPB    R0,#255
\   03AC   DF43                  JE      ?0047
```

397

Fuzzy Logic for Real World Design

```
\   03AE                ?0049:
\   03AE    A104001C        LD      R0,#Mbf_Degree
\   03B2    AF180420        LDBZE   R4,4[SP]
\   03B6    642020          ADD     R4,R4
\   03B9    64201C          ADD     R0,R4
\   03BC    A21C1C          LD      R0,[R0]
\   03BF    011E            CLR     R2
\   03C1    8B18021E        CMP     R2,2[SP]
\   03C5    D703            JNE     $+5
\   03C7    8A181C          CMP     R0,[SP]
\   03CA    DB15            JC      ?0051
\   03CC                ?0050:
\   03CC    A104001C        LD      R0,#Mbf_Degree
\   03D0    AF180420        LDBZE   R4,4[SP]
\   03D4    642020          ADD     R4,R4
\   03D7    64201C          ADD     R0,R4
\   03DA    A21C1C          LD      R0,[R0]
\   03DD    011E            CLR     R2
\   03DF    2007            SJMP    ?0052
\   03E1                ?0051:
\   03E1    A2181C          LD      R0,[SP]
\   03E4    A318021E        LD      R2,2[SP]
\   03E8                ?0052:
\   03E8    C2181C          ST      R0,[SP]
\   03EB    C318021E        ST      R2,2[SP]
\   03EF    279E            SJMP    ?0048
\   03F1                ?0047:
\   03F1                ?0054:
\   03F1    A18C001C        LD      R0,#RULE_BLOCK
\   03F5    A3180620        LD      R4,6[SP]
\   03F9    0720            INC     R4
\   03FB    C3180620        ST      R4,6[SP]
\   03FF    0520            DEC     R4
\   0401    64201C          ADD     R0,R4
\   0404    B21C1C          LDB     R0,[R0]
\   0407    C718041C        STB     R0,4[SP]
\   040B    99FF1C          CMPB    R0,#255
\   040E    DF34            JE      ?0053
\   0410                ?0055:
\   0410    A104001C        LD      R0,#Mbf_Degree
\   0414    AF180420        LDBZE   R4,4[SP]
\   0418    642020          ADD     R4,R4
\   041B    64201C          ADD     R0,R4
\   041E    A21C1C          LD      R0,[R0]
\   0421    011E            CLR     R2
\   0423    8B18021E        CMP     R2,2[SP]
\   0427    D703            JNE     $+5
\   0429    8A181C          CMP     R0,[SP]
\   042C    DB14            JC      ?0057
\   042E                ?0056:
\   042E    A104001C        LD      R0,#Mbf_Degree
\   0432    AF180420        LDBZE   R4,4[SP]
\   0436    642020          ADD     R4,R4
\   0439    64201C          ADD     R0,R4
\   043C    A21820          LD      R4,[SP]
\   043F    C21C20          ST      R4,[R0]
\   0442                ?0057:
\   0442    27AD            SJMP    ?0054
\   0444                ?0053:
\   0444    B3011E02        LDB     R0,j[0]
\   0448    1C
\   0449    171C            INCB    R0
```

398

Appendix B: Cross Compiling the Fuzzy Kernel

```
\    044B    C7011E02              STB       R0,j[0]
\    044F    1C
\    0450    2721                  SJMP      ?0044
\    0452                  ?0043:
\    0452    65080018              ADD       SP,#8
\    0456    F0                    RET


\    0457                  Defuzzify_Rulebase:
\    0457    69100018              SUB       SP,#16
\    045B    C7011E02              STB       0,j[0]
\    045F    00
\    0460                  ?0059:
\    0460    AF011E02              LDBZE     R0,j[0]
\    0464    1C
\    0465    8B010200              CMP       R0,Num_of_Outputs[0]
\    0469    1C
\    046A    D303                  JNC       $+5
\    046C    E7F600                LJMP      ?0058
\    046F                  ?0060:
\    046F    C3180400              ST        0,4[SP]
\    0473    C3180600              ST        0,6[SP]
\    0477    C3180800              ST        0,8[SP]
\    047B    C3180A00              ST        0,10[SP]
\    047F    A100001C              LD        R0,#CONFIG_BLOCK
\    0483    A3011A02              LD        R4,cfg_index[0]
\    0487    20
\    0488    0720                  INC       R4
\    048A    C3011A02              ST        R4,cfg_index[0]
\    048E    20
\    048F    0520                  DEC       R4
\    0491    642020                ADD       R4,R4
\    0494    64201C                ADD       R0,R4
\    0497    A21C1C                LD        R0,[R0]
\    049A    C318021C              ST        R0,2[SP]
\    049E    C21800                ST        0,[SP]
\    04A1                  ?0063:
\    04A1    A2181C                LD        R0,[SP]
\    04A4    8B18021C              CMP       R0,2[SP]
\    04A8    D303                  JNC       $+5
\    04AA    E78100                LJMP      ?0062
\    04AD                  ?0064:
\    04AD    A104001C              LD        R0,#Mbf_Degree
\    04B1    A3011C02              LD        R4,m_index[0]
\    04B5    20
\    04B6    0720                  INC       R4
\    04B8    C3011C02              ST        R4,m_index[0]
\    04BC    20
\    04BD    0520                  DEC       R4
\    04BF    642020                ADD       R4,R4
\    04C2    64201C                ADD       R0,R4
\    04C5    A21C1C                LD        R0,[R0]
\    04C8    011E                  CLR       R2
\    04CA    C3180C1C              ST        R0,12[SP]
\    04CE    C3180E1E              ST        R2,14[SP]
\    04D2    A3180C1C              LD        R0,12[SP]
\    04D6    A3180E1E              LD        R2,14[SP]
\    04DA    45040018              ADD       R4,SP,#4
\    04DE    20
\    04DF    66201C                ADD       R0,[R4]
\    04E2    C2211C                ST        R0,[R4]+
\    04E5    A6201E                ADDC      R2,[R4]
```

399

Fuzzy Logic for Real World Design

```
\   04E8   C2201E              ST      R2,[R4]
\   04EB   A100001C            LD      R0,#CONFIG_BLOCK
\   04EF   A3011A02            LD      R4,cfg_index[0]
\   04F3   20
\   04F4   0720                INC     R4
\   04F6   C3011A02            ST      R4,cfg_index[0]
\   04FA   20
\   04FB   0520                DEC     R4
\   04FD   642020              ADD     R4,R4
\   0500   64201C              ADD     R0,R4
\   0503   A21C1C              LD      R0,[R0]
\   0506   011E                CLR     R2
\   0508   A3180C20            LD      R4,12[SP]
\   050C   A3180E22            LD      R6,14[SP]
\   0510   EFEFEF              LCALL   ?UL_MUL_L02
\   0513   45080018            ADD     R4,SP,#8
\   0517   20
\   0518   66201C              ADD     R0,[R4]
\   051B   C2211C              ST      R0,[R4]+
\   051E   A6201E              ADDC    R2,[R4]
\   0521   C2201E              ST      R2,[R4]
\   0524   A2181C              LD      R0,[SP]
\   0527   071C                INC     R0
\   0529   C2181C              ST      R0,[SP]
\   052C   2773                SJMP    ?0063
\   052E                 ?0062:
\   052E   A318081C            LD      R0,8[SP]
\   0532   A3180A1E            LD      R2,10[SP]
\   0536   A3180420            LD      R4,4[SP]
\   053A   A3180622            LD      R6,6[SP]
\   053E   EFEFEF              LCALL   ?UL_DIV_SWAP_L02
\   0541   A1120220            LD      R4,#Crisp_Output
\   0545   C81C                PUSH    R0
\   0547   AF011E02            LDBZE   R0,j[0]
\   054B   1C
\   054C   641C1C              ADD     R0,R0
\   054F   64201C              ADD     R0,R4
\   0552   CC20                POP     R4
\   0554   C21C20              ST      R4,[R0]
\   0557   B3011E02            LDB     R0,j[0]
\   055B   1C
\   055C   171C                INCB    R0
\   055E   C7011E02            STB     R0,j[0]
\   0562   1C
\   0563   26FB                SJMP    ?0059
\   0565                 ?0058:
\   0565   65100018            ADD     SP,#16
\   0569   F0                  RET


\   0000                       RSEG    DATA
\   0000                Num_of_Inputs:
\   0000   0000                DCB     0,0
\   0002                Num_of_Outputs:
\   0002   0000                DCB     0,0
\   0004                Mbf_Degree:
\   0004                ?0000:
\   0202                       DSB     510
\   0202                ?0001:
\   0202                Crisp_Input:
\   0202   00000000            DCB     0,0,0,0,0,0,0,0,0,0,0,0,0,0
\   0206   00000000
```

400

Appendix B: Cross Compiling the Fuzzy Kernel

```
\   020A  00000000
\   020E  00000000
\   0212             Crisp_Output:
\   0212  00000000           DCB     0,0,0,0,0,0,0,0
\   0216  00000000
\   021A             cfg_index:
\   021A  0000               DCB     0,0
\   021C             m_index:
\   021C  0000               DCB     0,0
\   021E             j:
\   021E  00                 DCB     0
\   021F             Num_of_Rules:
\   021F  00                 DCB     0
\   0000                     RSEG    CONST
\   0000             CONFIG_BLOCK:
\   0000  0200               DCW     2
\   0002  0100               DCW     1
\   0004  0500               DCW     5
\   0006  0000               DCW     0
\   0008  0000               DCW     0
\   000A  CC2C               DCW     11468
\   000C  CC6C               DCW     27852
\   000E  FFFF               DCW     65535
\   0010  0300               DCW     3
\   0012  CC2C               DCW     11468
\   0014  CC6C               DCW     27852
\   0016  CC6C               DCW     27852
\   0018  FF7F               DCW     32767
\   001A  0300               DCW     3
\   001C  0D00               DCW     13
\   001E  CC6C               DCW     27852
\   0020  FF7F               DCW     32767
\   0022  FF7F               DCW     32767
\   0024  3393               DCW     37683
\   0026  0D00               DCW     13
\   0028  0D00               DCW     13
\   002A  FF7F               DCW     32767
\   002C  3393               DCW     37683
\   002E  3393               DCW     37683
\   0030  33D3               DCW     54067
\   0032  0D00               DCW     13
\   0034  0300               DCW     3
\   0036  3393               DCW     37683
\   0038  33D3               DCW     54067
\   003A  FFFF               DCW     65535
\   003C  FFFF               DCW     65535
\   003E  0300               DCW     3
\   0040  FFFF               DCW     65535
\   0042  0500               DCW     5
\   0044  0000               DCW     0
\   0046  0000               DCW     0
\   0048  3333               DCW     13107
\   004A  6666               DCW     26214
\   004C  FFFF               DCW     65535
\   004E  0500               DCW     5
\   0050  3333               DCW     13107
\   0052  6666               DCW     26214
\   0054  6666               DCW     26214
\   0056  FF7F               DCW     32767
\   0058  0500               DCW     5
\   005A  0A00               DCW     10
\   005C  6666               DCW     26214
```

Fuzzy Logic for Real World Design

```
\   005E    FF7F              DCW    32767
\   0060    FF7F              DCW    32767
\   0062    9999              DCW    39321
\   0064    0A00              DCW    10
\   0066    0900              DCW    9
\   0068    FF7F              DCW    32767
\   006A    9999              DCW    39321
\   006C    9999              DCW    39321
\   006E    CCCC              DCW    52428
\   0070    0900              DCW    9
\   0072    0500              DCW    5
\   0074    9999              DCW    39321
\   0076    CCCC              DCW    52428
\   0078    FFFF              DCW    65535
\   007A    FFFF              DCW    65535
\   007C    0500              DCW    5
\   007E    FFFF              DCW    65535
\   0080    0500              DCW    5
\   0082    D227              DCW    10194
\   0084    DD5D              DCW    24029
\   0086    0080              DCW    32768
\   0088    22A2              DCW    41506
\   008A    2DD8              DCW    55341
\   008C            RULE_BLOCK:
\   008C    19                DCB    25
\   008D    02                DCB    2
\   008E    05                DCB    5
\   008F    FF                DCB    255
\   0090    0D                DCB    13
\   0091    FF                DCB    255
\   0092    02                DCB    2
\   0093    06                DCB    6
\   0094    FF                DCB    255
\   0095    0D                DCB    13
\   0096    FF                DCB    255
\   0097    02                DCB    2
\   0098    07                DCB    7
\   0099    FF                DCB    255
\   009A    0C                DCB    12
\   009B    FF                DCB    255
\   009C    02                DCB    2
\   009D    08                DCB    8
\   009E    FF                DCB    255
\   009F    0B                DCB    11
\   00A0    FF                DCB    255
\   00A1    02                DCB    2
\   00A2    09                DCB    9
\   00A3    FF                DCB    255
\   00A4    0B                DCB    11
\   00A5    FF                DCB    255
\   00A6    00                DCB    0
\   00A7    05                DCB    5
\   00A8    FF                DCB    255
\   00A9    0E                DCB    14
\   00AA    FF                DCB    255
\   00AB    00                DCB    0
\   00AC    06                DCB    6
\   00AD    FF                DCB    255
\   00AE    0E                DCB    14
\   00AF    FF                DCB    255
\   00B0    00                DCB    0
\   00B1    07                DCB    7
```

Appendix B: Cross Compiling the Fuzzy Kernel

```
\   00B2   FF         DCB   255
\   00B3   0E         DCB   14
\   00B4   FF         DCB   255
\   00B5   00         DCB   0
\   00B6   08         DCB   8
\   00B7   FF         DCB   255
\   00B8   0E         DCB   14
\   00B9   FF         DCB   255
\   00BA   00         DCB   0
\   00BB   09         DCB   9
\   00BC   FF         DCB   255
\   00BD   0D         DCB   13
\   00BE   FF         DCB   255
\   00BF   01         DCB   1
\   00C0   05         DCB   5
\   00C1   FF         DCB   255
\   00C2   0E         DCB   14
\   00C3   FF         DCB   255
\   00C4   01         DCB   1
\   00C5   06         DCB   6
\   00C6   FF         DCB   255
\   00C7   0E         DCB   14
\   00C8   FF         DCB   255
\   00C9   01         DCB   1
\   00CA   07         DCB   7
\   00CB   FF         DCB   255
\   00CC   0D         DCB   13
\   00CD   FF         DCB   255
\   00CE   01         DCB   1
\   00CF   08         DCB   8
\   00D0   FF         DCB   255
\   00D1   0C         DCB   12
\   00D2   FF         DCB   255
\   00D3   01         DCB   1
\   00D4   09         DCB   9
\   00D5   FF         DCB   255
\   00D6   0C         DCB   12
\   00D7   FF         DCB   255
\   00D8   03         DCB   3
\   00D9   05         DCB   5
\   00DA   FF         DCB   255
\   00DB   0C         DCB   12
\   00DC   FF         DCB   255
\   00DD   03         DCB   3
\   00DE   06         DCB   6
\   00DF   FF         DCB   255
\   00E0   0C         DCB   12
\   00E1   FF         DCB   255
\   00E2   03         DCB   3
\   00E3   07         DCB   7
\   00E4   FF         DCB   255
\   00E5   0B         DCB   11
\   00E6   FF         DCB   255
\   00E7   03         DCB   3
\   00E8   08         DCB   8
\   00E9   FF         DCB   255
\   00EA   0A         DCB   10
\   00EB   FF         DCB   255
\   00EC   03         DCB   3
\   00ED   09         DCB   9
\   00EE   FF         DCB   255
\   00EF   0A         DCB   10
```

Fuzzy Logic for Real World Design

```
        \   00F0  FF            DCB   255
        \   00F1  04            DCB   4
        \   00F2  05            DCB   5
        \   00F3  FF            DCB   255
        \   00F4  0B            DCB   11
        \   00F5  FF            DCB   255
        \   00F6  04            DCB   4
        \   00F7  06            DCB   6
        \   00F8  FF            DCB   255
        \   00F9  0A            DCB   10
        \   00FA  FF            DCB   255
        \   00FB  04            DCB   4
        \   00FC  07            DCB   7
        \   00FD  FF            DCB   255
        \   00FE  0A            DCB   10
        \   00FF  FF            DCB   255
        \   0100  04            DCB   4
        \   0101  08            DCB   8
        \   0102  FF            DCB   255
        \   0103  0A            DCB   10
        \   0104  FF            DCB   255
        \   0105  04            DCB   4
        \   0106  09            DCB   9
        \   0107  FF            DCB   255
        \   0108  0A            DCB   10
        \   0109  FF            DCB   255
        \   0000                RSEG  ZVECT
        \   0000  0400          DCW   ?0000
        \   0002  0202          DCW   ?0001
        \   0004                END

Errors: none
Warnings: none
Code size: 1386
```

Glossary

This glossary includes terms and explanations used in this book. It also includes terms that have some relevance to fuzzy logic and fuzzy set theory.

Aggregation. Aggregation combines all rule consequents of a fuzzy output variable into a fuzzy output space. Two common aggregation algorithms are MAX and Additive. The resultant fuzzy space represents the result of the inference process for a specific combination of inputs. See Chapter 3.

Ambiguity. Ambiguity arises when a statement can have more than one meaning and we lack the information necessary to discriminate between the possibilities. For instance the phrase *that jacket is hot* may refer to thermodynamic temperature or trendy stylishness. Without further information it is not possible to determine which usage of the word *hot* was intended. Ambiguity is distinct from vagueness in that ambiguity results from an inability to distinguish between multiple possible meanings. See *Vagueness*.

AND. Fuzzy set intersection operator. The intersection of two fuzzy sets is achieved by taking the pointwise minimum of those two sets. The result is a third fuzzy set. See Chapter 2.

AND. Fuzzy proposition intersection operator. The AND operator performs the task of aggregating an overall truth value for the entire *IF-Side* of a rule from the individual degrees of membership of its constituent fuzzy propositions. See Chapter 3.

Antecedent. A fuzzy proposition found on the IF-side of a fuzzy rule. The degree to which an antecedent is true directly impacts the strength of that rule's consequent. Antecedents take the form *IF Fuzzy_Var IS Membership_Function*. See Chapter 3. See also *Consequent*.

Bayesian Probability. A probabilistic formulation of rational belief. *A priori* (before-the-fact) probabilities for events may be updated by future observations to yield *a posteriori* (after-the fact) probabilities. The domain of belief ranges continuously from complete disbelief (0) to complete belief (1).

Black, Max. (1909-1989) Early pioneer of continuously valued logics. Black felt that vagueness stemmed from a continuum, and that continuum implied degrees. Black interpreted the degrees of membership as probabilities. The fuzzy logic spoken of in this book is based on degrees of truth that are interpreted as possibilities rather than probabilities.

Bounded Intersection. The bounded intersection operator, whose mathematical form can be found in Chapter 3, is an alternate interpretation of the AND connective found in fuzzy IF/THEN rules.

Bounded Union. The bounded union operator, whose mathematical form can be found in Chapter 3, is an alternate interpretation of the OR connective found in fuzzy IF/THEN rules.

Centroid. The physicist will refer to this quantity as the center of mass, or the first moment of mass of particular object. In the context of fuzzy systems the centroid defuzzification algorithm takes its motivation from the physicists equation. This defuzzification method is influenced by all rules that possess a non-zero degree of truth. Centroid defuzzification produces a crisp value which corresponds to the center of area of the aggregated fuzzy consequent space.

Composite Maximum. This defuzzification method is influenced only by rules possessing the highest degree of truth. If more than one rule possesses the maximum degree of truth in the aggregated fuzzy consequent space, the defuzzified output is computed as the arithmetic average of all consequents having the maximum degree. See Chapter 3.

Composite Moment. See *Centroid*.

Composition. This term refers to the process of building up the consequent fuzzy space by combining the effects of all evaluated rules. The general compositional rules of inference specify how the consequent fuzzy space is built through the aggregation and correlation of fuzzy spaces associated with each rule.

Conjunction Fallacy. A contemporary of Aristotle expressed the following paradox: *Given a pile of sand, removing one grain does not destroy the pile.* The paradox is that if one grain is removed at a time, repeatedly, all grains of sand will have been removed while we still call it a pile of sand. Fuzzy logic provides the answer to this paradox in the following way: *Given a pile of sand, removing one grain reduces the degree to which it is considered a pile.* Removing grains repeatedly further reduces the degree to which the sand is considered a pile thus yielding the intuitively correct outcome.

Concentration. In the context of fuzzy logic, concentration refers to the operation of hedges like *very*, which act to decrease the degree of membership of those crisp values which lie furthest from the point of maximal truth in a given fuzzy set. For example, on a bell shaped fuzzy set a concentration operator would act in a way that squeezes the two sides of the bell closer together.

Consequent. A fuzzy proposition found on the *THEN-Side* of a fuzzy rule. The degree to which a consequent affects the system output depends upon the truth value of the *IF-Side*, and the inference, aggregation, and defuzzification methods. Consequents take the form ...THEN Fuzzy_Var IS Membership_Function. See Chapter 3. See also *Antecedent*.

Crisp Value A non-fuzzy quantity, which can be represented as a real number. *Crisp input* values are provided to a fuzzy engine which produces *crisp output* values.

Crisp set. We use this term to contrast against the *Fuzzy Set*. It is a collection of crisp values each of which belongs to the set with maximal degree. Like a Boolean set, a crisp value may be either a complete member or complete non-member of the set.

Deduction. This is a general description of a type of reasoning. Starting from propositions that are know to be true (called facts), the truth of new propositions is determined by the manipulation of the known facts according to the rules of logic (typically first order predicate logic).

Degree of Membership. See *Degree of Truth*.

Degree of Truth The degree to which a crisp input is a member of a given membership function. Degrees of truth range continuously from 0 (no membership at all) to 1 (the highest possible degree of membership). Also called *Degree of Membership*, or *Truth Value*.

Defuzzification. Defuzzification distills an entire fuzzy space into one crisp value. The fuzzy space typically operated on is the aggregated result of fuzzy rule evaluation. Since it is often impossible to convey the complexity of the aggregated fuzzy space in just one crisp number, there are a number of widely used defuzzification algorithms including, MAX, CENTROID, WEIGHTED AVERAGE, COMPOSITE MAXIMUM, LEFT and RIGHT MAXIMIZER, etc. See Chapter 3.

Dilation. In the context of fuzzy logic, dilation refers to the operation of hedges, like *somewhat*, that act to increase the degree of membership of those crisp values that lie furthest from the point of maximal truth in a given fuzzy set. For example, on a bell shaped fuzzy set a dilation operator would act in a way that pulls the two sides of the bell further apart.

Entropy. As a classical concept, entropy refers to the disorder of a system. Higher entropy implies higher disorder. In the context of an information system, entropy is a measure of information. High entropy implies a lack of information about the system. Gaining information about a system reduces entropy. Claude Shannon pioneered the entropy view of information in the 1950's.

Expert System. An expert system is one that possesses human expertise in a machine automatible form. The expert system accepts queries in its domain of expertise, performs inferences using its stored expert knowledge, and produces conclusions and observations, based on the query and its knowledge, as output.

Fuzzification. Fuzzification requires two things, a set of membership functions and a crisp input. Fuzzification describes the act of determining the degree of membership for that specific crisp input value in all relevant membership functions.

Fuzzy Cognitive Map(FCM). A method of modeling complex feedback dynamical systems developed by Rod Taber of Ring Technology Inc. and Bart Kosko of USC. An FCM consists of concepts and directional causal links between concepts. Causal links can inhibit (with a link strength between -1 and 0), reinforce (with a link strength between 0 and 1), or do nothing (with a link strength of 0). The map can be iterated to observe the effects of changing link strengths, adding concepts, etc. FCM have been used to model complex issues like the potential effects of health care reform and the economic trade regulations comprising GATT.

Fuzzy Model. A fuzzy model, through sets of membership functions and a collection of rules relating input variables to output variables, captures the relationship between input and output variables. A fuzzy model typically encodes knowledge that is approximate rather than exact. See Chapter 4. See also *Fuzzy Engine*.

Fuzzy Engine. Given a set of crisp inputs and the membership function definitions and rules of a *Fuzzy Model*, a fuzzy engine computes crisp outputs. The fuzzy engine processes the crisp inputs through fuzzification, rule evaluation, and defuzzification algorithms to produce crisp outputs. See Chapters 6 and 7.

Fuzzy Predicate. One or more fuzzy propositions on the left hand side (*IF-Side*) of a rule. The predicate is the entire left hand side considered in total. The degree of truth of the predicate may be evaluated if we know the degree of truth of each fuzzy proposition and we know how to combine the fuzzy propositions with *AND* and *OR* operators.

Fuzzy Set. See *Membership Function*.

Fuzzy Variable. An input or output variable of a fuzzy system, defined by a set of membership functions. Whereas a crisp variable is associated with crisp values, a fuzzy variable is associated with degrees of membership in the membership functions of that fuzzy variable.

Grade of Membership. See *Degree of Truth*.

Hedge. Hedges serve to further modify the basic definitions of fuzzy sets. For example, having defined a fuzzy set of temperatures labeled *Hot*, the hedge operator *Somewhat* may be applied to form a different, but related, fuzzy set representing the concept *Somewhat Hot*. Hedges that occur frequently in our daily speech patterns include *Very, Somewhat, Sort of, Nearly*, etc. See Chapter 2.

Induction. Induction is the process of obtaining general knowledge from specific examples. For instance, if you have a set of data, you may be able to generalize certain relationship that you see in the data. Such relationships can be thought of as knowledge, and the process used to generate that knowledge was induction.

Inference. The act of drawing conclusions based on premises. In fuzzy systems, inference is a rule driven process. The *IF-Side* of a rule contains the premise conditions and the *THEN-side* contains the conclusion to be drawn from that specific premise. This book illustrates correlation minimum and correlation product inference. See Chapter 3.

Intensification. In the context of Fuzzy Logic, intensification refers to the operation of hedges like *very*, which act to decrease the degree of membership of those crisp values which lie furthest from the point of maximal truth in a given fuzzy set. For example, on a bell shaped fuzzy set an intensification operator would act in a way that squeezes the two sides of the bell closer together.

Intersection. The intersection of two fuzzy sets is a third set, each element of which belongs to both initial sets to a non-zero degree. The *AND* operator specifies the precise mathematical action of the intersection operation.

Implication. Logical implication refers to the way that truth values flow from antecedents to consequents in a logical system. The most used method of implication in Fuzzy Logic systems is called *Modus Ponens*. See *Modus Ponens*.

Label. See *Linguistic Label*.

Law of Excluded Middle. This law is traced back to Aristotle's investigations of logic. The law says that a proposition is either true or untrue, but cannot be both. In the **Middle** would be a proposition that is both true and untrue at the same time. The law of excluded middle declares that no such proposition may exist in a logic system.

Law of Incompatibility. Zadeh's principle of incompatibility relates to the complexity of a system and precise knowledge about that system. It states that as the complexity of a system rises our ability to say something meaningful and precise about that system dwindles. Beyond some level of complexity meaningful statements and precise statements about the system become mutually exclusive.

Linguistic Label. The name of a membership function. The labeling of fuzzy sets usually proceeds from a human-centric understanding of the underlying classes. For example, if the variable is *Temperature*, the linguistic labels of *Hot*, *Warm*, *Cool*, and *Cold* make more sense than the labels *Set_A*, *Set_B*, *Set_C*, and *Set_D* which have no intuitive content. See Chapter 2.

Linguistic Variable. See *Fuzzy Variable*.

Lukasiewicz, Jan. (1878-1955) Logician and philosopher who pioneered work in multivalued logics. In addition to the truth values of 1, meaning *true*, and 0, meaning *false*, Lukasiewicz introduced 1/2 which meant *possible*. This simple addition allowed, for the first time in formal logic, partial truths and partial contradictions. Lukasiewicz also proposed that truth could take any value at all in the continuum between 1 and 0.

Mamdani, Ebrahim. In 1973 Mamdani constructed, with his student Sedrak Assilian, the first recognized fuzzy controller. Their controller simultaneously adjusted a steam boiler and steam piston to achieve constant boiler pressure and constant piston speed. Using four inputs, two outputs, and 24 fuzzy IF-THEN rules, their controller outperformed the conventional controllers they applied to the problem.

Membership Function. One classifier for a fuzzy variable. Each membership function spans some subset of the fuzzy variable's universe of discourse. Each crisp input will have some degree of membership in one or more of the fuzzy variable's membership functions. Also known as a *Fuzzy Set*.

Modus Ponens. The most frequently applied method of implication in Fuzzy Logic systems. This implication method allows us to infer the degree of truth of the consequent from the degree of truth of the corresponding antecedent. For instance, given the fuzzy rule *IF Turbine Temperature IS High THEN Turbine Lifetime IS Low*, if we know that the turbine temperature actually is high that knowledge implies that the turbine lifetime will be low.

Multivalued Logics. Logical systems that allow more than two discrete levels of truth. Boolean systems allow only true (1) and not-true (0) and thus are not considered multivalued. Fuzzy logic permits a continuum of truth values and hence may be considered a multi- or infinitely-valued logic.

Natural Classifications. These are the classes that enter our every day thoughts and speech. For example, natural classifications of temperature including *Hot, Warm, Cool,* and *Cold* are universal in their meanings. We express our knowledge and ideas to each other through a shared language laden with natural classifications. We further refine and shape those classes with hedges and other modifiers like *very, almost, somewhat,* etc. Natural classifications provide an intuitive basis for fuzzy set definition and labeling.

NORM. Normalization operator. If the maximum degree of membership of a fuzzy set is less than 1 or greater than 1, NORM will rescale that fuzzy set so that the highest possible degree of membership in that set is equal to 1. All other points of the fuzzy set are scaled proportionally so that the original shape is retained.

NOT. Fuzzy complement operator. Applying the NOT operator to a fuzzy set results in another fuzzy set, which contains all crisp values that did not belong to the original fuzzy set to degree 1. See Chapter 2.

OR. Fuzzy union operator. The union of two fuzzy sets is achieved by taking the pointwise maximum of those two sets. The result is a third fuzzy set. See Chapter 2.

Perspective Shift. This term describes the notion that our perception of a thing depends upon our specific perspective. For example, the notion of Tall Person will be quite different from a child's perspective than from an adult's perspective. Likewise, the notion of Middle-Aged from a young person's perspective may be different than from an older perspective.

Pi-Curve. This term describes a particular membership function shape. From a central peak with a 1.0 degree, degree of membership decreases to each side as one gets further from the central peak. The resulting shape looks a bit like the Greek letter representing Pi. The Pi-Curve can be constructed by placing an S-Curve and a Z-Curve back to back.

Possibility Theory. A calculus of possibility analysis based on fuzzy logic and approximate reasoning concepts. Whereas probability theory treats truth as a crisp quantity taking only the values 0 and 1, possibility theory treats truth as a fuzzy quantity thus permitting degrees of truth.

Quantifiers. Fuzzy quantifiers such as, *few, some, many* and *all* pervade our every day speech patterns. They each express a fuzzy quantity and convey information about proportion. Zadeh suggested a representation of quantifiers as fuzzy subsets of the unit interval.

Rule. This key element of fuzzy systems and approximate reasoning relates fuzzy propositions of input variables to fuzzy propositions of output variables. Using an IF-THEN construct, a rule combines fuzzy input and output variables with membership functions to capture the essence of human reasoning. The typical format of one rule is:

```
IF InVar1 IS Mbf1 [AND InVar2 IS.Mbf1..] THEN OutVar1 IS
Mbf1 [AND OutVar2 IS Mbf2...]
```

Rule Evaluation. In a fuzzy rule (see prior definition) the premise conditions specified on the *IF-Side* are evaluated individually and then combined to yield an overall degree of truth for the rule. The degree of truth of the rule is then applied to the consequent statements of the *THEN-Side* of the rule to determine the overall effect of the rule. See Chapter 3. See also *Fuzzification*, and *Defuzzification*.

Rulebase. A collection of fuzzy rules all relating to the same *Fuzzy Model*. A rulebase captures all of the relevant relations between the input fuzzy sets and output fuzzy sets of a particular system. See Chapter 3.

Scope. The range of crisp values for which a fuzzy set has non-zero degree of membership. Also known as the support set of a fuzzy set. See Chapter 2.

Semantic Shift. We may use the same words in classifying two completely different sets of objects. For example, we may describe both people and buildings as *Tall*. The meaning, or semantics, of *Tall* varies with the class of objects that we are describing. We are able to unambiguously apply terms like *large* and *small* to describe the members of a vast array of different classes because we inherently understand that the context shifts the meaning, or semantics, of those descriptors.

Singleton. A fuzzy set which is non-zero at only one point.

S-Curve. Smooth membership function with transition area from (left to right) complete non-membership to complete membership. When viewed on a graph, this shape looks like a stretched out 'S'. The *Z-Curve* is the complement to the S-Curve.

T-conorm. This is a family of mathematical functions that generalize the *Union* operation. See Chapter 3. The triangular conorm $S(a,b)$ must possess the following commutativity, associativity, monotonicity, and identity properties:

(1) $S(a,b) = S(b,a)$
(2) $S(a,S(b,c)) = S(S(a,b),c)$
(3) $S(a,b) \geq S(c,d) if (a \geq c) and (b \geq d)$
(4) $S(a,0) = a$

T-norm. This is a family of mathematical functions that generalize the *Intersection* operation. See Chapter 3. The triangular norm $T(a,b)$ must possess the following commutativity, associativity, monotonicity, and identity properties:

(1) $T(a,b) = T(b,a)$
(2) $T(a,T(b,c)) = T(T(a,b),c)$
(3) $T(a,b) \geq T(c,d) if (a \geq c) and (b \geq d)$
(4) $T(a,1) = a$

Truth Value. See *Degree of Truth*.

Uncertainty. The condition of being in doubt as to the exact state or nature of a thing. Uncertainty does not mean that we know nothing, rather it refers to the incompleteness of our knowledge. Fuzzy logic accommodates uncertainty by allowing for degrees of truth when describing our knowledge. The general state or nature of a thing can be known and expressed while its precise nature or state is still uncertain.

Union. The union of two fuzzy sets is a third set, each element of which belongs to at least one of the initial sets to a non-zero degree. The *OR* operator specifies the precise mathematical action of the union operation. See Chapters 2 and 3.

Universe of Discourse. Set of values that are of interest to a fuzzy variable, usually represented by the X-axis of a fuzzy variable graph. At one end of the universe of discourse is the minimum value of interest. At the other end is the maximum value of interest. For example, the universe of discourse for fuzzy variable *Temperature* might be -10 to 100 degrees.

Vagueness. Any classification or statement with an imprecise meaning can be described as vague. Vagueness is distinct from ambiguity in that vagueness describes the imprecision inherent in a particular meaning. For instance, the phrase *the orchestra gave an unsatisfying performance* leaves much unsaid about the actual nature or caliber of the performance and is thus a vague description of that performance. See *Ambiguity* and *Uncertainty*.

Zadeh, Lotfi. Considered the father of Fuzzy Logic. Zadeh, of U.C. Berkeley, introduced the notion of a *Fuzzy Set* to the world in his 1965 paper simply titled "Fuzzy Sets".

Z-Curve. Smooth membership function with transition area from (left to right) complete membership to complete non-membership. When viewed on a graph, this shape looks like a stretched out 'Z'. The *S-Curve* is the complement to the Z-Curve.

Index

—A—

absolute error, 153
accuracy, 17, 77
adaptive control, 19
adaptive fuzzy control systems, 262
adaptive fuzzy systems, 139
additive, 405
additive aggregation, 94, 224, 317
aggregation, 79, 94, 126, 135, 192, 198, 224, 226, 245, 260, 261, 317, 360, 384, 391, 405, 407
alarms, 103, 138
algorithm design, 226
algorithms, 304, 408
alpha-cut, 45, 103, 138
ambiguity, 405, 415
analog architectures, 240
analog implementation, 238
analog-to-digital conversion, 197, 209
AND, 41, 81, 83, 86, 93, 126, 316, 405, 406, 409, 410
ANSI C, 229
antecedent, 71, 79, 89, 223, 246, 251, 259, 264, 406, 410, 412
antecedent evaluation, 198
antecedent processing, 181
anti-lock braking, 237
application class hierarchy, 256
Apply_Crisp_Outputs, 229
approximate reasoning, 8, 15, 27, 72, 75, 128, 413

architecture, 191, 197, 238
Aristotle, 3, 410
arithmetic processing units, 238
artificial intelligence, 1, 5, 213
assembly code, 214, 237
assertion, 71, 102, 223
assignment operator, 321
Assilian, Sedrak, 411
association, 245
associativity, 47, 414
asymmetric membership function, 118
asymmetric response, 175
attributes, 245, 246
automatic traction control, 237
automobile cruise control, 161
automotive applications, 237
average error, 153

—B—

BASIC, 214
Bayesian probability, 406
BBS, 355
belief, domain of, 406
bell curve, 57
benchmarking, 201
bibliography, 342
BIDS, 299
binary trees, 299
BINTER, 84, 86, 198
bipolar fuzzy variable, 118
bipolar junction transistors, 241
bipolar variables, 113

417

bivalued logic, 2
Black, Max, 7, 406
Boole, George, 4
Boolean operators, 40
Boolean set, 407
Boolean systems, 412
Boolean truth values, 10
Borland, 265, 272, 299, 320
bounded intersection, 84, 406
bounded union, 85, 406
Buchanan, 5
building fuzzy application, 318
bulletin board servers, 355
BUNION, 85, 86, 198

—C—

C, 214, 229, 237, 359, 384, 390
C++, 267, 272, 299, 301
C++ classes, 321
C++ data structures, 304
causal links, 409
Centroid, 199, 406
centroid algorithm, 226
centroid defuzzification, 97, 100, 317
chaining, 196
class, 29, 301
class hierarchies, 255
class reuse, 262
classes, 75, 243, 318, 412
classification, 33
classifier, 411
classifying, 29
closed loop feedback, 150
code enhancing, 320
code generation tools, 346
code reuse, 262
collecting inputs, 319
collection iterators, 301
collection managers, 300, 302
commutativity, 48, 414
companion disk, 28, 229, 265, 267, 321, 323
compensatory operators, 81, 83
compilation, 237, 320
compilers, 214

complement operator, 40, 412
complex systems, 19
complexity, 13, 191, 196, 210, 411
component reusability, 266
composite maximum, 99, 199, 406
composite moment, 406
composition, 407
computation speed, 100
computational cost, 263
computational efficiency, 317
computational power, 201
computer simulation, 131
concentration, 407
concentration operators, 50
conclusion, 410
concurrent engineering, 189
conditional statement, 81
CONFIG_BLOCK, 215, 222, 223
configuration file, 270, 320
conflicting rules, 120
conjunction fallacy, 407
consequent, 71, 90, 223, 264, 406, 407, 412, 413
consequent aggregation, 199
consequent fuzzy set, 157
consequent fuzzy space, 407
consequent space, 406
consequents, 89, 207, 246, 405, 410
consistent rule set, 207
constraints, 190
constructor, 305
consulting, 347
consumer products, 24, 105
container classes, 299
continuously valued logic, 406
continuum, 6
continuum of truth, 2, 7, 412
contradiction, 3
control, 114
control applications, 114
control flow, 224, 255
control functions, 188
control loop, 189, 258, 269, 312
control loop timing, 132, 138, 192
control rules, 146
control space, 153

Index

control surface, 133
control system, 111, 138, 148, 150
controllability, 207
controller, 160
controller design, 169
controller loop, 144
controller tuning, 162
convergence, 148
copy semantics, 321
correlation, 407
Correlation Minimum, 198
correlation minimum inference, 90, 93, 99, 224, 410
Correlation Product, 198
correlation product inference, 90, 100, 224, 317, 410
cost, 242
coverage map, 124
crisp input, 34, 70, 201, 217, 228, 259, 313, 408, 411
crisp input tuple, 312
crisp input, output, 93, 197
crisp output, 99, 228, 253, 261, 409
crisp rules, 180
crisp set, 407
crisp value, 33, 96, 118, 407
crisp values, 110, 137, 408, 410, 412, 414
cross compilers, 229
cross compiling, 237
cross compiling fuzzy kernel, 359
cruise control, 161
customizing your application, 319

—D—

damping, 153
data block, 222
data path controls, 238
data structures, 199, 215, 299
data tuple, 255, 258
databases, 137
DDE, 346
De Morgan laws, 48
debugging, 211
debugging tools, 345
decision aid, 179

decision making elements, 108
decision points, 224
decision support system, 180
deduction, 407
defuzzification, 94, 96, 97, 99, 100, 135, 192, 198, 199, 215, 224, 226, 238, 261, 317, 360, 384, 391, 406, 407, 408, 409
defuzzification strategy, 126
defuzzifications per second, 201
degree of control, 114
degree of fit, 31
degree of freedom, 142
degree of membership, 31, 33, 217, 407, 408, 409, 410, 411, 414
degree of truth, 70, 90, 259, 406, 408, 409, 412, 413
degree of truth calculation, 314
degrees of membership, 405
degrees of truth, 181, 406, 415
DeMorgan conjugates, 86
DENDRAL, 5
design decisions, 224
design services, 347
design tools, 242
development cost, 190
development effort, 211
development process, 120
development time, 242
development tools, 341, 344
diagnostics, 139
differential equation, 167
digital architectures, 238
digital implementation, 238
digital signal processing, 189
digital-to-analog conversion, 197
digitizing, 194, 197
dilation operator, 52, 408
distributive property, 48
DLL, 346
domain, 35
domain of belief, 406
dominance properties, 47
DOS, 262, 265, 268
DSP, 189
Duda, 5
dynamic system, 193

—E—

education, 347
Einstein, 6
Email servers, 355
embedded application, 237
embedded control, 114, 266
embedded expert system, 179
embedded fuzzy models, 199
embedded microcontroller, 214
embedded system, 63, 187, 210, 224, 347
empirical data, 165
empirical methods, 206
entropy, 408
equations of motion, 142
equilibrium point, 147
error, 153
error flags, 136, 138
error rate of change, 160
Euler's method, 167
exception conditions, 103
exception handling, 136, 138
excluded middle, 3
execution speed, 127
execution time restrictions, 63
expert system, 126, 179, 210, 408
extreme values, 110

—F—

facts, 408
failure analysis, 137
FAM, 122
fcFuzzyEngine, 318
fcFuzzyProposition construction, 309
FCM, 409
file transfer repository, 356
first order predicate logic, 5, 408
flexibility, 199, 210, 213, 223, 236, 238, 242, 262, 266
FLIPS, 201
floating point, 210
FORTH, 214
Fourier expansion, 76
Frege, 5
ftp sites, 356

function approximation, 75
function call, 257
function partitioning, 145
function shapes, 56
fuzzification, 71, 79, 198, 201, 215, 238, 259, 313, 408, 409
fuzzification algorithm, 217, 220
fuzzy algorithms, 210
fuzzy application, 254
fuzzy application object, 305
fuzzy associative memory, 122
fuzzy chips, 134
fuzzy classes, 113, 114
fuzzy cognitive map, 409
fuzzy control system, 323
fuzzy controller, 142, 144, 150, 161, 169, 188, 207, 209, 305, 411
fuzzy engine, 243, 253, 255, 261, 264, 265, 267, 313, 318, 346, 409
fuzzy engine code, 320
fuzzy engine design, 213
fuzzy engineering, 105, 107, 136, 185
fuzzy function approximation, 75
fuzzy inference engine, 210
fuzzy information models, 210
fuzzy input, 195, 210
fuzzy input, output, 223
fuzzy intersection, 192
fuzzy kernel, 192, 215, 224, 226, 228, 359, 390
fuzzy kernel in C, 215
fuzzy logic, 7, 23
fuzzy logic classes, 243
Fuzzy Logic Inferences Per Second, 201
fuzzy logical operators, 40
fuzzy model, 76, 77, 93, 114, 127, 133, 138, 196, 198, 199, 209, 215, 237, 269, 409, 413
fuzzy model development, 341
fuzzy model generation tools, 346
fuzzy model implementation, 134
fuzzy model integration, 136
fuzzy model simulation, 130
fuzzy modeling, 120
fuzzy models, 108, 190, 195, 213, 346
fuzzy output set, 224

Index

fuzzy output space, 99
fuzzy output variable, 405
fuzzy outputs, 67
fuzzy predicate, 409
fuzzy processing, 201
fuzzy processing chip, 238
fuzzy processing hardware, 238
fuzzy processor, 211, 239
fuzzy proposition, 70, 251, 264, 406
fuzzy relation, 73
fuzzy rule, 18, 69, 71, 79, 89, 120, 180, 246, 271, 406
fuzzy rule aggregation, 245
fuzzy rulebased engineering, 139
fuzzy set, 1, 7, 11, 18, 29, 32, 36, 90, 103, 118, 216, 238, 407, 408, 410, 411, 412, 414, 415
fuzzy set definition, 412
fuzzy set representation, 220
fuzzy set shapes, 198
fuzzy singleton, 66
fuzzy system, 36, 69, 96
fuzzy variable, 36, 56, 110, 113, 210, 215, 217, 246, 250, 251, 253, 259, 261, 269, 271, 305, 310, 318, 409, 411, 415
fuzzy variable parsing, 320
FuzzyAppl, 265
FuzzyEngine, 265
FuzzyLab, 28, 146, 254, 265, 319
FuzzyLab primer, 323
FuzzyVar construction, 306
FZKERN.C, 229, 237

—G—

gaussian curves, 62
Gaussian shapes, 114, 193
generality, 199
generalization, 410
generic base class, 319
Get_Crisp_Inputs, 229
glossary, 405
goal constraints, 180
goals, 190
grade of membership, 409

—H—

hand coding, 210
handwriting recognition, 237
hardware, 192, 197, 211, 213, 237, 242
hardware acceleration, 347
hardware platforms, 201
hedge, 49, 264, 407, 410
hedging, 89
heuristics, 179
high speed architecture, 238
human judgment, 31
Hungarian notation, 272
hybrid, 213, 242
hypertext, 354

—I—

identity, 414
IF/THEN rule construction, 120
IfSide, 246
IfSide construction, 308
IF-side evaluation, 224
implementation, 213, 238
implementation constraints, 126, 134
implementation cost, 24, 190
implication, 410, 412
imprecise knowledge, 184
induction, 410
industrial control, 237
inference, 89, 90, 224, 260, 317, 407, 410
inference algorithms, 236
inference engine, 215
inference loop, 201
inference loop timings, 236
inference method, 100, 126
inference process, 405
inferencing speed, 137, 211, 237, 238
infinite series, 76
information preservation, 129
information system, 111, 114, 137, 243
information system models, 348
inheritance, 245, 255
initial conditions, 142
initial rule set, 171
initialization, 319

421

Fuzzy Logic for Real World Design

Initialize_Fuzzy_Engine, 228
initializing, 228
input membership function, 170
input range, 110
input sensors, 108, 129, 136
input space, 123
input tuple, 253, 312, 313
inputs, 22, 215
integrated tools, 348
Intel 80386DX, 236
Intel 80486, 199
Intel 80C196, 237, 390
intensification, 410
interface to databases, 136
interfaces, 137
Internet sites, 355
interpolation, 189, 315
interpolative properties, 75
intersection, 41, 81, 84, 85, 89, 135, 137, 182, 210, 224, 360, 384, 391, 405, 410, 414
intractability, 14
inventory forecasting link, 137
inverted pendulum, 17, 108, 131, 142, 229
iterator, 301
iterator class, 303

—K—

kernel performance, 236
knowledge, 410
Kosko, Bart, 409

—L—

label, 36, 410
labeling, 412
Law of Excluded Middle, 410
Law of Incompatibility, 411
law of non-contradiction, 40
law of the excluded middle, 40
Left- or Right-most maximizer, 199
limits, 110
Lindsay, 5
linear control systems, 194
linear systems, 21, 132

linguistic hedge, 49
linguistic label, 36, 49, 411
linguistic rule, 19
linguistic variable, 38, 69, 411
link, 320
linked lists, 299
links, 255
local variables, 228
logic gates, 241
logical operators, 40
loop timing, 138, 236
Lukasiewicz, Jan, 7, 411

—M—

machine intelligence, 27
macro, 223
main control loop, 255
maintainability, 134
maintenance, 24
Mamdani, Ebrahim, 411
mapping, 181
mapping classes, 301
mathematical approximation, 76
mathematical model, 129
mathematical operations, 198
mathematical operator, 81
mathematical simulation, 185
mathematics of simulation, 168
matrix of rules, 122
MAX, 93, 198, 224, 226, 240, 405
max aggregation, 96, 129, 360, 384, 391
MAX defuzzification, 360
MAX union operator, 83
maximizer defuzzification, 99
Mbf construction, 307
measurement uncertainty, 129
medical diagnosis, 137
membership, 10
membership function, 29, 32, 46, 56, 70, 77, 93, 111, 123, 128, 130, 133, 135, 151, 186, 199, 215, 250, 251, 259, 269, 271, 300, 305, 310, 313, 317, 318, 320, 408, 409, 411, 413
membership function domain, 35
membership function labels, 207

422

Index

membership function overlap, 116
membership function partitioning, 118, 141
membership function partitions, 175
membership function placement, 114
membership function shape, 113, 137, 193, 413
membership functions, number of, 112
memory, 127
memory map, 237
memory requirements, 63
message, 304, 313
message passing, 257
methods, 245
microcomputer, 187
microcontroller, 24, 114, 134, 187, 192, 195, 198, 199, 209, 210, 214, 220, 229, 237, 242, 359, 384, 390
microprocessor, 236
migration pathing, 190
MIN, 198, 224, 240
MIN intersection, 83, 90, 129, 193, 360, 384, 391
min-max inference, 93
MIN-MAX rule, 197
mirror image partitioning, 118
model semantics, 127
model-free, 19
model-free estimation, 262
modifiers, 12
Modus Ponens, 410, 412
monotonicity, 414
Motorola 68HC11, 237, 359
Motorola 68HC16, 237, 384
multiple rules, 73
multiplicity relationship, 249
multivalued logic, 7, 412
MYCIN, 5

—N—

natural classifications, 12, 412
natural language, 36, 69
NeuraLogix, 211
noisy input data, 47
non-conflicting rules, 120, 125

non-linear control, 161
non-linear interpolation, 189
non-linear stability, 132
non-linear systems, 21, 194, 207
NORM, 44, 90, 412
normal distribution, 62
normalization operator, 44
NOT, 40, 412
number of rules, 120, 125, 210
numerical integration, 167

—O—

object construction, 304
object hierarchy, 304
Object Modeling Technique, 244
object-oriented design, 257, 260, 301
object-oriented paradigm, 243
objects, 267, 318
observability, 207
OMT, 244
on-line resources, 354
OO design, 301
open loop response, 146, 171
operating points, 207
operator alerting, 136
operator selection, 126
optimization, 130
optimizing compilers, 214
OR, 43, 82, 83, 86, 406, 409, 412, 415
oscillation, 148, 155, 173
output actuators, 136
output effectors, 108
output fuzzy set, 114
output range, 110
output scaling, 229
output sets, 157
output singleton, 145, 170
output tuple, 253
output variable, 405
outputs, 22, 215
overlap, 116
overlap ratio, 116
overlapping rules, 123
overloaded operators, 321
overshoot, 148, 153, 174

423

—P—

parameter memory, 238
parsing, 320
partitioning, 114
pattern recognition, 237
performance, 130, 190, 242
performance comparisons, 203
performance data, 139
performance objectives, 153
personal computers, 214
perspective shift, 412
PhaseView, 146
PI, 159
PI-control, 169
Pi-Curve, 61, 413
Pi-curves, 114
PID, 159
piecewise linear functions, 63
piecewise linear membership function, 262
piecewise linear shapes, 114
pointwise minimum, 405
pole and cart, 19
portability, 134, 213, 214
possibilities, 406
possibility theory, 413
precision, 13, 16
premise conditions, 90, 410
principle of continuity, 207
Principle of Incompatibility, 14
probabilistic sum, 86
probability, 406
problem complexity, 19
problem solving, 13
problem-solving techniques, 105
process control, 113, 118
processing power, 201
processing speed, 191, 192, 200
processing time, 210
PROD, 198
PRODUCT, 224
product evaluation method, 264
product inference, 261
PRODUCT operator, 85
program branching, 64
properties, 47

proportional-integral control, 159
proportional-integral-derivative control, 159
proposition, 70, 79
propositions, 4
PROSPECTOR, 5
PSUM, 198
pulse width modulation, 197

—Q—

qualifiers, 12
quantifiers, 413

—R—

RAM, 228
RAM space, 192
range of interest, 36
real-time control, 192, 237
reasoning, 407
register allocation table, 237
reliability, 190
resolution, 191, 194, 209
response time, 155, 241
reusability, 134, 262
reuse of code, 262, 268
rewriting, 264
ROM, 220
Rosch, 30
rule, 69, 73, 77, 79, 157, 171, 200, 207, 210, 215, 222, 238, 246, 300, 310, 317, 405, 406, 407, 409, 410, 413
rule consequents, 405
Rule construction, 308
rule coverage map, 124
rule evaluation, 181, 193, 198, 201, 215, 224, 238, 264, 408, 409, 413
rule evaluation methods, 264
rule inference, 198
rule matrix, 208
rule parsing, 320
rule set continuity, 207
rule set, complete, 207
rule structure, 223
rule transition, 72

rule writing, 120, 141
RULE_BLOCK, 215, 222, 224
rulebase, 111, 122, 129, 249, 304, 318, 319, 413
Rulebase construction, 308
rulebase evaluation, 93, 259, 316
rules, number of, 120
Rumbaugh, 244
Runge-Kutta, 168
running the fuzzy engine, 313

—S—

scaling, 264
scaling factors, 160
scaling inputs, 229
scope, 35, 46, 71, 414
S-Curve, 57, 114, 413, 414, 415
selection of operators, 126
self-tuning, 139
semantic operations, 49
semantic shift, 414
sensor resolution, 138
serial operation, 196
set-point, 113, 146, 174, 269
set-point control, 159
sets, 10
Shannon, Claude, 408
Shortliff, 5
simple association, 255
simulation, 131, 135, 141, 144, 150, 207, 243, 323
simulation loop, 146
simulation mathematics, 169
simulation model, 162
simulation tools, 348
singleton, 66, 100, 114, 216, 414
singleton output, 226
singleton output membership function, 193
size, 134, 266
Slightly, 54
SMALLC, 359
smoothness, 124
software, 197, 210, 213, 242
software architecture, 199
Somewhat, 52

Sort of, 55
source code, 267, 272, 304
specificity, 191, 196, 210
speed, 127, 134, 266
speed of development, 24
spreadsheet interface tools, 349
stability, 132, 193, 206, 209
state space evolution, 153, 185
state space graph, 148
state space performance, 171
state space region, 157
state space trajectory, 143
state variables, 142, 144
state vector, 188
static data blocks, 215
stock trading, 126
strength, 260, 316, 406
strength attribute, 246
summarizing, 29
support set, 414
supportability, 190
syllogisms, 4
symmetric behavior, 124
symmetric membership function, 118
symmetry, 113, 118, 147
syntax, 69
system behavior, 130
system complexity, 411
system description, 107
system model, 22
system theory, 21
system variables, 22

—T—

Taber, Rod, 409
tails, 46
target system, 162
Taylor power series expansion, 76
T-conorm, 414
Tempcntl, 265
TEMPCNTL.CFG, 270
TEMPCNTL.EXE, 268
temperature compensation, 241
templates, 299
terminology, 29

testing, 206, 207
ThenSide, 246
ThenSide construction, 309
thermostatic control, 244, 267
thresholding, 103
time response, 148
T-norm, 414
tools, 214, 348
tractability, 16
trade-offs, 127
transforming, 264
translating, 264
trapezoidal membership function, 63, 114, 262, 271
trapezoidal shapes, 216
traveling salesman problem, 15
triangular conorm, 414
triangular membership function, 63, 65, 114, 193
triangular norm, 414
triangular shapes, 216
truncate, 93
truncation, 90
truth value, 33, 94, 99, 195, 259, 405, 414
tuning, 114, 130, 135, 139, 141, 152, 162, 169, 175, 185
tuple, 253, 255, 258, 312, 318

—U—

uncertainty, 31, 415
union, 82, 85, 86, 89, 210, 414, 415
union operator, 43, 412
unipolar variables, 113
universe of discourse, 36, 110, 114, 150, 173, 195, 250, 271, 313, 411, 415
universes of discourse, 207
unsharp boundary, 29
user-interface, 265

—V—

vague set, 7
vagueness, 6, 405, 415
variable parsing, 320
variables, 228, 409
vectors, 299
vehicle simulation, 162
Very, 50
VLSI chips, 213
voice recognition, 237
voltage mode circuits, 241

—W—

weighted average defuzzification, 384, 391
weighted average intersection operator, 183
whole-part, 245, 255
Windows, 262, 265, 272, 323, 346
workspace variables, 228
World Wide Web, 354
World Wide Web sites, 357
writing rules, 120
www, 354
www sites, 357

—Y—

Yager functions, 86, 198
Yager intersection, 193, 210
Yamaichi Securities, 126

—Z—

Zadeh, ix, xvi, xviii, 1, 7, 8, 14, 38, 50, 52, 57, 58, 72, 79, 81, 82, 179, 411, 413, 415
Zadeh intersection, 81
Zadeh union, 82
Z-Curve, 60, 114, 413, 414, 415
zero point, 113

You are welcome to send us comments or questions concerning this Annabooks product, or to request a catalog of our other products, services, conferences, and classes.

Annabooks
11838 Bernardo Plaza Court
San Diego, CA 92128-2414

800-462-1042   619-673-0870
619-673-1432 FAX
73204.3405 @ compuserve.com

Companion Disk Information

The companion disk for Fuzzy Logic for Real World Design contains the following software:

- FuzzyLab 1.1
- Fuzzy Kernel
- Fuzzy Application Executable and C++ Code

Setup Instructions:
Insert disk into drive A (or B).
Run MS Windows Program Manager.
From File menu select Run.
Enter A:Setup (or B:Setup) and hit ENTER.

Licensing Agreement

By opening this package, you are agreeing to be bound by the following agreement:

COPYRIGHT:
Copyright ©1995 Synerdaptix, Inc. Permission to copy all or part of this software is granted, provided that the copies are not made or distributed for resale, and provided that the NO WARRANTY, AUTHOR CONTACT, and COPYRIGHT notice are retained verbatim and are displayed conspicuously. If anyone needs other permissions that are not covered by the above, please contact the authors.

NO WARRANTY:
All of the software on this Companion Disk is provided on an "as is" basis. The authors, publisher, and their dealers and distributors provide no warranty whatsoever, either express or implied, regarding the work, including warranties with respect to its merchantability or fitness for any particular purpose.

AUTHOR CONTACT:
Synerdaptix, Inc.
P.O. Box 775
Suwanee, GA 30174
CompuServe 75104,3236
         Internet 75104.3236@compuserve.com